CLASSICAL MECHANICS

CLASSICAL MECHANICS

H. C. CORBEN
Professor Emeritus
UNIVERSITY OF TORONTO

PHILIP STEHLE
Professor Emeritus
UNIVERSITY OF PITTSBURGH

2ND EDITION

DOVER PUBLICATIONS, INC.
NEW YORK

Copyright

Bibliographical Note

This Dover edition, first published in 1994, is an unabridged, slightly corrected and enlarged, republication of the work first published by John Wiley & Sons, New York, 1960 (fourth printing, 1966; original edition, 1950).

Library of Congress Cataloging-in-Publication Data

Corben, H. C. (Herbert Charles)
Classical mechanics / H.C. Corben, Philip Stehle.—2nd ed.
p. cm.
Originally published: 2nd ed. New York : Wiley, 1960.
Includes bibliographical references and index.
ISBN 0-486-68063-0 (pbk.)
1. Mechanics, Analytic. I. Stehle, Philip. II. Title.
QA805.C75 1994
531—dc20 94-17085
CIP

Manufactured in the United States of America
Dover Publications, Inc., 31 East 2nd Street, Mineola, N.Y. 11501

PREFACE

"Classical mechanics" denotes the theory of the motion of particles and particle systems under conditions in which Heisenberg's uncertainty principle has essentially no effect on the motion and therefore may be neglected. It is the mechanics of Newton, Lagrange, and Hamilton and it is now extended to include the mechanics of Einstein. When coupled with classical electromagnetism, its principles become the basis for more accurate and more general physical theories, and its applications provide the structure for almost all modern developments in technology outside of nuclear and solid state phenomena.

In the ten years since the first edition of this book was published, the subject has been applied with enormous effort to problems in space technology, accelerator design, plasma theory, and magnetohydrodynamics as well as to the older but currently very active fields of vacuum and gaseous electronics, aerodynamics, elasticity, and so forth. In this second edition we have therefore included some applications to problems not usually taught in physics departments, for example, the theory of space-charge limited currents, atmospheric drag, the motion of meteoritic dust, variational principles in rocket motion, transfer functions, and dissipative systems. Some special applications which are of current interest in more basic physics research are also treated, for instance, spin motion and rotating coordinate systems, noncentral forces, the Boltzmann and Navier-Stokes equations, the inverted pendulum, Thomas precession, and the motion of particles in high energy accelerators, a chapter which had to be completely rewritten to give some account of recent work in this field. Since so many treatises on relativity theory are available, the emphasis in our discussion of this subject is on aspects not treated in detail elsewhere.

At the same time this book is an attempt to present classical mechanics in a way that shows the underlying assumptions and that as a consequence indicates the boundaries beyond which its uncritical extension is dangerous. The presentation is designed to make the transition from classical mechanics to quantum mechanics and to relativistic mechanics smooth so that the reader will be able to sense the continuity in physical thought as the change is made.

The place of classical mechanics in the scheme of present-day physics is discussed and the idealizations made in describing a system in terms of classical concepts are analyzed. The problem of central motion is treated with more than the usual emphasis on scattering problems such as the nature of a cross section, the Rutherford formula, and the transformation between center-of-mass and laboratory coordinate systems. The transformation theory of mechanics has been based more directly on the variational principles, and the section on perturbation theory has been amplified.

The simplified index notation adopted does not distinguish between convariant and contravariant indices since this is not necessary for most of the development. Where such a distinction is useful, as in demonstrating the connection between Newton's and Lagrange's form of the equations of motion, the analysis is given in an appendix. Other purely mathematical and specialized topics, such as group theory and molecular vibrations, linear vector spaces, and quaternions, have also been removed from the body of the text to appendices.

Complementing the theory of the motion of particles under the influence of applied forces is the theory of classical fields, such as electricity, magnetism, and gravitation, in which the applied forces are seen to arise from a distribution of charges, currents, and masses. The techniques used in classical field theory are described here because they are of general interest, being applicable to any situation in which the transport of energy and momentum may be described in terms of one or more functions of position and time. Even the wave functions of modern physics satisfy this condition, so that they could be included in classical field theory even though wave mechanics is the very antithesis of classical mechanics. However, such developments lie outside of the scope of this book.

A number of exercises, many of them not previously published, are included. Some of these emphasize the need for visualization of the basic phenomena involved in solving problems in theoretical physics, whereas others stress the more formal aspects of theory. At the end of nearly every chapter is given a list of references to books and other published literature where additional relevant material may be found.

H. C. CORBEN
PHILIP STEHLE

August, 1960

CONTENTS

Chapter 1. Kinematics of Particles 1

1. Introduction 1
2. Definition and Description of Particles 6
3. Velocity 11
4. Acceleration 12
5. Special Coordinate Systems 13
6. Vector Algebra 16
7. Kinematics and Measurement 22
   Exercises 25

Chapter 2. The Laws of Motion 27

8. Mass 27
9. Momentum and Force 28
10. Kinetic Energy 31
11. Potential Energy 32
12. Conservation of Energy 35
13. Angular Momentum 36
14. Rigid Body Rotating about a Fixed Point 38
15. A Theorem on Quadratic Functions 40
16. Inertial and Gravitational Masses 43
    Exercises 44

Chapter 3. Conservative Systems with One Degree of Freedom 47

17. The Oscillator 47
18. The Plane Pendulum 50
19. Child-Langmuir Law 54
    Exercises 56

Chapter 4. Two-Particle Systems 58

20. Introduction 58
21. Reduced Mass 59
22. Relative Kinetic Energy 61
23. Laboratory and Center-of-Mass Systems 62
24. Central Motion 63
Exercises 65

Chapter 5. Time-Dependent Forces and Nonconservative Motion 66

25. Introduction 66
26. The Inverted Pendulum 67
27. Rocket Motion. 70
28. Atmospheric Drag. 71
29. The Poynting-Robertson Effect 73
30. The Damped Oscillator 74
Exercises 76

Chapter 6. Lagrange's Equations of Motion 77

31. Derivation of Lagrange's Equations. 77
32. The Lagrangian Function 81
33. The Jacobian Integral 82
34. Momentum Integrals 83
35. Charged Particle in an Electromagnetic Field 85
Exercises 88

Chapter 7. Applications of Lagrange's Equations 90

36. Orbits under a Central Force 90
37. Kepler Motion. 93
38. Rutherford Scattering. 100
39. The Spherical Pendulum 103
40. Larmor's Theorem. 107
41. The Cylindrical Magnetron 110
Exercises 111

Chapter 8. Small Oscillations 113

42. Oscillations of a Natural System 113
43. Systems with Few Degrees of Freedom. 117
44. The Stretched String, Discrete Masses 124

45. Reduction of the Number of Degrees of Freedom . . . 129
46. Laplace Transforms and Dissipative Systems 131
Exercises 134

Chapter 9. Rigid Bodies **136**

47. Displacements of a Rigid Body 136
48. Euler's Angles 140
49. Kinematics of Rotation 141
50. The Momental Ellipsoid 146
51. The Free Rotator 148
52. Euler's Equations of Motion 149
Exercises 152

Chapter 10. Hamiltonian Theory **154**

53. Hamilton's Equations 154
54. Hamilton's Equations in Various Coordinate Systems . 159
55. Charged Particle in an Electromagnetic Field 162
56. The Virial Theorem 164
57. Variational Principles 166
58. Contact Transformations 172
59. Alternative Forms of Contact Transformations . . . 177
60. Alternative Forms of the Equations of Motion . . . 179
Exercises 181

Chapter 11. The Hamilton-Jacobi Method **183**

61. The Hamilton-Jacobi Equation 183
62. Action and Angle Variables—Periodic Systems . . . 188
63. Separable Multiply-Periodic Systems 196
64. Applications 200
Exercises 213

Chapter 12. Infinitesimal Contact Transformations **215**

65. Transformation Theory of Classical Dynamics . . . 215
66. Poisson Brackets 220
67. Jacobi's Identity 228
68. Poisson Brackets in Quantum Mechanics 230
Exercises 231

Chapter 13. Further Development of Transformation Theory 232

69. Notation 232
70. Integral Invariants and Liouville's Theorem 233
71. Lagrange Brackets 237
72. Change of Independent Variable 238
73. Extended Contact Transformations. 241
74. Perturbation Theory 244
75. Stationary State Perturbation Theory 245
76. Time-Dependent Perturbation Theory 251
77. Quasi Coordinates and Quasi Momenta 254
Exercises 257

Chapter 14. Special Applications 258

78. Noncentral Forces. 258
79. Spin Motion 262
80. Variational Principles in Rocket Motion 265
81. The Boltzmann and Navier-Stokes Equations. 268

Chapter 15. Continuous Media and Fields 273

82. The Stretched String 273
83. Energy-Momentum Relations 277
84. Three-Dimensional Media and Fields 280
85. Hamiltonian Form of Field Theory 284
Exercises 286

Chapter 16. Introduction to Special Relativity Theory 287

86. Introduction 287
87. Space-Time and the Lorentz Transformation 288
88. The Motion of a Free Particle 294
89. Charged Particle in an Electromagnetic Field. 296
90. Hamiltonian Formulation of the Equations of Motion . 298
91. Transformation Theory and the Lorentz Group . . . 302
92. Thomas Precession 304
Exercises 312

Chapter 17. The Orbits of Particles in High Energy Accelerators 314

93. Introduction 314
94. Equilibrium Orbits 315

95. Betatron Oscillations 318
96. Weak Focusing Accelerators 323
97. Strong Focusing Accelerators 324
98. Acceleration and Synchrotron Oscillations. 329

Appendix I **Riemannian Geometry** **335**

Appendix II **Linear Vector Spaces** **345**

Appendix III **Group Theory and Molecular Vibrations** **364**

Appendix IV **Quaternions and Pauli Spin Matrices** **373**

Index **383**

1 KINEMATICS OF PARTICLES

1. INTRODUCTION

The sciences of physics and chemistry are concerned with the interactions between inanimate material systems. Although chemistry is related chiefly to the interactions between atoms to form molecules, physics is primarily the study of the motion induced in one system of matter by the presence of other systems and of the characterization and measurement of such states of motion by a human observer.

The situations of interest to the physicist are among the least complicated of all known phenomena. This fundamental simplicity of the subject matter allows the science to reach and even demand a degree of precision in both theory and experiment not known elsewhere. Sometimes this precision can be realized, however, only by the introduction of complexities of theory and experiment which mask the very simplicity that made it possible.

The aim of theoretical physics is to provide and develop a self-consistent mathematical structure which runs so closely parallel to the development of physical phenomena that, starting from a minimum number of hypotheses, it may be used to accurately describe and even predict the results of all carefully controlled experiments. The desire for accuracy, however, must be tempered by the need for reasonable simplicity, and the theoretical description of a physical situation is always simplified for convenience of analytical treatment. Such simplifications may be thought of as arising both from physical approximations, i.e., the neglect of certain physical effects which are judged to be of negligible importance, and from mathematical approximations made during the development of the analysis. However, these two types of approximations are not really distinct, for usually each may be discussed in the language of the other. Representing as they do an economy rather than an ignorance, such approximations may be refined by a series of increasingly accurate calculations, performed either algebraically or with an analog or digital computer.

More subtle approximations appear in the laws of motion which are assumed as a starting point in any theoretical analysis of a problem. At present the most refined form of theoretical physics is called quantum field theory, and the theory most accurately confirmed by experiment is a special case of quantum field theory called quantum electrodynamics. According to this discipline, the interactions between electrons, positrons, and electromagnetic radiation have been computed and shown to agree with the results of experiment with an over-all accuracy of one part in 10^9. Unfortunately, analogous attempts to describe the interactions between mesons, hyperons, and nucleons are at present unsuccessful.

These recent developments are built on a solid structure which has been developed over the last three centuries and which is now called classical mechanics. Figures 1-1 and 1-2 illustrate how classical mechanics is related to other basic physical theories. A line connecting two boxes indicates that the branch of theoretical physics to which the arrow points is a particular case of the branch from which the line originates. The solid lines indicate special cases from a mathematical point of view; i.e., they lead from a theory to one of its applications. The dashed lines, on the other hand, refer to particular cases that may be obtained by neglecting one or the other of the three basic effects that have been discovered since the development of nonrelativistic classical theory—radiative effects, quantum effects, and relativistic effects. For some situations (e.g., quantum effects in atomic theory) one of these may be of dominant importance, whereas for others (e.g., celestial mechanics) the same effect may be completely negligible.

A theory that describes the motion of a particle at any level of approximation must eventually reduce to nonrelativistic classical mechanics when conditions are such that relativistic, quantum, and radiative corrections can be neglected. This fact makes the subject basic to the student's understanding of the rest of physics, in the same way that over the centuries it has been the foundation of human understanding of the behavior of physical phenomena.

Classical mechanics accurately describes the motion of a material system provided that the angular momentum of the system with respect to the nearest system which is influencing its motion is large compared with the quantum unit of angular momentum $\hbar = 1.054 \times$

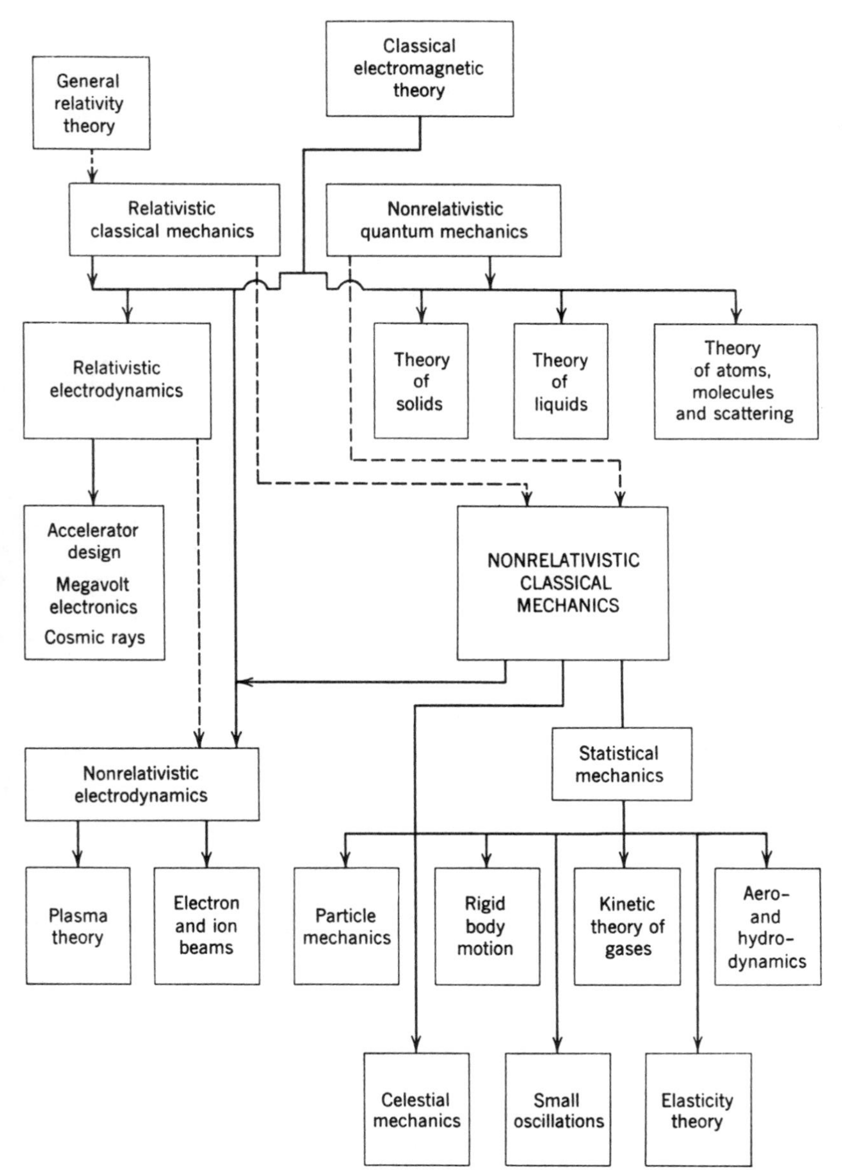

Fig. 1-1. The basic physical theories known in 1926 with some applications made possible within their framework.

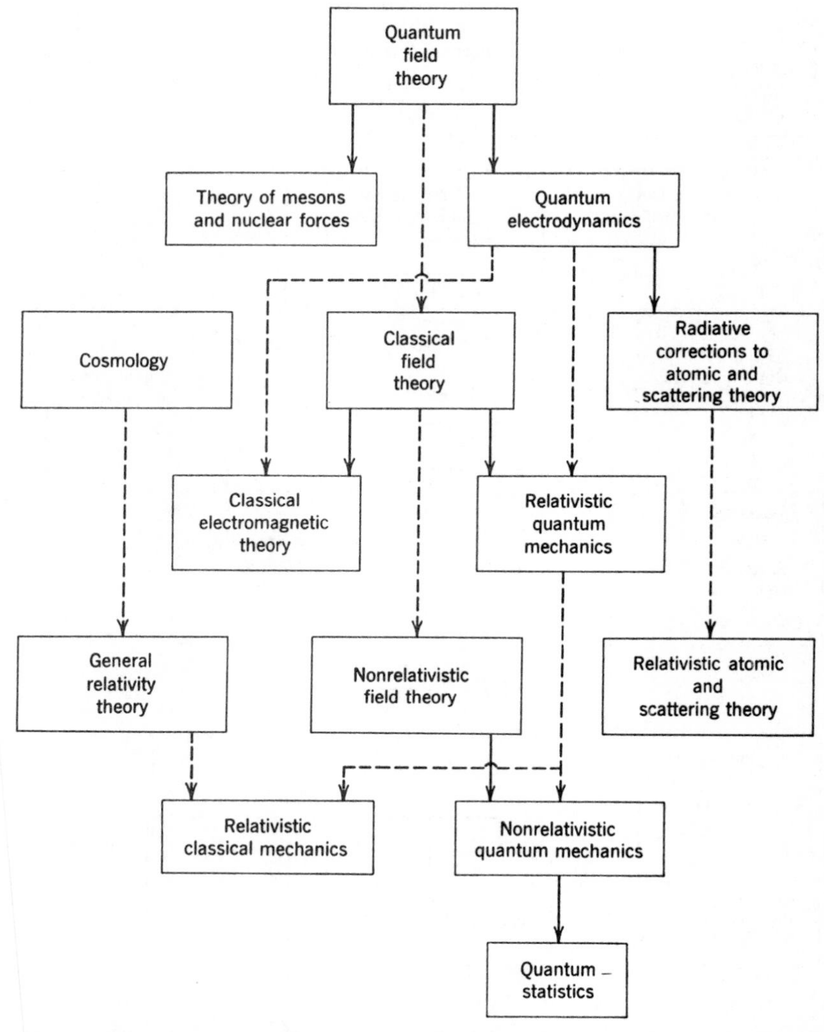

Fig. 1-2. The basic physical theories discovered, apart from classical electromagnetic theory, since 1904.

10^{-27} gram-cm^2/sec. Some examples are given in Table 1-1. Clearly, in all but the last case the existence of a smallest unit of angular momentum is irrelevant, and the error introduced by using the approximation of classical mechanics will be small compared with both unavoidable experimental errors and other errors and approximations made in describing theoretically the actual physical situation. However, classical mechanics should not be studied only as an introduction to the more refined theories, for despite advances made during this century it continues to be the mechanics used to describe the motion of directly observable macroscopic systems. Although an old subject, the mechanics of particles and rigid bodies is finding new applications in a number of areas, including the fields of vacuum and gaseous electronics, accelerator design, space technology, plasma physics, and magnetohydrodynamics. Indeed, more effort is being put into the development of the consequences of classical mechanics today than at any time since it was the only theory known.

TABLE 1-1

System	Approximate Angular Momentum (Units of $\hbar$)
Earth moving around the sun	10^{64}
Steel ball 1 cm radius rolling at 10 cm/sec along a plane	10^{29}
Electron moving in a circle of radius 1 cm at 10^8 cm/sec	10^8
Electron moving in an atom	$0, 1, 2 \cdots$

Throughout all but the last chapters of this book we are concerned with nonrelativistic classical mechanics. This theory is applicable to the motion of systems which not only possess an angular momentum large compared with $\hbar$ but are moving with relative velocities small compared with that of light ($c = 2.99776 \times 10^{10}$ cm/sec). In Chapter 16, however, we give an introduction to the special theory of relativity, which is capable of describing situations in which this restriction on the velocity does not necessarily hold.

In addition, none of the above theories is applicable to the description of the interaction between systems separated by distances comparable to Gm/c^2, where m is the largest mass and G is the gravitational constant ($G = 6.670 \times 10^{-8}$ dyne-cm^2/gram^2). For a proton and an electron this distance is extremely small ($\sim 10^{-52}$ cm) and many unknown properties of the interaction between these particles already occur at very much larger distances ($\sim 10^{-13}$ cm). For a planet interacting with the sun, the critical distance is 1.47 km, so that although Mercury is much farther from the sun than this, there exists in its motion a very small anomaly that cannot be described even by relativistic classical mechanics but which is adequately explained by the general theory of relativity.

2. DEFINITION AND DESCRIPTION OF PARTICLES

The simplest mechanical system is one which may be represented in the mathematical scheme of mechanics by a point. Such a system is called a *particle*. How good a representation this is for a particular system can be determined only when we come to consider a more complicated representation of the system and can estimate the effects due to features neglected in the particle treatment. Intuitively it is plausible that it will be a good approximation for a given body provided that this body is small compared with its distance from other bodies and provided that the motion of the body as a whole is only slightly affected by its internal motion. Thus a particle is a body whose internal motion is irrelevant to its motion as a whole, so that it may be represented by a mass point having no extension in space.

A particle is described when its position in space is given and when the values of certain parameters such as mass, electric charge, and magnetic moment are given. By our definition of a particle, these parameters must have constant values because they describe the internal constitution of the particle. If these parameters do vary with time, we are not dealing with a simple particle. The position of a particle may, of course, vary with time.

We may specify the position of a single particle in space by giving its distances from each of three mutually perpendicular planes. These three numbers are called the *cartesian coordinates* of the particle. Since three mutually perpendicular planes meet in three mutually perpendicular lines, we may also consider the cartesian coordinates of a particle as the displacements in the directions of these three lines needed to move the particle from the point of intersection of the three lines to its actual position. These coordinates will be denoted by x_1, x_2, x_3, or in general by x_i $(i = 1, 2, 3)$. To avoid the use of indices in particular problems, we shall also use the letters x, y, z as coordinates.

The coordinate system introduced in this way plays the role of the observer of the particle in the mathematical scheme. The unessential features of the observer have been completely removed, only an extreme simplification of the observer's measuring implements in the form of a set of coordinates being left. The very complex relationship between the particle and the observer which is to be described mathematically is reduced to the relationship of a point to a set of coordinates. In this way we tend almost to forget the observer entirely. In quantum mechanics and relativity theory the role of the observer himself is stressed considerably more.

The coordinates x_i may be considered as the components of a vector, the radius vector $\mathbf{r}$ of the particle

$$\mathbf{r} = \mathbf{e}_1 x_1 + \mathbf{e}_2 x_2 + \mathbf{e}_3 x_3 \tag{2.1}$$

where $\mathbf{e}_i$ is the unit vector along the i coordinate axis. Sometimes these vectors are denoted by $\mathbf{i}$, $\mathbf{j}$, $\mathbf{k}$ respectively.

We now introduce a notation that makes many formulas simpler in appearance and easier to understand. Whenever an italic index appears twice, it is to be summed over the entire range of the index. Greek indices are not to be summed unless a summation is explicitly indicated. We shall usually reserve such indices for numbering individual particles in many-particle systems. In this notation (2.1) becomes simply

$$\mathbf{r} = \mathbf{e}_i x_i \tag{2.2}$$

If N particles are present, the position of each one is specified by its radius vector, that of the ρth particle being denoted by $\mathbf{r}_\rho$. The $3N = f'$ quantities $x_{\rho i}$ ($i = 1, 2, 3$; $\rho = 1, 2, \cdots, N$) specify the configuration of the entire system. For many purposes it is more convenient to regard the f' quantities $x_{\rho,i}$ as the cartesian coordinates of a single particle which is located in an f'-dimensional space. (For reasons which will be explained in Sec. 10, coordinates $\bar{x}_{\rho,i} = \sqrt{m_\rho}\, x_{\rho,i}$ are frequently used, m_ρ being the mass of the ρth particle.) This f'-dimensional space is called the *unconstrained configuration space* of the system.

Very often it is not convenient to describe the position of even a single particle in terms of rectangular cartesian coordinates referred to a particular set of coordinate axes. If, for example, the particle moves in a plane under the influence of a force which is directed toward a fixed point in the plane and which is independent of the azimuthal angle θ, it is usually more convenient to use the *plane polar coordinates*

$$\begin{aligned} q_1 &= r = (x^2 + y^2)^{1/2} \\ q_2 &= \theta = \tan^{-1} \frac{y}{x} \end{aligned} \tag{2.3}$$

or if the force is spherically symmetrical it is natural to use the *spher-*

ical coordinates

$$q_1 = r = (x^2 + y^2 + z^2)^{1/2}$$

$$q_2 = \theta = \cot^{-1} \frac{z}{(x^2 + y^2)^{1/2}} \tag{2.4}$$

$$q_3 = \phi = \tan^{-1} \frac{y}{x}$$

These coordinates are also used if the particle is constrained to move on a fixed circle or fixed sphere, respectively.

Sometimes it is useful to look at the motion of the particle from the point of view of a set of coordinates in uniform motion in the x direction with velocity v

$$\begin{aligned} q_1 &= x - vt \\ q_2 &= y \qquad\qquad (v = \text{const}) \\ q_3 &= z \end{aligned} \tag{2.5}$$

or from a uniformly accelerated system

$$\begin{aligned} q_1 &= x - \tfrac{1}{2}gt^2 \\ q_2 &= y \qquad\qquad (g = \text{const}) \\ q_3 &= z \end{aligned} \tag{2.6}$$

In general, each transformation of the coordinate system x_i to a new set q_m may be expressed as a set of three equations of the form

$$x_i = x_i(q_1, q_2, q_3, t) \tag{2.7}$$

For the stationary coordinate systems (2.3) and (2.4), the relations between x_i and q_i do not involve the time t.

If equations (2.7) are such that the three coordinates q_m can be expressed as functions of the x_i, we have

$$q_m = q_m(x_1, x_2, x_3, t) \tag{2.8}$$

The q_m are as effective as the x_i in describing the position of the particle. The q_m are called *generalized coordinates* of the particle. The generalized coordinates may themselves be rectangular cartesian coordinates [as in (2.5), (2.6)] or they may be a set of any three variables, not necessarily with the dimension of length, which between them specify unambigously the position of the particle relative to some set of axes.

To obtain the necessary and sufficient conditions that equations (2.7) may be solved for the q_m at a fixed value of time to give a set of equations of the form (2.8), we consider a small displacement of the particle defined by changes δx_i in the x_i, with t held fixed:

$$\delta x_i = \frac{\partial x_i}{\partial q_m} \delta q_m \tag{2.9}$$

(This is, according to our notation, the sum of three terms!) Equation (2.9) is a set of inhomogeneous linear equations for the δq_m which can be solved if and only if the determinant of the coefficients does not vanish:

$$\left| \frac{\partial x_i}{\partial q_m} \right| \neq 0 \tag{2.10}$$

This determinant is called the Jacobian determinant of the x's with respect to the q's and is denoted by $[\partial(x_1, x_2, x_3)]/[\partial(q_1, q_2, q_3)]$. The determinant $|\partial q_m/\partial x_i|$ is the reciprocal of that appearing in (2.10).

If the inequality (2.10) is satisfied, we may write

$$\delta q_m = \frac{\partial q_m}{\partial x_i} \delta x_i \tag{2.11}$$

and the integration of these equations yields the q's as functions of the x's. It is not necessary to make this integration, as (2.7) can be solved to yield (2.8) directly provided that (2.10) holds.

The idea of the last few paragraphs can be generalized to the case of N particles. The $3N = f'$ coordinates $x_{\rho,i}$ may be considered as functions of f' parameters q_m and perhaps time:

$$x_{\rho,i} = x_{\rho,i}(q_1, q_2, \cdots, q_{f'}, t) \tag{2.12}$$

If the Jacobian determinant does not vanish, i.e.,

$$\frac{\partial(x_{1,1} \cdots x_{N,3})}{\partial(q_1 \cdots q_{f'})} \neq 0 \tag{2.13}$$

the q_m can be expressed as functions of the $x_{\rho,i}$ and time:

$$q_m = q_m(x_{1,1} \cdots x_{N,3}, t) \tag{2.14}$$

The q's are now as effective as the x's in describing the configuration of the system.

In the case of a system of many particles it is often necessary to introduce the generalized coordinates because of the presence in a particular problem of a number—sometimes a very large number—of constraints which prevent the coordinates from changing independently of each other. These constraints may take the form of k relations:

$$\phi_s(x_{1,1} \cdots x_{N,3}, t) = 0 \qquad (s = 1, 2, \cdots, k) \tag{2.15}$$

Thus, for example, the coordinates of the N particles in a rigid structure are not independent, there existing $3N - 6$ such relations between them (see Sec. 47). Any relation of the form (2.15) between the particle coordinates and, possibly, the time is called a *holonomic constraint,* whether it be due to the properties of the system itself (as in the case of a rigid body) or to the conditions under which the system is interacting with its environment (e.g., a bead moving along a smooth wire). If the time t appears explicitly in the equation describing a given constraint, that constraint is said to be *moving* (as when, in the case of the bead on the wire, the wire is forced to move in some specified manner relative to the observer). If t does not appear explicitly in one of the equations (2.15), that constraint is said to be *fixed,* or *workless.*

The advantage of generalized coordinates now appears in the fact that it is possible to introduce these coordinates in such a way that the constraints (2.15) become trivial. There are f' coordinates and k constraints. Let the last k of the generalized coordinates q_{f+1}, $q_{f+2}, \cdots, q_{f+k}$ ($f + k = f' = 3N$) be defined by the equations

$$q_{f+s} = \phi_s(x_{1,1} \cdots x_{N,3}, t) \qquad (s = 1, 2, \cdots, k) \tag{2.16}$$

and let the other f generalized coordinates be defined by

$$q_m = q_m(x_{1,1} \cdots x_{N,3}, t) \qquad (m = 1, 2, \cdots, f) \tag{2.17}$$

Again the Jacobian determinant of the transformation from the $3N = f'$ variables $x_{\rho,i}$ to the f' variables q_{f+s}, q_m must not vanish. The equations of constraint are now simply

$$q_{f+s} = 0 \qquad (s = 1, 2, \cdots, k) \tag{2.18}$$

and the configuration of the system depends in a nontrivial way only on the rest of the q's, i.e., the f variables $q_1, q_2, \cdots, q_f$.

The number $f = 3N - k$ is called the number of *degrees of freedom* of the system. The configuration of the system may be represented by the position of a point in an f-dimensional space which is called the *configuration space* of the system.

There exists another type of constraint which involves conditions only on infinitesimal changes in the coordinates in a way which cannot be integrated to yield relations such as (2.15) between the coordinates themselves. The classic example of this kind of constraint is a sphere constrained to roll without slipping on a plane. Five coordinates are needed to specify the configuration of this system, two to locate the center of the sphere and three to give the orientation of the sphere about its center (Sec. 47). There are two constraints: both components of the velocity of the point of contact of the sphere with the plane must vanish. These constraints are of a differential character. Their nonintegrability means that the sphere can be given any orientation at any position of the center by a displacement which does not violate the constraints, and so no relation between the coordinates can exist. We shall have no occasion to discuss this *nonholonomic* type of constraint.

3. VELOCITY

The coordinates describing a system of particles may vary with time. Consider a system consisting of a single particle. It is described by the values of its cartesian coordinates x_i at each value of the time t. The rate at which the coordinates change with time gives the velocity of the particle. Denoting the cartesian components of velocity by v_i, we have

$$v_i = \frac{dx_i}{dt} \equiv \dot{x}_i \tag{3.1}$$

This may be written in vector notation as

$$\mathbf{v} = \frac{d\mathbf{r}}{dt} \equiv \dot{\mathbf{r}} \tag{3.2}$$

If the system consists of N particles, the cartesian components of the velocity of each particle may be defined as above:

$$v_{\rho,i} = \dot{x}_{\rho,i} \tag{3.3}$$

The quantities $v_{\rho,i}$ may be considered the cartesian components of a vector in the f'-dimensional unconstrained configuration space of the system.

Velocity may be described in terms of the generalized coordinates of Sec. 2. From (2.12) we see that

$$x_{\rho,i} = x_{\rho,i}(q_1 \cdots q_f, t) \tag{3.4}$$

where f replaces f' because we assume the constraints to have been eliminated in the manner of Sec. 2. Then

$$\dot{x}_{\rho,i} = \frac{\partial x_{\rho,i}}{\partial q_m}\dot{q}_m + \frac{\partial x_{\rho,i}}{\partial t} \tag{3.5}$$

If desired, these equations can be solved for the $\dot{q}_m$ in terms of the $\dot{x}_{\rho,i}$ even though the number of equations may be greater than the number of unknowns, since the equations are not independent but must satisfy the constraints.

The cartesian components of velocity are seen to be linear functions of the generalized velocity components no matter how the generalized coordinates are defined. This means that it is easy to express velocities in generalized coordinates. The term $\partial x_{\rho,i}/\partial t$ appears only when there are moving constraints on the system or in the rare cases where it is convenient to introduce moving coordinate axes.

4. ACCELERATION

The velocity components of a system of particles may vary with time. The rate of change of the velocity components gives the acceleration components only when the coordinates are cartesian:

$$a_{\rho,i} = \dot{v}_{\rho,i} = \ddot{x}_{\rho,i} \tag{4.1}$$

If there is only one particle in the system, the acceleration is represented by a vector:

$$\mathbf{a} = \dot{\mathbf{v}} = \ddot{\mathbf{r}} \tag{4.2}$$

If there is more than one particle, the $a_{\rho,i}$ may be considered the cartesian components of a vector in the unconstrained configuration space of the system.

Acceleration components may be given in terms of the generalized coordinates. From (3.5) we obtain

$$\ddot{x}_{\rho,i} = \frac{\partial x_{\rho,i}}{\partial q_m}\ddot{q}_m + \frac{\partial^2 x_{\rho,i}}{\partial q_m\,\partial q_n}\dot{q}_m\dot{q}_n + 2\frac{\partial^2 x_{\rho,i}}{\partial q_m\,\partial t}\dot{q}_m + \frac{\partial^2 x_{\rho,i}}{\partial t^2} \tag{4.3}$$

If the x's do not depend explicitly on the time, which is the usual situation, (4.3) reduces to

$$\ddot{x}_{\rho,i} = \frac{\partial x_{\rho,i}}{\partial q_m}\ddot{q}_m + \frac{\partial^2 x_{\rho,i}}{\partial q_m\,\partial q_n}\dot{q}_m\dot{q}_n \tag{4.4}$$

The cartesian acceleration components are not linear functions of the second derivatives of the generalized coordinates alone, but depend quadratically on the generalized velocity components as well. This quadratic dependence on the generalized velocity components disappears only if all the second derivatives of the x's with respect to the q's vanish, i.e., only if the x's are linear functions of the q's. The terms quadratic in the velocity components, which enter whenever the coordinate curves q_m = const are not straight lines, represent effects like the centripetal and Coriolis accelerations.

Higher derivatives of the coordinates with respect to time could be named and discussed. This proves unnecessary because the laws of mechanics are stated in terms of the acceleration. Even the computation of the components (4.3) of the acceleration in terms of generalized coordinates can become tedious for relatively simple problems. The advantage of introducing generalized coordinates would then seem to be counterbalanced by the algebraic complexity of the acceleration components which are to be inserted in the dynamic law $\mathbf{F} = m\mathbf{a}$. Fortunately, a method due to Lagrange, and discussed in Chapter 6, makes it possible to avoid this difficulty and to write down equations of motion in terms of generalized coordinates without ever having to compute the second time derivatives of these coordinates.

5. SPECIAL COORDINATE SYSTEMS

We may apply the results of the last section to the commonly used special coordinate systems defined by (2.3) and (2.4). In order to describe the motion of a particle in a plane, it may be convenient to introduce the plane polar coordinates r, θ defined by (2.3), r being the length of the radius vector to the particle and θ the angle between the radius vector and the x axis. If the particle is not confined to a plane, but is subjected to forces which possess cylindrical symmetry, i.e., are independent of θ, it is convenient to supplement (2.3) with the extra coordinate

$$q_3 = z$$

The set of coordinates q_1, q_2, q_3 is then called a *cylindrical coordinate* system (see Fig. 5-1). (The symbols r, θ are often replaced by ρ, ϕ to avoid confusion with the quantities r, θ used for spherical polar coordinates.)

When solved for x and y, equations (2.3) give

$$x = r\cos\theta \qquad y = r\sin\theta \tag{5.1}$$

The velocity components are found by differentiating (5.1):

$$\dot{x} = \dot{r}\cos\theta - r\dot{\theta}\sin\theta \qquad \dot{y} = \dot{r}\sin\theta + r\dot{\theta}\cos\theta \tag{5.2}$$

The terms in $\dot{r}$ give the velocity toward or away from the origin, and those in $\dot{\theta}$ give the velocity around the origin. Equations (5.2) may be solved for $\dot{r}$ and $\dot{\theta}$:

$$\dot{r} = \frac{\dot{x}x + \dot{y}y}{(x^2 + y^2)^{1/2}} \tag{5.3}$$
$$\dot{\theta} = \frac{x\dot{y} - y\dot{x}}{x^2 + y^2}$$

The acceleration components are found by differentiating (5.2). The result is

$$\ddot{x} = (\ddot{r} - r\dot{\theta}^2)\cos\theta - (r\ddot{\theta} + 2\dot{r}\dot{\theta})\sin\theta \tag{5.4}$$
$$\ddot{y} = (\ddot{r} - r\dot{\theta}^2)\sin\theta + (r\ddot{\theta} + 2\dot{r}\dot{\theta})\cos\theta$$

This is seen to involve terms quadratic in the velocity components, as was shown would be the case in (4.4). The various terms can be identified. Those in the first parentheses give the acceleration in the

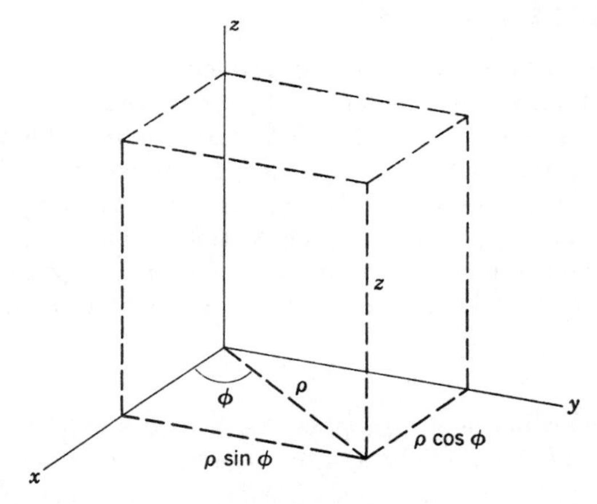

Fig. 5-1. Cylindrical coordinates.

r direction, and those in the second give the acceleration in the θ direction. The r acceleration consists of two parts, that due to the second derivative of r and that due to the centripetal acceleration. The latter arises because the curves of r = const are not lines but circles, and the motion of a particle along a circle with constant speed is an accelerated motion since the direction of the velocity changes continually. The θ acceleration also consists of two parts. The first is due to the second derivative of θ, and the second is the Coriolis acceleration. The latter arises for two reasons: because the linear speed is given by $r\dot{\theta}$, which changes as r changes even if $\dot{\theta}$ is constant, and because even a constant velocity vector has a component in the θ direction which changes as θ changes.

For a particle moving under the influence of a force which possesses spherical symmetry, i.e., is directed towards a fixed point and depends only on the distance of the particle from the point, it is convenient to introduce the spherical coordinates defined by (2.4). When solved for the x's, these yield (see Fig. 5-2)

$$\begin{aligned} x_1 &\equiv x = r \sin \theta \cos \phi \\ x_2 &\equiv y = r \sin \theta \sin \phi \\ x_3 &\equiv z = r \cos \theta \end{aligned} \tag{5.5}$$

The geometrical significance of these coordinates is the following: r is the length of the radius vector of the particle; θ is the angle between

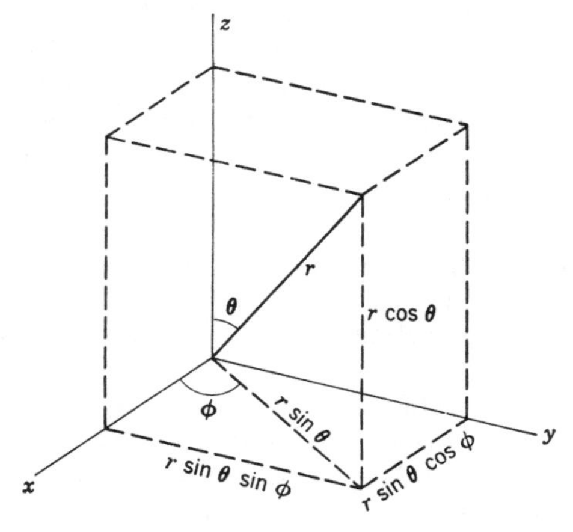

Fig. 5-2. Spherical polar coordinates.

the radius vector and the z axis, or the complement of the latitude of the particle; ϕ is the angle between the plane containing the radius vector and the z axis and the plane containing the x axis and the z axis, or the longitude of the particle.

The velocity components are found by differentiating (5.5):

$$\begin{aligned} \dot{x} &= \dot{r} \sin \theta \cos \phi + r\dot{\theta} \cos \theta \cos \phi - r\dot{\phi} \sin \theta \sin \phi \\ \dot{y} &= \dot{r} \sin \theta \sin \phi + r\dot{\theta} \cos \theta \sin \phi + r\dot{\phi} \sin \theta \cos \phi \\ \dot{z} &= \dot{r} \cos \theta - r\dot{\theta} \sin \theta \end{aligned} \tag{5.6}$$

Equations (5.6) may be solved for $\dot{r}$, $\dot{\theta}$, $\dot{\phi}$ in terms of the coordinates

x, y, z:

$$\dot{r} = \frac{x\dot{x} + y\dot{y} + z\dot{z}}{(x^2 + y^2 + z^2)^{1/2}}, \qquad \dot{\theta} = \frac{(x\dot{x} + y\dot{y})\dfrac{z}{\rho} - \dot{z}\rho}{(x^2 + y^2 + z^2)^{1/2}}$$

$$\dot{\phi} = \frac{x\dot{y} - y\dot{x}}{x^2 + y^2}, \qquad \rho^2 = x^2 + y^2$$

The acceleration components could be found by differentiating (5.6), but this would lead to a multitude of terms; as has been pointed out, it will not be necessary to perform this differentiation.

On rare occasions the analysis is simplified by the introduction of elliptical coordinates or parabolic coordinates to specify the position of a particle.

Elliptical coordinates in the plane are defined by a one-parameter family of confocal ellipses and a one-parameter family of confocal hyperbolas, the latter having the same focuses as the ellipses. Through any point in the plane there pass one and only one ellipse and one and only one hyperbola. The values of the parameters for these two curves are the coordinates of the point. Parabolic coordinates in the plane are defined by two one-parameter families of parabolas with the origin as common focus. One family opens to the right and one to the left. The values of the parameters which give the two parabolas intersecting at a point give the parabolic coordinates of that point.

The equations describing these coordinates are given in Exercises 4, 5, and 6 at the end of this chapter and the coordinate systems are discussed in more detail in Sec. 64, where we consider the problems to which they are especially applicable—the motion of a particle under the influence of two fixed centers of force, and the motion of a particle near one such attracting center in the presence of an external uniform field.

6. VECTOR ALGEBRA

In this section we are concerned with coordinate transformations of a simpler type than those discussed in Sec. 5. We consider a rectangular cartesian set of coordinate axes relative to which the position of a point is (x_1, x_2, x_3) and imagine the axes rotated about the origin to a new orientation, the coordinates of the point relative to this new set of axes being $(\bar{x}_1, \bar{x}_2, \bar{x}_3)$. Once the rotation is specified, the relations between the x_i and the $\bar{x}_j$ become known:

$$x_i = x_i(\bar{x}_1, \bar{x}_2, \bar{x}_3) \qquad (i = 1, 2, 3) \tag{6.1}$$

For this type of coordinate transformation, which is of the general form (2.7), the x_i are linear functions of the $\bar{x}_j$, and vice versa; i.e., equations (6.1) are of the form

$$\bar{x}_i = S_{ij}x_j \tag{6.2}$$

where the S_{ij} are a set of constants. Thus, for example, the reader may verify that if the rotation from the x's to the $\bar{x}$'s were through an

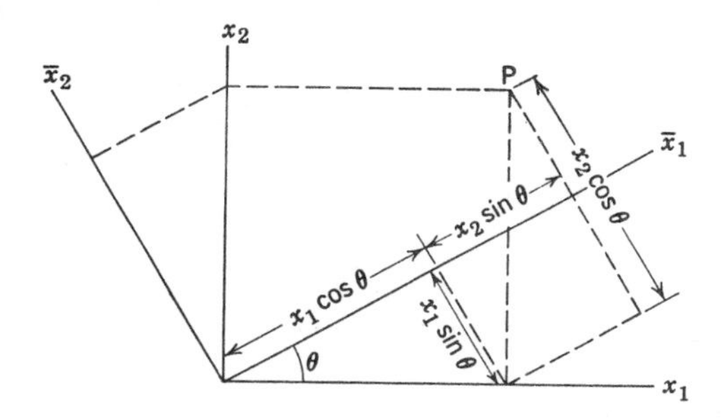

Fig. 6-1. Rotation of the coordinate axes through an angle θ in the (1-2) plane.

angle θ about the 3 axis, the transformation equations would be (see Fig. 6-1)

$$\begin{aligned} \bar{x}_1 &= x_1 \cos\theta + x_2 \sin\theta \\ \bar{x}_2 &= -x_1 \sin\theta + x_2 \cos\theta \\ \bar{x}_3 &= x_3 \end{aligned} \tag{6.3}$$

Although a detailed discussion of the transformation (6.2) is given in Appendix II, we should draw attention here to the fact that (6.2) is of basic significance in defining a vector. If three quantities have the values a_1, a_2, a_3 when referred to the x_i axes, and the values $\bar{a}_1$, $\bar{a}_2$, $\bar{a}_3$ when referred to the $\bar{x}_i$ axes, and if

$$\bar{a}_i = S_{ij}a_j \tag{6.4}$$

i.e., if the three quantities transform under a rotation of the axes in the same way that the coordinates x_i transform, then the set of quantities is called a *vector*.

It is usual to represent a vector by a boldface symbol, **a**, or by a symbol with an arrow over it, $\vec{a}$. Any relation of the form

$$\mathbf{a} = \mathbf{b}$$

between two vectors, if true in one coordinate system, is automatically true in any other coordinate system obtained from the first by a rotation about the origin. We may therefore write relations between vectors without specifying the orientation of the axes to which the vectors are referred. We give below a review of vector algebra with which the student should be familiar.

Given two vectors **a** and **b**, their *scalar product* is defined to be

$$\mathbf{a} \cdot \mathbf{b} = |\mathbf{a}|\,|\mathbf{b}| \cos \theta$$

where $|\mathbf{a}|$ is read "length of **a**," and where θ is the angle between the positive senses of the two vectors. In cartesian components,

$$\mathbf{a} \cdot \mathbf{b} = \delta_{ij} a_i b_j = a_1 b_1 + a_2 b_2 + a_3 b_3 \tag{6.5}$$

Scalar multiplication is commutative and distributive; i.e.,

$$\begin{aligned} \mathbf{a} \cdot \mathbf{b} &= \mathbf{b} \cdot \mathbf{a} \\ \mathbf{a} \cdot (\mathbf{b} + \mathbf{c}) &= \mathbf{a} \cdot \mathbf{b} + \mathbf{a} \cdot \mathbf{c} \end{aligned} \tag{6.6}$$

As the name implies, the result of the scalar multiplication of two vectors is a scalar. The scalar product is sometimes known as the *inner product.*

The *vector product* of two vectors, **a** and **b**, is defined by

$$\mathbf{a} \times \mathbf{b} = |\mathbf{a}|\,|\mathbf{b}| \sin \theta \mathbf{e} \tag{6.7}$$

where the symbols are as above and **e** is a unit vector in the direction perpendicular to the plane of **a** and **b**, drawn in the sense which makes the triple of vectors **a**, **b**, **e** a positive or a right-handed one.

From the definition it follows that the vector product is not commutative. Rather

$$\mathbf{a} \times \mathbf{b} = -\mathbf{b} \times \mathbf{a} \tag{6.8}$$

The distributive law holds:

$$\mathbf{a} \times (\mathbf{b} + \mathbf{c}) = \mathbf{a} \times \mathbf{b} + \mathbf{a} \times \mathbf{c} \tag{6.9}$$

The associative law does not hold, for in general

$$\mathbf{a} \times (\mathbf{b} \times \mathbf{c}) \neq (\mathbf{a} \times \mathbf{b}) \times \mathbf{c}$$

In terms of cartesian coordinates

$$(\mathbf{a} \times \mathbf{b})_1 = a_2 b_3 - a_3 b_2 \tag{6.10}$$

and the other components are obtained by cyclically permuting the indices 1, 2, 3.

Two kinds of triple products occur. The first is

$$\mathbf{a} \cdot (\mathbf{b} \times \mathbf{c}) = (\mathbf{a} \times \mathbf{b}) \cdot \mathbf{c} = \begin{vmatrix} a_1 & a_2 & a_3 \\ b_1 & b_2 & b_3 \\ c_1 & c_2 & c_3 \end{vmatrix} \tag{6.11}$$

This is a scalar, and it gives the volume of the parallelepiped with **a, b, c** as edges. It is clearly unchanged by a cyclic permutation of **a, b, c**. It is positive if the vectors taken in this order form a right-handed triple, and negative if they form a left-handed one.

The second kind of triple product is

$$\mathbf{a} \times (\mathbf{b} \times \mathbf{c}) = (\mathbf{a} \cdot \mathbf{c})\mathbf{b} - (\mathbf{a} \cdot \mathbf{b})\mathbf{c} \tag{6.12}$$

The result is a vector in the plane of **b** and **c**. The order of multiplication is important.

Many formulas of vector analysis become simpler if one introduces the symbols ϵ_{ijk}. These quantities are defined in cartesian coordinates by

$$\epsilon_{ijk} = \begin{Bmatrix} +1 \\ -1 \\ 0 \end{Bmatrix} \tag{6.13}$$

when i, j, k is $\begin{Bmatrix} \text{an even} \\ \text{an odd} \\ \text{no} \end{Bmatrix}$ permutation of 1, 2, 3.

Interchanging any pair of indices changes the sign of the quantity, so that the value of the symbol is zero if any two indices are alike. There are two identities which the reader should verify:

$$\epsilon_{ikl}\epsilon_{imn} = \delta_{mk}\delta_{nl} - \delta_{nk}\delta_{ml} \tag{6.14}$$

$$\epsilon_{ikl}\epsilon_{ikn} = 2\delta_{nl} \tag{6.15}$$

The vector product of two vectors can be expressed in terms of the ϵ's. Thus, if

$$\mathbf{c} = \mathbf{a} \times \mathbf{b}$$

then

$$c_i = \epsilon_{imn}a_m b_n \tag{6.16}$$

This shows that c_i changes sign when the order of **a** and **b** is reversed, since this is equivalent to interchanging m and n on ϵ.

The first of the two triple products becomes

$$\mathbf{a} \cdot (\mathbf{b} \times \mathbf{c}) = a_i(\mathbf{b} \times \mathbf{c})_i = a_i\epsilon_{imn}b_mc_n = \epsilon_{imn}a_ib_mc_n \tag{6.17}$$

The second triple product becomes

$$[\mathbf{a} \times (\mathbf{b} \times \mathbf{c})]_k = a_j\epsilon_{kji}\epsilon_{imn}b_mc_n = (\delta_{mk}\delta_{nj} - \delta_{nk}\delta_{mj})a_jb_mc_n = a_nc_nb_k - a_mb_mc_k \tag{6.18}$$

which coincides with (6.12).

We shall also be concerned with vector quantities, such as electric and magnetic field strengths, which depend on position and time in such a manner that the first and second partial derivatives of their components with respect to x_i, t may be defined. In cartesian coordinates we define the differential operator

$$\boldsymbol{\nabla} \equiv \frac{\partial}{\partial \mathbf{x}}$$

by its components

$$\partial_i \equiv \frac{\partial}{\partial x_i}$$

and use (6.5) and (6.16) to define

$$\boldsymbol{\nabla} \cdot \mathbf{A} \equiv \operatorname{div} \mathbf{A} = \partial_i A_i$$

$$\boldsymbol{\nabla} \times \mathbf{A} \equiv \operatorname{curl} \mathbf{A} = \epsilon_{ikl}\partial_k A_l$$

The identities (6.14) and (6.15) may then be used to deduce relations such as the following. Operating on $\partial_k(a_nb_m)$ by both sides of (6.14), we have

$$\epsilon_{ikl}\epsilon_{imn}\partial_k(a_nb_m) = \partial_k(a_lb_k) - \partial_k(a_kb_l)$$

i.e.,

$$-\epsilon_{ikl}\partial_k(a \times b)_i = \epsilon_{lki}\partial_k(a \times b)_i = [\boldsymbol{\nabla} \times (\mathbf{a} \times \mathbf{b})]_l = a_l\partial_kb_k - b_l\partial_ka_k + b_k\partial_ka_l - a_k\partial_kb_l = [\mathbf{a}\boldsymbol{\nabla} \cdot \mathbf{b} - \mathbf{b}\boldsymbol{\nabla} \cdot \mathbf{a} + \mathbf{b} \cdot \boldsymbol{\nabla}\mathbf{a} - \mathbf{a} \cdot \boldsymbol{\nabla}\mathbf{b}]_l$$

Vector analysis is useful in physics because it de-emphasizes the unimportant question as to which way the axes are pointing, and makes it possible to formulate the laws of physics in a manner which is automatically independent of any preferred direction in space. We should also expect these laws to be independent of whether we choose a right-handed or a left-handed system of axes, since such a choice is purely a matter of convention. However, under a transformation from right- to left-handed axes (e.g., a simple interchange of x_1 and x_2), not all vectors transform in the same manner. Vectors which transform in the way that the displacement vector transforms are called *polar vectors* (e.g., if $\bar{x}_1 = x_2$, $\bar{x}_2 = x_1$, $\bar{x}_3 = x_3$, then, for a polar vector, $\bar{P}_1 = P_2, \bar{P}_2 = P_1, \bar{P}_3 = P_3$). If, then, **a** and **b** are two polar vectors, the relation $\mathbf{a} = \mathbf{b}$, if true for a right-handed set of axes, is automatically true if left-handed axes are used.

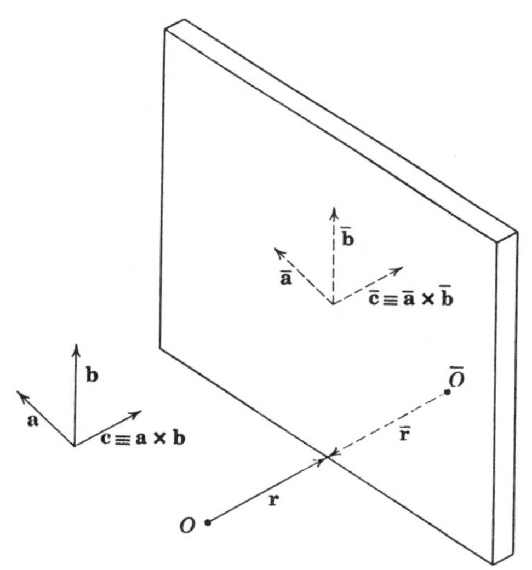

Fig. 6-2. Reflection of vectors in a mirror. The polar vector **r** perpendicular to the mirror is reversed in direction but the axial vector $\mathbf{c} = \mathbf{a} \times \mathbf{b}$ is unchanged, if **c** is defined by (6.7).

There is another class of vectors, called *axial vectors*, which do not transform under reflection in the same way that polar vectors transform. This may be seen by referring to (6.7), in which an asymmetry between left- and right-handedness was introduced in the definition of the vector product

$$\mathbf{c} = \mathbf{a} \times \mathbf{b}$$

To illustrate the difference in transformation properties between the vector **c** defined in this way and the vector **r** denoting the position of a point, we imagine a mirror placed in a plane parallel to that defined by the vectors **a** and **b** and such that the vectors **c** and **r** point toward the mirror. When examined in such a mirror, a right-handed system of axes becomes a left-handed system (see Fig. 6-2). The mirror image

$\bar{\mathbf{r}}$ of $\mathbf{r}$ is then a vector antiparallel to $\mathbf{r}$, but since $\mathbf{a}$ and $\mathbf{b}$ are parallel to the mirror they are unchanged by reflection. If defined as in (6.7), the vector product of the reflections of $\mathbf{a}$ and $\mathbf{b}$ is therefore the same as $\mathbf{a} \times \mathbf{b}$ itself; hence, unlike $\mathbf{r}$, the axial vector $\mathbf{c}$ does not change when examined in the mirror. We should therefore not expect to encounter equations in which a polar vector is equated to an axial vector, for the meaning of the equation would depend on whether we were using a right- or left-handed set of axes. Examples of axial vectors which we shall encounter later include the angular velocity vector and magnetic field strength.

The vector product $\mathbf{a} \times \mathbf{b}$ could alternatively be defined by (6.7), with $\mathbf{a}$, $\mathbf{b}$, $\mathbf{c}$ forming a right-handed triple if right-handed coordinates were used and a left-handed triple if left-handed coordinates were used. This definition would allow (6.16) to hold with ϵ_{imn} given by (6.13) in either case. With this definition, however, the relation $\mathbf{c} = \mathbf{a} \times \mathbf{b}$ between three vectors could not hold if $\mathbf{c}$ were a polar vector, for the validity of the equation would depend on the "handedness" of the coordinate system that was used.

The set of nine quantities that may be formed by taking the products in pairs of the components of two vectors

$$T_{ij} = a_i b_j \tag{6.19}$$

is called the *outer product* of the vectors $\mathbf{a}$ and $\mathbf{b}$. Such a set transforms according to the rule

$$\begin{aligned} \bar{T}_{ij} &= \bar{a}_i \bar{b}_j \\ &= S_{ik} S_{jl} a_k b_l \end{aligned}$$

or

$$\bar{T}_{ij} = S_{ik} S_{jl} T_{kl} \tag{6.20}$$

Any set of quantities T_{ij} which transforms according to (6.20) is called a tensor, even if these quantities cannot be expressed as in (6.19) as the outer product of two vectors. For further discussion of the transformation properties of tensors, the reader is referred to Appendix I.

7. KINEMATICS AND MEASUREMENT

Physics is a science of measurable quantities, and whereas the experimentalist is primarily concerned with making such measurements the theoretician must shape his calculations so that they ultimately lead to a quantitative description of phenomena that can be observed. One important way of developing a theory so that it begins to run parallel to experiment is to define some of the variables which appear in the theory in terms of experiments which could be performed, at least conceptually. This operational viewpoint, emphasized chiefly by Mach and Bridgman, already applies to discussions of kinematics, or the description of motion, before we even concern ourselves with the forces causing that motion. Use of the words "distance" and "time" implies an understanding of how these quantities are to be measured.

Although the explicit recognition of this idea was not necessary for the growth of nonrelativistic classical mechanics, it is generally thought to be essential for an understanding of the reasons why this theory is unable to describe atomic phenomena or the motion of fast particles. Indeed, we take the measurement of time and distance so much for granted that a discussion of how such measurements are made would appear at first sight to be pedantic. It is well known that in order to measure distances we agree on a standard length, give it a name, and compare all other lengths and distances to this standard. Thus, for example, a particular bar of 90% platinum and 10% iridium has been set up at Sèvres so that at 0°C the separation of two marks on this bar is called the international prototype meter. More accurately, the meter has been defined as 1,533,164.13 times the wavelength of a particular red line in the spectrum of cadmium at normal temperature and pressure.

In comparing a length to a copy of the standard bar we say that two objects at rest relative to each other are of the same length when their end points can be brought into coincidence, each to each. Of course if the object to be measured is moving parallel to the measuring stick, it is necessary to add that they are of the same length if their end points can be brought into coincidence *at the same time*. (We cannot measure the length of a moving train by noting the position of the engine and, some time later, that of the caboose!) Thus, in general, the measurement of length is intimately related to the problem of determining whether two events at different places are simultaneous. In nonrelativistic kinematics we do not raise this question, but assume an absolute time by which simultaneity is automatically defined.

Similarly, in comparing two lengths, or measuring the distance between two objects, it is necessary for some signal to pass between the two objects and the observer. Although the signal need not be sufficiently energetic to materially disturb a macroscopic object, the effect such a signal could produce during the measurement of the position of an electron may not always be ignored, as it is in classical mechanics.

Although a standard interval of time cannot be preserved in the same way as a space interval, similar considerations apply to the measurement of time. The basis of such measurements is the occurrence of some regular periodic process, such as the rotation of the earth. Thus the second may be defined by the statement that the earth rotates through 2π radians relative to the stars in 86,162 · · · sec. As long as we are satisfied with an accuracy of one part in 10^5, this definition is sufficient and within this limit the regularity of the earth's rotation agrees with that of other periodic processes—the oscillation of a crystal, or some frequency characteristic of a particular atom. As time develops, however, tidal forces will cause the earth to slow down relative to a clock based on atomic vibrations so that time intervals are more accurately defined in terms of an atomic clock, which is not subject to such vagaries. Measurement of time intervals with an accuracy of one part in 10^{11} is presently possible by such high stability oscillators, and frequency differences of very narrow line-width gamma rays are being measured with an over-all accuracy that is even greater.

As in the case of length measurements, it is assumed in nonrelativistic classical theory that the magnitude of a time interval is independent of the motion of the observer who measures it, and that signals used in measuring time intervals between two events do not disturb those events. Again these assumptions are denied by the theories of relativity and quantum mechanics respectively. It may be that there are other assumptions about the nature of measurement made throughout all of theoretical physics that are not valid in the realm of high energy and nuclear phenomena.

GENERAL REFERENCES

P. W. Bridgman, *The Logic of Modern Physics.* Macmillan, New York (1927).

A. d'Abro, *The Evolution of Scientific Thought.* Dover, New York (1950).

H. Goldstein, *Classical Mechanics.* Addison-Wesley, Reading (1950).

R. B. Lindsay, *Concepts and Methods of Theoretical Physics.* Van Nostrand, New York (1951).

R. B. Lindsay and H. Margenau, *The Foundations of Physics.* Wiley, New York (1936).

E. Mach, *The Science of Mechanics.* Open Court, Chicago (1907).

C. Weatherburn, *Elementary Vector Analysis. Advanced Vector Analysis.* G. Bell, London (1924).

E. T. Whittaker, *Analytical Dynamics.* Dover, New York (1944).

EXERCISES

1. A particle is constrained to move with constant speed in the circle $r = a$. Find the cartesian and polar coordinates of its velocity and acceleration.

2. A particle is constrained to move with constant speed on the ellipse

$$a_{ij}x_ix_j = 1 \qquad (i, j = 1, 2)$$

Find the cartesian and polar components of its acceleration.

3. In Exercise 2 introduce as a generalized coordinate the angle θ between the radius vector of the particle and the major axis of the ellipse, and find the velocity and acceleration in terms of θ.

4. A possible pair of coordinates of a point with cartesian coordinates (x, y) is furnished by the two roots of the equation

$$\frac{x^2}{q - e_1} + \frac{y^2}{\bar{q} - e_2} = 1 \qquad (e_1 > e_2)$$

Show that these roots q_1, q_2 satisfy the inequalities

$$q_1 \geq e_1 \geq q_2 \geq e_2$$

Also show that the curves $q_1 =$ const form a family of confocal ellipses, that the curves $q_2 =$ const form a family of hyperbolas confocal with the ellipses, and that curves of these two families form an orthogonal net.

5. A possible pair of coordinates of a point with cartesian coordinates (x, y) is furnished by q_1 and q_2 defined by the equations

$$x = \cosh q_1 \cos q_2, \qquad y = \sinh q_1 \sin q_2$$

Show that these are also a set of elliptic coordinates, and find q_1, q_2 as functions of $\bar{q}_1$, $\bar{q}_2$.

6. A possible pair of coordinates of a point with cartesian coordinates (x, y) is furnished by

$$x = \tfrac{1}{2}(q_1{}^2 - q_2{}^2), \qquad y = q_1q_2$$

Show that the curves $q_1 =$ const form a family of parabolas with the origin as focus and opening to the left, and that the curves $q_2 =$ const form a similar family opening to the right.

7. Prove that

$$(\mathbf{a} \times \mathbf{b}) \cdot (\mathbf{c} \times \mathbf{d}) = (\mathbf{a} \cdot \mathbf{c})(\mathbf{b} \cdot \mathbf{d}) - (\mathbf{a} \cdot \mathbf{d})(\mathbf{b} \cdot \mathbf{c})$$

8. Show that the component of a vector **a** in the direction orthogonal to a vector **b** is

$$\mathbf{a} - \frac{(\mathbf{a} \cdot \mathbf{b})\mathbf{b}}{b^2} = \frac{1}{b^2}[\mathbf{b} \times (\mathbf{a} \times \mathbf{b})]$$

9. Prove the following identities:
(*a*) $\nabla(\mathbf{a} \cdot \mathbf{b}) = (\mathbf{a} \cdot \nabla)\mathbf{b} + (\mathbf{b} \cdot \nabla)\mathbf{a} + \mathbf{a} \times (\nabla \times \mathbf{b}) + \mathbf{b} \times (\nabla \times \mathbf{a})$
(*b*) $\nabla \cdot (\mathbf{a} \times \mathbf{b}) = \mathbf{b} \cdot (\nabla \times \mathbf{a}) - \mathbf{a} \cdot (\nabla \times \mathbf{b})$

(c) $\nabla \times \nabla\phi = 0$
(d) $\nabla \times (\nabla \times \mathbf{a}) = \nabla(\nabla \cdot \mathbf{a}) - \nabla^2\mathbf{a}$

10. A system of orthogonal axes $\bar{x}_i$ rotates about the $\bar{x}_3$ axis with constant angular velocity ω relative to the orthogonal system x_j, the x_3 and $\bar{x}_3$ axes being coincident, and the two systems having a common origin. Determine relations of the form (2.7) between measurements made in these coordinate systems of the components of the position of a particle relative to the origin.

11. The transformation $\bar{x}_j = -x_j$ $(j = 1, 2, 3)$ describes a reflection in the origin. Draw a diagram to illustrate how two radius vectors $\mathbf{r}_1$, $\mathbf{r}_2$ and their vector product $\mathbf{r}_1 \times \mathbf{r}_2$ transform under this reflection.

12. Show that under double reflection in two mirrors, one in the x-z plane and the other in the x-y plane, an axial vector transforms in the same way as a polar vector.

13. Show that the vector product of two axial vectors is an axial vector, and that the vector product of an axial vector with a polar vector is a polar vector.

14. The maximum acceleration of a train is α and its maximum retardation β. Show that it cannot run a distance a from rest to rest in a shorter time than

$$\left[\frac{2a(\alpha + \beta)}{\alpha\beta}\right]^{1/2}$$

15. A body moving in a straight line with uniform acceleration passes two consecutive equal spaces, each of length a, in times t_1, t_2. Show that its acceleration is

$$\frac{2a(t_1 - t_2)}{t_1t_2(t_1 + t_2)}$$

16. A car is moving with constant velocity v. Find the velocities of the points on the tread of the tire which are at distance y from the road ($y \leq 2r$; r = radius of wheel).

2 THE LAWS OF MOTION

8. MASS

In order to understand the motion of a system of particles it is necessary to consider the environment of that system—potentially all of the other particles in the universe—and learn how that environment influences the motion of the system in question. We begin by considering two particles which influence each other's motion but which move in such a manner that we may reasonably expect all the other matter in the universe to have a negligible effect on their relative motion. Thus, for example, we may imagine two masses connected by a small spring and free to move on a smooth horizontal table. We should expect that the matter in the earth would not affect the motion of the masses in the plane of the table, and that extraterrestial matter would be too far away to have anything but a negligible effect. It is found that under such conditions, if $\mathbf{a}_A$ and $\mathbf{a}_B$ are the accelerations of the two particles A, B, then these vectors are parallel and in the opposite sense, and that the ratio of the magnitudes of $\mathbf{a}_A$ and $\mathbf{a}_B$ is a constant for a given pair of particles. This ratio is called the ratio of the masses of the two particles

$$\frac{m_A}{m_B} = \frac{|\mathbf{a}_B|}{|\mathbf{a}_A|} \tag{8.1}$$

It is also found that, if mass C is allowed to interact with A in the absence of B, not only is it true that $\mathbf{a}_A$ and $\mathbf{a}_C$ are parallel and in the opposite sense but that $\frac{m_A}{m_C} = \frac{|\mathbf{a}_C|}{|\mathbf{a}_A|}$ is identical with the ratio $\frac{m_A/m_B}{m_C/m_B}$ determined by comparing A and B in the absence of C and C and B in the absence of A.

Thus with each particle there may be associated a mass that has a unique meaning no matter how many stages it goes through in being compared to another and which may therefore eventually be compared with a standard mass of platinum called the international prototype kilogram, which is preserved in Sèvres. If in an experiment the vectors $\mathbf{a}_A$ and $\mathbf{a}_B$ were found not to be parallel, the two particles would be considered as not acting on each other alone, and the discrepancy would be attributed to the influence of another particle or system.

Thus, somewhat arbitrarily, the mks (meter-kilogram-second) system of units for length, mass, and time has come into being, with its alternatives the cgs system (centimeter-gram-second) and the British system based on feet, slugs, and seconds. It is of course meaningless to write down a physical quantity without noting the units in terms of which it is expressed.

In many problems it is convenient to abandon these arbitrary units and to introduce new units which are especially suited to the problem and which are sometimes called *natural units.* Thus we may take one mass appearing in the problem as the unit of mass, and similarly one length and one interval of time may be taken as the units of length and time respectively. Other physical quantities appearing in the problem may then be expressed in terms of these units, with a consequent simplification of many of the equations. At the end of the calculation it is then usually necessary to express the result in terms of one of the standard systems of units.

9. MOMENTUM AND FORCE

The *momentum* of a particle acted upon by mechanical, gravitational, or electrostatic forces is defined to be the product of its mass and its velocity

$$\mathbf{p} = m\mathbf{v} \tag{9.1}$$

The force $\mathbf{F}$ acting on a particle is defined by the rate of change of momentum it produces:

$$\mathbf{F} = \dot{\mathbf{p}} \tag{9.2}$$

Although introduced here as a definition, this statement is usually referred to as *Newton's second law of motion.* If the mass of the particle is constant in time, then

$$\dot{\mathbf{p}} = m\dot{\mathbf{v}} = m\mathbf{a} \tag{9.3}$$

where $\mathbf{a}$ is the acceleration vector. Thus in this case the force on the particle may be defined by

$$\mathbf{F} = m\mathbf{a} = m\ddot{\mathbf{r}} \tag{9.4}$$

Thus if $\mathbf{F} = 0$, the velocity of the particle is constant, a result known as *Newton's first law of motion.*

From the definition (9.2) of $\mathbf{F}$, it follows that since $\mathbf{p}$ is a vector, so also is $\mathbf{F}$. Thus, if the force $\mathbf{F}$ is the sum of two forces $\mathbf{F}_1$ and $\mathbf{F}_2$, this sum must be understood as a vector sum. This constitutes the parallelogram law for the composition of forces.

As stated above, the parallelogram law is a trivial mathematical fact. It acquires physical significance in those cases where the force between two particles is independent of the presence of other particles. Consider a system of three particles ρ, σ, τ. The force on particle ρ when σ is present and τ is absent we denote by $\mathbf{F}^{(\sigma\rightarrow\rho)}$, and similarly for $\mathbf{F}^{(\tau\rightarrow\rho)}$. Then, with both σ and τ present,

$$\mathbf{F} = \mathbf{F}^{(\sigma\rightarrow\rho)} + \mathbf{F}^{(\tau\rightarrow\rho)} \tag{9.5}$$

the usual parallelogram law. The law is valid only when the various forces are independent. This independence does exist for most of the forces met with in mechanics, such as gravitation and the forces between charged particles. It does not exist between polarizable molecules moving in electric fields, for the induced electric moment of a molecule depends on the field at the location of the molecule, and this depends on the location of other molecules and the fields at those locations. The forces between nuclear particles may be of this *many-body* character rather than simple two-body forces.

The *equations of motion* of a particle, as (9.2) and (9.4) are called, are a set of ordinary differential equations of the second order. If the forces are given as functions of position and time, the values of the coordinates and of the velocity components at a given time t_0 determine the solution of the equations uniquely and thus determine the whole future course of the motion. The combination of initial coordinates and initial velocity components is termed the *initial state* of the system. The future states of a system are determined by the state at any given time and by the equations of motion.

When two particles exert forces on each other, as they are made to do in the measurement of their mass ratio, we have,

$$m_A\mathbf{a}_A = -m_B\mathbf{a}_B \tag{9.6}$$

Thus when two particles exert forces on each other, these forces are equal in magnitude and opposite in direction. This is the *third law of Newton* that *action is equal and opposite to reaction.*

The total momentum of a system of particles is defined to be the sum of the momenta $\mathbf{p}_\rho$ of the individual particles constituting the system:

$$\mathbf{p} = \sum_\rho \mathbf{p}_\rho \tag{9.7}$$

The total or resultant force $\mathbf{F}$ is then given by

$$\mathbf{F} = \dot{\mathbf{p}} = \sum_\rho \dot{\mathbf{p}}_\rho = \sum_\rho \mathbf{F}_\rho \tag{9.8}$$

where $\mathbf{F}_\rho$ is the force acting on the ρth particle. *If the resultant force* $\mathbf{F}$ *vanishes, the total linear momentum is constant.*

If the forces acting on the particles of the system are additive, i.e., if (9.5) is valid, we may divide the force $\mathbf{F}_\rho$ acting on the ρth particle into two parts: an internal force due to other particles in the system, ($\mathbf{F}_\rho^{\text{int}}$), and an external force due to influences outside the system, ($\mathbf{F}_\rho^{\text{ext}}$):

$$\mathbf{F}_\rho = \mathbf{F}_\rho^{\text{int}} + \mathbf{F}_\rho^{\text{ext}} \tag{9.9}$$

From Newton's third law of motion it follows that the forces $\mathbf{F}_\rho^{\text{int}}$, if summed over all the particles, cancel in pairs, so that

$$\sum_\rho \mathbf{F}_\rho^{\text{int}} = 0$$

and (9.8) may be written

$$\dot{\mathbf{p}} = \sum_\rho \mathbf{F}_\rho^{\text{ext}} = \mathbf{F}^{\text{ext}} \tag{9.10}$$

If the forces are not additive, we may *define* the total external force acting by (9.10). Thus the *rate of change of the total linear momentum of a system is equal to the total* external *force acting on the system.*

If a system consists of N particles, the ρth particle having mass m_ρ and radius vector $\mathbf{r}_\rho$, then the radius vector $\mathbf{r}$ of the *center of mass* of the system is defined by

$$\mathbf{r} = \sum_\rho \frac{m_\rho \mathbf{r}_\rho}{m} \tag{9.11}$$

where

$$m = \sum_\rho m_\rho$$

is the total mass. If the masses do not vary with time, it follows from (9.7), (9.10), and (9.11) that

$$\mathbf{F}^{\text{ext}} = \dot{\mathbf{p}} = \sum_{p} m_p \ddot{\mathbf{r}}_p = m\ddot{\mathbf{r}} \tag{9.12}$$

The total momentum of a system of particles is its total mass multiplied by the velocity of its center of mass. From (9.12) *the center of mass of the system moves as though it were a particle of mass equal to the total mass of the system, acted on by the resultant of the external forces acting on the system.* A special case of this result is that the center of mass of an isolated system moves with constant velocity.

These results are absolutely essential if the theory is to be consistent. We have defined a particle as a system whose internal constitution is irrelevant to its "motion as a whole." We now define the "motion as a whole" to be the motion of the center of mass. Under circumstances where the system can be treated as a particle, it must still be possible to treat it as a composite system without changing the conclusions. Equation (9.12) shows that this is indeed the case.

10. KINETIC ENERGY

The resultant force acting on a particle is defined by (9.2), the mass of the particle being assumed constant:

$$\mathbf{F} = m\ddot{\mathbf{r}} \tag{10.1}$$

Scalar multiplication of both sides of (10.1) by $d\mathbf{r}$ yields

$$\begin{aligned} \mathbf{F} \cdot d\mathbf{r} &= m\ddot{\mathbf{r}} \cdot d\mathbf{r} \\ &= m\dot{\mathbf{r}} \cdot d\dot{\mathbf{r}} \\ &= d(\tfrac{1}{2}m\dot{r}^2) \\ &= dT \end{aligned} \tag{10.2}$$

The quantity on the left of (10.2) is called the *increment of work* done by the force $\mathbf{F}$ in the displacement $d\mathbf{r}$. The quantity on the right is the differential of a function of the velocity called the *kinetic energy* of the particle. Thus between two points A and B the work done on the particle by the resultant force acting on it may be said to appear as kinetic energy of the particle:

$$\begin{aligned} \int_A^B \mathbf{F} \cdot d\mathbf{r} &= (\tfrac{1}{2}m\dot{\mathbf{r}}^2)_B - (\tfrac{1}{2}m\dot{\mathbf{r}}^2)_A \\ &= T_B - T_A \end{aligned} \tag{10.3}$$

If $\mathbf{F} \cdot d\mathbf{r}$ is negative, i.e., if the displacement takes place against the force, the particle is said to do work and the kinetic energy of the particle decreases.

The kinetic energy of a system of particles is the sum of the individual kinetic energies of the particles. The total kinetic energy of an isolated system of particles is not necessarily constant. Even though the sum of the forces acting on the system is zero, in finding the change in kinetic energy the forces are multiplied by various $d\mathbf{r}$, and the sum of these products does not vanish in general.

The kinetic energy of a system of particles can be represented in the unconstrained configuration space of the system:

$$T = \frac{1}{2} \sum_{\rho,i} m_\rho (\dot{x}_{\rho,i})^2 \tag{10.4}$$

This expression becomes considerably simpler if we introduce new rectangular coordinates $\bar{x}_{\rho,i}$ by the equation

$$\bar{x}_{\rho,i} = \sqrt{m_\rho}\, x_{\rho,i} \tag{10.5}$$

Then

$$T = \frac{1}{2} \sum_{\rho,i} (\dot{\bar{x}}_{\rho,i})^2 \tag{10.6}$$

and the masses have been eliminated. Thus we can say that our system is represented by a single particle of unit mass whose position is described by cartesian coordinates $\bar{x}_{\rho,i}$. The separation of two neighboring points in this configuration space ds is given by

$$ds^2 = \sum_{\rho} (d\bar{x}_{\rho,i})^2 \tag{10.7}$$

This is related to the kinetic energy, and we may write it as

$$ds^2 = 2T\, dt^2 \tag{10.8}$$

This equation is independent of the coordinate system used and can be taken as the definition of the arc length in the configuration space of the generalized coordinates q_m (cf. the remarks in Sec. 2).

11. POTENTIAL ENERGY

If the value of the force which a particle would experience at any point of a region of space is given, a force field is said to exist in that region. The value of the force at a given point may vary with time. If it does, we speak of a varying or an unsteady field.

In many physical problems the force has the characteristic that, at a given value of time, the line integral of the force around a closed path vanishes:

$$\oint \mathbf{F}(\mathbf{r}, t) \cdot d\mathbf{r} = 0 \tag{11.1}$$

Because this integral must be evaluated at a given time, it says nothing about what happens in the actual displacement of a particle around a closed path unless the field is a steady one and the time does not appear explicitly in (11.1).

An immediate consequence of (11.1) is that the integral

$$\int_{r_0}^{r} \mathbf{F}(\mathbf{r}, t) \cdot d\mathbf{r}$$

evaluated at a fixed time is independent of the path of integration and therefore defines a function of $\mathbf{r}$, the potential energy V.

$$\int_{r_0}^{r} \mathbf{F}(\mathbf{r}, t) \cdot d\mathbf{r} = -V(\mathbf{r}, t) + V(\mathbf{r}_0, t) \tag{11.2}$$

Here $V(\mathbf{r}_0, t)$ may be considered as a constant of integration; it is an arbitrary number, and $\oint \mathbf{F}(\mathbf{r}, t) \cdot d\mathbf{r} = 0$, where the integration is taken over any closed path at constant t.

Differentiating (11.2) with respect to $\mathbf{r}$, remembering that the time is treated as a constant, we obtain

$$\begin{aligned} \mathbf{F}(\mathbf{r}, t) \cdot d\mathbf{r} &= -dV(\mathbf{r}, t) \Big|_{t=\text{const}} \\ &= -\frac{\partial V}{dx_i} dx_i \end{aligned} \tag{11.3}$$

We introduce the vector operation $\boldsymbol{\nabla}$ with cartesian components defined by

$$\boldsymbol{\nabla}_i = \partial_i = \frac{\partial}{\partial x_i} \tag{11.4}$$

Thus

$$\frac{\partial V}{\partial x_i} = \partial_i V \tag{11.5}$$

Then (11.3) can be written as

$$\mathbf{F} \cdot d\mathbf{r} = -\boldsymbol{\nabla} V \cdot d\mathbf{r} \tag{11.6}$$

Since this holds for arbitrary $d\mathbf{r}$, we can conclude that

$$\mathbf{F} = -\nabla V \tag{11.7}$$

The vector ∇V is called the gradient of V.

Vector fields which can be expressed as the gradient of a scalar function are termed *irrotational* or *lamellar* fields. A lamellar field which is independent of the time is called a *conservative* field. The reason for this name is that in such a field the quantity $T + V$, the kinetic energy plus the potential energy, is constant in time or is conserved.

$$\begin{aligned} \frac{d}{dt}(T + V) &= \frac{dT}{dt} + \frac{dV}{dt} \\ &= \mathbf{F} \cdot \dot{\mathbf{r}} + \nabla V \cdot \dot{\mathbf{r}} + \frac{\partial V}{\partial t} \\ &= \frac{\partial V}{\partial t} \end{aligned} \tag{11.8}$$

Thus if V does not contain the time explicitly, the quantity on the left in (11.8) is constant in time.

The argument just given for the case of a single particle can be generalized to fit the case of many particles. If the sum of the integrals around closed paths at a particular value of t vanishes, there exists a function of the coordinates of the various particles such that

$$\sum_\rho \mathbf{F}_\rho \cdot d\mathbf{r}_\rho = -\sum_\rho \nabla_\rho V \cdot d\mathbf{r}_\rho \tag{11.9}$$

where ∇_ρ is the gradient operator operating on $\mathbf{r}_\rho$ only, leaving all the other $\mathbf{r}$'s unchanged. Since (11.9) is to hold for any set of $d\mathbf{r}_\rho$, it follows that

$$\mathbf{F}_\rho = -\nabla_\rho V \tag{11.10}$$

In the case of many particles also, if V does not contain the time explicitly, then

$$\frac{d}{dt}\Big(\sum_\rho T_\rho + V\Big) = 0 \tag{11.11}$$

and the sum is constant. If V does contain the time explicitly, then as in (11.8)

$$\frac{d}{dt}\left(\sum_{\rho} T_{\rho} + V\right) = \frac{\partial V}{\partial t} \tag{11.12}$$

The remaks made about many-body forces in Sec. 9 may be illustrated by use of the potential function for a system of particles. If the force on a given particle of the system can be represented as the sum of the forces that would act on the particle if the other particles were to act one at a time, and if each of these forces is derivable from a potential, the potential function of the system can be represented as a sum of functions each of which depends on the coordinates of two particles only:

$$V = \sum_{\rho \neq \sigma} V_{\rho\sigma}(\mathbf{r}_{\rho}, \mathbf{r}_{\sigma}, t) \tag{11.13}$$

From the point of view of general dynamic theory there is no need to restrict the potential to be of this form. Any differentiable function of the coordinates is admissible, and even the condition of differentiability can be weakened if the utmost generality is desired. Thus a function $V(\mathbf{r}_1, \mathbf{r}_2, \cdots, r_N, t)$ is a possible potential function even when it cannot be expressed in the form (11.13). With a potential of this more general form it is impossible to give a unique meaning to "the force on the ρth particle due to the σth."

12. CONSERVATION OF ENERGY

In an isolated system the sum $T + V$ of the kinetic and potential energies is not necessarily constant. There may be forces acting which do not fulfill (11.1) or its analog for the case of many particles. Among such forces we can name friction and the radiation damping of charged particles. Some examples of this type of force are discussed in Chapter 5.

If, however, the system is isolated, it is believed that the *total* energy of the system is constant. This total energy may include the thermal motion of particles (which is connected with the mechanical motion through friction), light waves and other electromagnetic fields (connected with the mechanical motion through the electric charge), and mechanical mass (which represents a form of energy according to the theory of relativity). If the hitherto known forms of energy are not conserved in an isolated system, it is usually taken as an indication of a new form of energy rather than as a violation of the principle of the conservation of energy. The neutrino in the theory of beta decay is an outstanding example of this interpretation of conservation laws.

13. ANGULAR MOMENTUM

If $\mathbf{A}_\rho$ is a vector passing through the point $\mathbf{r}_\rho$, the vector $\mathbf{r}_\rho \times \mathbf{A}_\rho$ is called the *moment* of the vector $\mathbf{A}_\rho$ about the origin of the vector $\mathbf{r}_\rho$.

The moment of the momentum of a system of particles about a given point is called its *angular momentum* about that point. We show below that, *if the total momentum is zero, the angular momentum is independent of the point about which the moments are taken.*

Let us take the reference point to be the origin, and let $\mathbf{r}_\rho$ and $\mathbf{p}_\rho$ be the radius vector and the momentum of the ρth particle respectively. Then the angular momentum of the ρth particle about the origin is

$$\mathbf{j}_\rho = \mathbf{r}_\rho \times \mathbf{p}_\rho \tag{13.1}$$

The total angular momentum of a system of particles about a point is the sum of the individual angular momenta taken about that common point:

$$\mathbf{j} = \sum_\rho \mathbf{j}_\rho \tag{13.2}$$

Let us denote by a prime moments taken about a fixed point with radius vector $\mathbf{a}$ (= const) rather than about the origin. Then

$$\mathbf{j}_\rho' = (\mathbf{r}_\rho - \mathbf{a}) \times \mathbf{p}_\rho \tag{13.3}$$

and

$$\begin{aligned} \mathbf{j}' &= \sum_\rho (\mathbf{r}_\rho - \mathbf{a}) \times \mathbf{p}_\rho \\ &= \sum_\rho \mathbf{r}_\rho \times \mathbf{p}_\rho - \mathbf{a} \times \Sigma \mathbf{p}_\rho \end{aligned} \tag{13.4}$$

which proves the theorem since $\mathbf{j}$ is independent of $\mathbf{a}$ if the total momentum vanishes.

From the definition (13.2) of angular momentum it follows that

$$\begin{aligned} \frac{d}{dt}\mathbf{j} &= \frac{d}{dt}\sum_\rho \mathbf{r} \times \mathbf{p}_\rho \\ &= \frac{d}{dt}\Sigma m_\rho \mathbf{r}_\rho \times \dot{\mathbf{r}}_\rho \\ &= \sum_\rho \mathbf{r}_\rho \times m_\rho \ddot{\mathbf{r}}_\rho \end{aligned}$$

since $\dot{\mathbf{r}}_\rho \times \dot{\mathbf{r}}_\rho = 0$. Thus

$$\frac{d}{dt}\mathbf{j} = \sum_\rho \mathbf{r}_\rho \times \mathbf{F}_\rho \tag{13.5}$$

or the rate of change of the angular momentum about a point is equal to the sum of the moments of the forces about that point.

If, as in (9.9), we may break up the force $\mathbf{F}_\rho$ on the ρth particle into two parts,

$$\frac{d}{dt}\mathbf{j} = \sum_\rho \mathbf{r}_\rho \times \mathbf{F}_\rho^{\text{ext}} + \sum_\rho \mathbf{r}_\rho \times \mathbf{F}_\rho^{\text{int}} \tag{13.6}$$

we may show that under certain conditions the second term vanishes; i.e., *the rate of change of the angular momentum of a system about a point is equal to the sum of the moments of the* external *forces about that point.*

We shall prove this theorem here only where the force on a particle of the system is the sum of the forces the particle would experience if the other particles acted one at a time, and where the force between a pair of particles acts along the line joining them.

If the force on particle ρ due to particle σ acts in the line joining them, it can be represented as

$$\mathbf{F}^{(\sigma\to\rho)} = \alpha_{\sigma\rho}(\mathbf{r}_\sigma - \mathbf{r}_\rho) \tag{13.7}$$

since the vector $\mathbf{r}_\sigma - \mathbf{r}_\rho$ is the line segment joining the particles. Furthermore, since

$$\mathbf{F}^{(\sigma\to\rho)} = -\mathbf{F}^{(\rho\to\sigma)} \tag{13.8}$$

we must have

$$\alpha_{\rho\sigma} = \alpha_{\sigma\rho} \tag{13.9}$$

Thus, since $\mathbf{F}_\rho^{\text{int}} = \sum_\sigma \mathbf{F}^{(\sigma\to\rho)}$,

$$\begin{aligned} \sum_\rho \mathbf{r}_\rho \times \mathbf{F}_\rho^{\text{int}} &= \sum_\rho \sum_\sigma \mathbf{r}_\rho \times \alpha_{\sigma\rho}(\mathbf{r}_\sigma - \mathbf{r}_\rho) \\ &= \sum_\rho \sum_\sigma \alpha_{\sigma\rho}\mathbf{r}_\rho \times \mathbf{r}_\sigma \end{aligned} \tag{13.10}$$

since $\mathbf{r}_\rho \times \mathbf{r}_\rho = 0$. By (13.9) this may also be written

$$\begin{aligned} \sum_\rho \mathbf{r}_\rho \times \mathbf{F}_\rho^{\text{int}} &= \sum_\rho \sum_\sigma \alpha_{\rho\sigma}\mathbf{r}_\rho \times \mathbf{r}_\sigma \\ &= \sum_\sigma \sum_\rho \alpha_{\sigma\rho}\mathbf{r}_\sigma \times \mathbf{r}_\rho \end{aligned}$$

by relabeling the indices. This is the same as (13.10) except for a change in sign; hence

$$\sum_\rho \mathbf{r}_\rho \times \mathbf{F}_\rho^{\text{int}} = -\Sigma \mathbf{r}_\rho \times \mathbf{F}_\rho^{\text{int}} = 0$$

thus proving the theorem. The extension of this result to more general conditions will be considered later (Sec. 34).

A particular consequence of this theorem, when supplemented by (9.12), is that the necessary conditions for the *equilibrium* of a system are

$$\sum \mathbf{F}_\rho^{\text{ext}} = 0 \tag{13.11}$$

and

$$\sum_\rho \mathbf{r}_\rho \times \mathbf{F}_\rho^{\text{ext}} = 0 \tag{13.12}$$

More generally, if these conditions are satisfied the total momentum and total angular momentum of the system are both constants of the motion. Thus if the external force $\mathbf{F}_\rho^{\text{ext}}$ acting on the ρth particle is always parallel to the vector $\mathbf{r}_\rho$ joining that particle to a fixed origin (as, for example, the force which the sun exerts on the individual particles of the earth), then the total angular momentum of the system about that origin is a constant. It also follows that if a system is completely *isolated*, i.e., $\mathbf{F}_\rho^{\text{ext}} = 0$ for each particle in the system, the total momentum and total angular momentum of the system are both constants of the motion.

14. RIGID BODY ROTATING ABOUT A FIXED POINT

If a system of particles were to move subject to the constraint that *each particle remained at a fixed distance from the origin,* it would be required that

$$\frac{d}{dt}(r_\rho)^2 = 2\mathbf{r}_\rho \cdot \dot{\mathbf{r}}_\rho = 0$$

for each value of ρ, i.e., the velocity of each particle would be at right angles to its radius vector. This condition may be expressed thus:

$$\dot{\mathbf{r}}_\rho = \boldsymbol{\omega}_\rho \times \mathbf{r}_\rho \tag{14.1}$$

where the axial vector $\boldsymbol{\omega}_\rho$ so introduced is called the *angular velocity* of particle ρ about the origin, although it is not possible to define it uniquely for a single particle.

If the system is such that the distance between *any* two particles ρ, σ is constant, i.e.,

$$(14.2) \qquad (\mathbf{r}_\rho - \mathbf{r}_\sigma) \cdot (\dot{\mathbf{r}}_\rho - \dot{\mathbf{r}}_\sigma) = 0$$

then the system is called a *rigid body*. The definition is usually restricted to the cases for which there are at least three non-collinear particles present.

For a rigid body rotating around a fixed point, it follows from (14.1) and (14.2) that

$$(14.3) \qquad (\boldsymbol{\omega}_\rho - \boldsymbol{\omega}_\sigma) \cdot (\mathbf{r}_\rho \times \mathbf{r}_\sigma) = 0$$

In order to satisfy this equation for each pair of particles it is necessary that

$$(14.4) \qquad \boldsymbol{\omega}_\rho = \boldsymbol{\omega} + \epsilon_\rho \mathbf{r}_\rho$$

where $\boldsymbol{\omega}$ *is independent of* ρ, and ϵ_ρ is an undetermined number. From (14.1) we may then write

$$(14.5) \qquad \dot{\mathbf{r}}_\rho = \boldsymbol{\omega} \times \mathbf{r}_\rho$$

since the extra term in (14.4) involving the unknown ϵ_ρ gives no contribution. The vector $\boldsymbol{\omega}$ is now the angular velocity of the rigid body as a whole.

We shall denote the angular momentum of a rigid body rotating with one point fixed by the vector $\boldsymbol{\sigma}$, defined as in (13.1) and (13.2):

$$(14.6) \qquad \boldsymbol{\sigma} = \sum_\rho m_\rho (\mathbf{r}_\rho \times \dot{\mathbf{r}}_\rho)$$

Thus, from (14.5),

$$(14.7) \qquad \boldsymbol{\sigma} = \sum_\rho m_\rho [\mathbf{r}_\rho \times (\boldsymbol{\omega} \times \mathbf{r}_\rho)]$$

$$= \sum_\rho m_\rho [r_\rho{}^2 \boldsymbol{\omega} - (\mathbf{r}_\rho \cdot \boldsymbol{\omega}) \mathbf{r}_\rho]$$

This equation may be alternatively written in the form

$$(14.8) \qquad \sigma_i = J_{ij} \omega_j$$

where

$$(14.9) \qquad J_{ij} = \sum_\rho m_\rho (x_{\rho,k} x_{\rho,k} \delta_{ij} - x_{\rho,i} x_{\rho,j})$$

In (14.8) and (14.9), summation on the singly repeated indices (j and k respectively) is automatically implied.

The quantities J_{ij} depend on the distribution of mass in the body and the set of them is called the *inertia tensor* of the body with respect to the origin. The elements $J_{11} = A$, $J_{22} = B$, $J_{33} = C$ are called the *moments of inertia* of the body about the three axes, for example,

$$J_{11} = A = \Sigma m_\rho(x_{\rho,2}^2 + x_{\rho,3}^2)$$

and the negatives of the other components $F = -J_{23}$, $G = -J_{31}$, $H = -J_{12}$ are called the *products of inertia* with respect to the axes, for example,

$$F = -J_{23} = -J_{32} = \sum_\rho m_\rho x_{\rho,2} x_{\rho,3}$$

We note from (9.12) that if the point in the rigid body which is held fixed is also the center of mass of the body, then the resultant of the external forces acting on the body is zero and the total momentum of the body is zero. Thus, from Sec. 13, the angular momentum $\boldsymbol{\sigma}$ is independent of the point about which the moments are taken.

The kinetic energy of a rigid body rotating with one point (the origin) held at rest is

$$\begin{aligned}
T &= \frac{1}{2}\sum_\rho m_\rho(\dot{\mathbf{r}}_\rho \cdot \dot{\mathbf{r}}_\rho) \\
&= \frac{1}{2}\sum_\rho m_\rho(\boldsymbol{\omega} \times \mathbf{r}_\rho) \cdot (\boldsymbol{\omega} \times \mathbf{r}_\rho) \\
&= \tfrac{1}{2}\boldsymbol{\omega} \cdot \sum_\rho m_\rho[\mathbf{r}_\rho \times (\boldsymbol{\omega} \times \mathbf{r}_\rho)] \\
&= \tfrac{1}{2}\boldsymbol{\omega} \cdot \boldsymbol{\sigma} \\
&= \tfrac{1}{2}J_{ij}\omega_i\omega_j
\end{aligned} \tag{14.10}$$

Written out in detail, this becomes

$$T = \tfrac{1}{2}(A\omega_1^2 + B\omega_2^2 + C\omega_3^2 - 2F\omega_2\omega_3 - 2G\omega_3\omega_1 - 2H\omega_1\omega_2) \tag{14.11}$$

15. A THEOREM ON QUADRATIC FUNCTIONS

In terms of a set of rectangular cartesian coordinates, the kinetic energy, angular momentum, and inertia tensor of a system of particles may be written thus:

$$
\begin{aligned}
T &= \frac{1}{2} \sum_{\rho} m_{\rho} \dot{x}_{\rho,k} \dot{x}_{\rho,k} \\
j_i &= \epsilon_{ijk} \sum_{\rho} m_{\rho} x_{\rho,j} \dot{x}_{\rho,k} \qquad (15.1) \\
J_{ij} &= \delta_{ij} \sum_{\rho} m_{\rho} x_{\rho,k} x_{\rho,k} - \sum_{\rho} m_{\rho} x_{\rho,i} x_{\rho,j}
\end{aligned}
$$

These important physical variables have in common the property that any one of them may be written as a quadratic function of the position or velocity components:

$$
X = \sum_{\alpha\beta} A_{\alpha\beta} \sum_{\rho} m_{\rho} Q_{\rho,\alpha} Q_{\rho,\beta} \tag{15.2}
$$

where α, β assume the range of values 1 to 6. For a given ρ, the first three of the quantities $Q_{\rho,\alpha}$ are the position coordinates $x_{\rho,i}$ of the ρth particle, and the second three of the $Q_{\rho,\alpha}$ are the corresponding velocities:

$$
\left.
\begin{aligned}
Q_{\rho,i} &= x_{\rho,i} \\
Q_{\rho,i+3} &= \dot{x}_{\rho,i}
\end{aligned}
\right\} \qquad (i = 1, 2, 3) \tag{15.3}
$$

The set of quantities $A_{\alpha\beta}$ are *independent of* ρ, and are chosen to yield whichever of the quantities (15.1) X is to represent. Thus, for example, for $X = T$ we require $A_{ij} = 0$ (i or $j = 1, 2, 3$) and $A_{i+3,j+3} = \frac{1}{2}\delta_{ij}$.

We now write

$$
\begin{aligned}
x_{\rho,i} &= x_i + r_{\rho,i} \\
\dot{x}_{\rho,i} &= \dot{x}_i + \dot{r}_{\rho,i}
\end{aligned} \tag{15.4}
$$

or in abbreviated notation

$$
Q_{\rho,\alpha} = X_{\alpha} + R_{\rho,\alpha} \tag{15.5}
$$

where the X_{α} denote the position ($\mathbf{r} \equiv x_i$) and velocity of the center of mass and the $R_{\rho,\alpha}$ denote the position ($\mathbf{r}_{\rho} \equiv r_{\rho,i}$) and velocity of particle ρ relative to the center of mass. Thus from (9.11)

$$
\Sigma m_{\rho} R_{\rho,\alpha} = 0 \tag{15.6}
$$

so that (15.2) becomes

$$
X = m \sum_{\alpha\beta} A_{\alpha\beta} X_{\alpha} X_{\beta} + \sum_{\alpha\beta} A_{\alpha\beta} \sum_{\rho} m_{\rho} R_{\rho\alpha} R_{\rho\beta} \tag{15.7}
$$

Any one of the quantities (15.1) may therefore be broken into two parts, one proportional to the total mass m of the system and determined by the position and velocity of the center of mass, the other specifying the distribution and velocity of the particles with respect to the center of mass. Thus the kinetic energy becomes

$$T = \tfrac{1}{2}m\dot{r}^2 + \tfrac{1}{2}\Sigma m_\rho \dot{r}_\rho{}^2 \tag{15.8}$$

the kinetic energy that would be attributed to a particle of mass m moving with the center of mass, together with a term giving the kinetic energy of motion relative to the center of mass. Similarly,

$$\mathbf{j} = \mathbf{r} \times \mathbf{p} + \Sigma m_\rho(\mathbf{r}_\rho \times \dot{\mathbf{r}}_\rho) \tag{15.9}$$

where the first term denotes the angular momentum about the origin again associated with a particle of mass m moving with the center of mass. The second term is identical with $\boldsymbol{\sigma}$ defined in (14.6) provided that the origin is taken at the center of mass. In that case, as already noted, $\boldsymbol{\sigma}$ is independent of the point about which moments are taken. In this manner the total angular momentum $\mathbf{j}$ of a system of particles may be broken unambigously into two parts:

$$\mathbf{j} = \boldsymbol{l} + \boldsymbol{\sigma} \tag{15.10}$$

where

$$\boldsymbol{l} = \mathbf{r} \times \mathbf{p} \tag{15.11}$$

is called the *orbital angular momentum*, and

$$\boldsymbol{\sigma} = \Sigma m_\rho(\mathbf{r}_\rho \times \dot{\mathbf{r}}_\rho) \tag{15.12}$$

is the *spin angular momentum* or angular momentum about the center of mass.

If the resultant external force on the system passes through the center of mass, the angular momentum $\boldsymbol{\sigma}$ about the center of mass will remain constant throughout the motion. If, in addition, the external force on each particle of the system is directed toward a fixed point and the internal forces are along the lines joining the particles, then, as we have seen at the end of Sec. 13, $\mathbf{j}$ is a constant. Under the combined conditions it then follows that $\boldsymbol{l}$ and $\boldsymbol{\sigma}$ are separately constants of the motion. In the case of the earth's motion, these independent constants of the motion relate respectively to the year and to the day. In the case of the motion of an electron in a hydrogen atom it is only approximately true that $\boldsymbol{l}$ and $\boldsymbol{\sigma}$ are separately constants since a moving electron has a small electric dipole moment which allows the electric field of the proton to exert a torque on it (see Sec. 92). However, the spin of the electron cannot be represented in terms of its structure by a classical expression such as (15.12).

Returning to (15.7), we may allow X to denote the moment of inertia C of a system about the z axis. In this case $A_{11} = A_{22} = 1$ and all other $A_{\alpha\beta}$ vanish. Thus

$$C = m(x_1^2 + x_2^2) + C_0 \tag{15.13}$$

where C_0 is the moment of inertia about the axis in the z direction which passes through the center of mass and $m(x_1^2 + x_2^2)$ is the moment of inertia about the z axis itself that a particle of mass m would have if placed at the center of mass. The moment of inertia about an axis through the center of mass is therefore less than that about any parallel axis.

16. INERTIAL AND GRAVITATIONAL MASSES

The familiar condition (13.12) for the static equilibrium of a system allows two weights w_A, w_B to be compared by a balance with arms of unequal lengths l_A, l_B:

$$\frac{w_A}{w_B} = \frac{l_B}{l_A} \tag{16.1}$$

Since the weights are the forces exerted by gravity on the masses ($w_A = m_A g, \cdots$) and gravity is assumed not to vary across the balance, we have

$$\frac{m_A}{m_B} = \frac{l_B}{l_A} \tag{16.2}$$

The masses compared in this manner are sometimes referred to as *gravitational masses*, in contrast to the *inertial masses* defined by (8.1). According to *Newton's law of gravitation*, the gravitational attractive force between two masses m_A, m_B separated by a distance r is

$$F = \frac{G m_A m_B}{r^2} \tag{16.3}$$

where G is the gravitational constant, and m_A, m_B are the gravitational masses. If, however, the masses of two particles attracting each other by gravity are compared by the method of Sec. 8—i.e., taking the ratio of their accelerations—then it is the masses appearing in the equation $\mathbf{F} = m\mathbf{a}$ which are compared. A priori there is no reason for these to be identical with the masses appearing in (16.3).

However, it has not proved possible to distinguish experimentally* between these two apparently different types of masses. In Newtonian theory we accept this result as an empirical fact and refer to the mass of a body without specifying which method is to be used to measure it. One important feature of the general theory of relativity is that from this point of view the distinction between the two types of masses loses its meaning so that they become automatically identical.

GENERAL REFERENCES

P. G. Bergman, *Basic Theories of Physics. Mechanics and Electrodynamics.* Prentice-Hall, New York (1949).

K. E. Bullen, *An Introduction to the Theory of Mechanics.* Science Press, Sydney (1949).

J. C. Slater and N. H. Frank, *Mechanics.* McGraw-Hill, New York (1947).

R. J. Stephenson, *Mechanics and Properties of Matter.* Wiley, New York (1952).

J. L. Synge and B. A. Griffith, *Principles of Mechanics.* McGraw-Hill, New York (1942).

EXERCISES

1. A gun is mounted on a hill of height h above a level plain. Assuming that the path of the projectile is a parabola, find the angle of elevation α for greatest horizontal range and given muzzle speed V.

$$\operatorname{cosec}^2 \alpha = 2\left(1 + \frac{gh}{V^2}\right)$$

What physical effects are neglected in the above approximation?

2. With the same assumptions as in Exercise 1, show that if a projectile is thrown over a double inclined plane from one end of the horizontal base to the other and if it just grazes the summit in its flight, its angle of projection is

$$\tan^{-1}(\tan\theta + \tan\phi)$$

where θ, ϕ are the slopes of the faces and the motion is in a vertical plane through the line of greatest slope.

3. Again with the same assumptions as in Exercise 1, if a bomb bursts on contact with level ground and pieces of it fly off in all directions with speeds up to v ft/sec, find at what distance from the bomb a man is in danger from flying metal.

* R. V. Eötvös, *Ann. phys.*, 59, 354 (1896). L. Southerns, *Proc. Roy. Soc. (London)*, *A*, 84, 325 (1910). P. Zeeman, *Proc. Amst.*, 20, 542 (1917). R. Dicke, Princeton University (1959), unpublished.

4. Show that for a satellite of mass m moving with velocity v in a circular orbit of radius r about an attracting center of mass M,

$$v = \sqrt{\frac{GM}{r}} = \frac{v_e}{\sqrt{2}}, \qquad l = m\sqrt{GMr}$$

where v_e is the escape velocity, and l is the orbital angular momentum.

5. Show that the acceleration of a thin circular ring, rolling without sliding down a plane of inclination α to the horizontal, is $\frac{1}{2}g \sin \alpha$ and that the least coefficient of friction necessary to prevent sliding is $\frac{1}{2} \tan \alpha$.

6. A reel of thread whose rim and spindle are of radii a and b respectively rests on a rough horizontal table. The loose end of the thread passes under the spindle and leads off at an angle α above the horizontal ($\alpha < \frac{1}{2}\pi$). Show that the least tension in it will, in general, wind or unwind the thread according as α is less or greater than a certain value. When α has this critical value, show that there will be no motion unless the tension exceeds a critical value.

7. A garden roller of external radius a is pulled along a rough horizontal path by a force $\mathbf{F}$ which acts at a point on its axle and is inclined at an angle α with the horizontal. Find the relation that exists between $\mathbf{F}$ and the weight of the roller if the resultant force on the path is at right angles to $\mathbf{F}$, and show that the acceleration of the roller is then

$$\frac{ga^2 \sin \alpha \cos \alpha}{k^2 + a^2 \sin^2 \alpha}$$

where k is the radius of gyration of the roller about its axis.

8. Show that three particles can be found which, when placed one at each end and one at the center of a nonuniform rod, form a system equimomental with it.

9. Find the center of mass of each of the following bodies: (*a*) a hemisphere of radius a; (*b*) a hollow right circular cone of height h, half angle θ; (*c*) a solid right circular cone of height h, half angle θ; (*d*) the half of an ellipsoid lying on one side of a plane containing two principal axes; (*e*) that part of a parabola bounded by the *latus rectum;* (*f*) that part of a paraboloid of revolution bounded by the plane containing the *latus rectums.*

10. A system of particles moves in a uniform gravitational field g in the z direction. Show that g can be eliminated from the equations of motion by a transformation of coordinates given by

$$\bar{x} = x, \qquad \bar{y} = y, \qquad \bar{z} = z - \tfrac{1}{2}gt^2$$

(This is an example which forms the basis for the "principle of equivalence" in general relativity. The principle states that the gravitational field at any point in space and time can be eliminated by a suitable coordinate transformation.)

11. A particle moves under the influence of gravity in a curve lying in a vertical plane. Find the shape of curve which makes the potential energy proportional to the square of the arc length from a level point of the curve.

This gives the path of a truly isochronous pendulum. Find the maximum amplitude of a seconds pendulum if $g = 980$ cm/sec^2.

12. A particle of mass m confined to the x axis experiences a force $-kx$. Find the motion resulting from a given initial displacement x_0 and initial velocity v_0. Show that the period is independent of the initial conditions, that a potential energy function exists, and that the energy of the system is constant.

13. A mass approaches the solar system with a velocity v_0, and if it had not been attracted toward the sun it would have missed the sun by a distance d. Use the laws of conservation of energy and angular momentum and the law of gravitation to compute its closest distance of approach a to the sun. Neglect the gravitational attractions of the planets and assume the sun fixed.

(*Ans.* $a = (d^2 + d_0^2)^{1/2} - d_0$, where $d_0 = GM/v_0^2$ and M is the mass of the sun.)

3 CONSERVATIVE SYSTEMS WITH ONE DEGREE OF FREEDOM

17. THE OSCILLATOR

A system with a single degree of freedom has its configuration specified by the value of a single coordinate and its velocity specified by the time derivative of that coordinate. (For the present we assume that only one value of the coordinate corresponds to a given configuration. This is not so when the system consists of a particle confined to a closed curve if the arc length along the curve is chosen as the coordinate, a choice that we shall make later.) We call such a system an *oscillator* when in the course of the motion the initial configuration and the initial velocity (i.e., the initial state of the system) recur after a finite time interval. Since the future course of the motion is completely determined by the initial state and the equations of motion, the initial state, if it recurs once, will recur an infinite number of times separated by equal time intervals. The time interval between two successive occurrences of the same state we call the *period* of the motion. We have assumed that the forces entering the equations of motion do not depend explicitly on the time.

We investigate oscillators in the case where the force is derivable from a potential which is independent of the time.

The particle which, having only one degree of freedom, is constrained to move on a curve may be located by its distance along the curve away from some fixed point. Distances to one side of the point will be positive, those to the other side negative. Calling this coordinate x, we see that the kinetic energy of the particle is

$$T = \tfrac{1}{2}m\dot{x}^2 \tag{17.1}$$

and that the potential energy is some function of x alone:

$$V = V(x) \tag{17.2}$$

According to (11.8) the sum of the kinetic energy and the potential energy is a constant E:

$$T + V = E \tag{17.3}$$

Equation (17.3) alone is enough to find the coordinate x as a function of the time. For

$$\tfrac{1}{2}m\dot{x}^2 + V(x) = E \tag{17.4}$$

and so

$$t - t_0 = \int \frac{dx}{\{(2/m)[E - V(x)]\}^{1/2}} \tag{17.5}$$

If $V(x)$ is known the integral can be evaluated, so we have here the general solution of the one-dimensional conservative problem.

The question we wish to discuss is: When is the motion given by (17.5) an oscillatory one? On physical grounds the answer is clear. The particle starts from x_0 with velocity $\dot{x}_0$. If it is to return to x_0, the velocity must be reversed at some point, and so $\dot{x}$ must vanish once. This is not sufficient, for the original value of $\dot{x}_0$ must also be repeated; thus another reversal of $\dot{x}$ and another vanishing of $\dot{x}$ are called for. There is one other requirement which is also obvious. At neither of the places where $\dot{x}$ vanishes may the force vanish. Otherwise the particle would remain at rest at that place.

The places where $\dot{x}$ vanishes we call the *turning points* of the motion. Let these points be $x = x'$ and $x = x''$. From (17.4) we then obtain

$$V(x') = V(x'') = E \tag{17.6}$$

These two equations serve to determine x' and x''. For $x' < x < x''$, $V(x)$ must be less than E since T is never negative. If there are more than two values of x satisfying (17.6), two adjacent ones must be chosen which satisfy the condition just stated. If there are not two values of x satisfying (17.6), there is no oscillation. The requirement that the force does not vanish at the turning points means that dV/dx must not vanish there. This is equivalent to saying that the roots of (17.6) must be single roots.

The absence of oscillation when a double root occurs can be seen analytically from (17.5). For oscillations to take place, any value of x in the interval $x' < x < x''$ must be reached in a finite time interval; i.e., the integral in (17.5) must converge for all such points. Now the integrand becomes infinite at $x = x'$ and $x = x''$. This infinity of the integrand is integrable provided that the infinity is not too severe a one, i.e., if in the neighborhood of the turning point, say x', the integrand becomes infinite as $(x - x')^{-p}$, where $p < 1$. For $p \geq 1$ the integral becomes infinite also. If dV/dx is finite for $x = x'$, then $p = \frac{1}{2}$; whereas, if dV/dx vanishes at x', p is at least unity and may be greater, depending on how many other derivatives vanish also.

Nearly all the above information can be obtained from an inspection of a graph of $V(x)$. Three cases are shown in Fig. 17-1. In (a) the

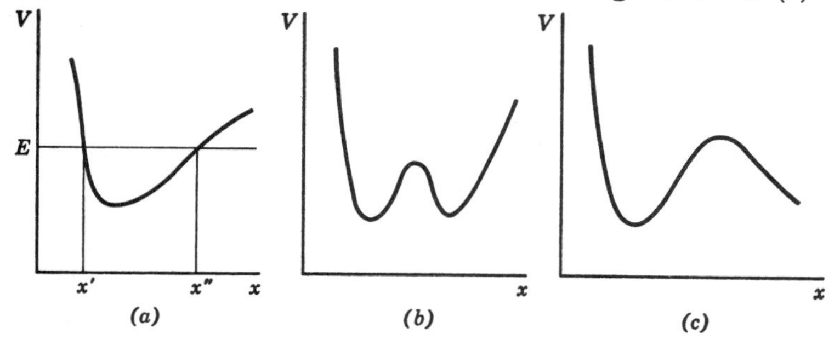

Fig. 17-1. Oscillator potentials.

potential has a single minimum. Oscillations will take place with an amplitude which increases as the energy E increases. In (b) the potential has a double minimum. In this case, if the energy E is below the barrier between the two minima, oscillations may take place on either side of the barrier. For E greater than the barrier height there will be oscillations across the whole width of the potential well. If E has exactly the value corresponding to the top of the barrier, the motion will not be oscillatory but the particle will stick at the top. An infinitesimal energy change will make the motion oscillatory. In (c), if the energy E is less than the barrier height and the particle is to the left of the barrier, there will be oscillation. Otherwise there will not be.

The character of the motion in the potential with a double minimum shows up one unsatisfactory feature of classical mechanics. An infinitesimal change in the energy can cause a large change in the character of the motion. This sensitivity is due to the fact that a classical particle moves under the influence of the potential only in its immediate vicinity. In quantum mechanics this is not so, the form of the potential everywhere being effective in determining the state of the system.

Suppose that we have a classical particle oscillating about one of the two potential minima. We perturb the system in such a way that the energy changes only a very small amount per period. This is possible as long as the energy stays below the barrier energy. If the transfer of energy is sufficiently slow, the exact manner in which it takes place is unimportant. Such a perturbation is called *adiabatic*. As the barrier energy is approached, however, no finite rate of energy transfer is possible which does not need to be specified exactly, since the period becomes infinitely long, and thus a finite amount of energy must be given in an arbitrarily small fraction of a period. In the neighborhood of the barrier energy, then, we cannot make a change in the energy in an arbitrary way, no matter how slowly, without the result being dependent on the manner of transfer. It is impossible to go from oscillation about one minimum to oscillation across the entire width of the potential well by an adiabatic change.

The situation is perhaps even more clear if the energy is initially above the barrier energy. As the energy decreases, the particle will be confined to one side or the other of the barrier. To which side it is confined depends on the details of the energy transfer at the time the energy decreases below the barrier energy.

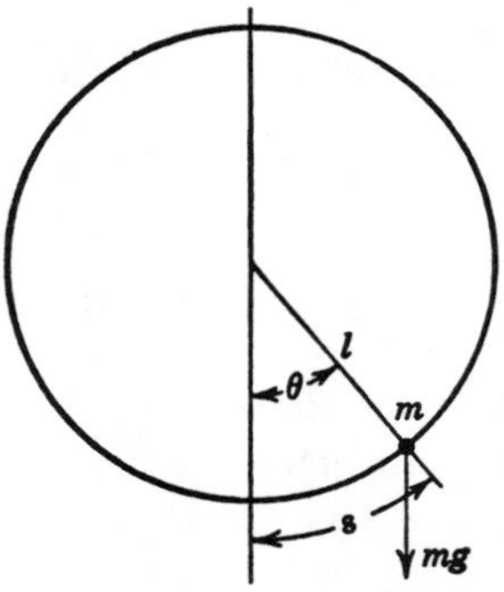

Fig. 18-1. Plane pendulum.

18. THE PLANE PENDULUM

An important example of an oscillator is the harmonic oscillator of Exercise 12, Chapter 2. It has the property that its period is independent of the energy, or of the amplitude. Such an oscillator is called isochronous. Another important example is the plane pendulum. Its motion is isochronous for infinitesimal amplitudes and very nearly so for finite but small amplitudes.

The plane pendulum (Fig. 18-1) consists of a particle constrained to move in a vertical circle of radius l. (The physical pendulum which consists of many particles constrained to move in vertical circles around a common center is always representable by an equivalent simple pendulum of the kind described here.) We choose as coordinate the angle θ between the downward vertical and the radius vector of the particle measured from the center of the circle. The arc length

s along the circle is

$$s = l\theta \tag{18.1}$$

The kinetic energy of the system is given by

$$T = \tfrac{1}{2}m\dot{s}^2 = \tfrac{1}{2}ml^2\dot{\theta}^2 \tag{18.2}$$

The potential energy is the gravitational potential energy which depends only on the height of the bob of the pendulum above the bottom of the circle:

$$V = mgl(1 - \cos\theta) \tag{18.3}$$

It is left to the reader to show that for vanishing amplitudes the motion is harmonic.

The analysis of the problem for finite amplitudes is simplified by introducing natural units. We measure all masses in terms of m, all lengths in terms of l, and time in terms of $(l/g)^{1/2}$. Thus m, l, and g can be made equal to unity, giving

$$T = \tfrac{1}{2}\dot{\theta}^2, \qquad V = 1 - \cos\theta \tag{18.4}$$

The conservation of energy, which applies because V does not contain the time explicitly, tells us that

$$\tfrac{1}{2}\dot{\theta}^2 + 1 - \cos\theta = E \tag{18.5}$$

or

$$\tfrac{1}{2}\dot{\theta}^2 + 2\sin^2\tfrac{1}{2}\theta = E \tag{18.6}$$

so that

$$\dot{\theta} = (2E)^{1/2}\left(1 - \frac{2}{E}\sin^2\frac{\theta}{2}\right)^{1/2} \tag{18.7}$$

For the motion to be oscillatory this quantity must have two simple zeros, as it does if $E < 2$. Now let $y = \sin\theta/2$, so that

$$\dot{\theta} = \frac{2\dot{y}}{(1 - y^2)^{1/2}} \tag{18.8}$$

Inserting this in (18.7) and simplifying, we obtain

$$\dot{y} = \left[(1 - y^2)\left(\frac{E}{2} - y^2\right)\right]^{1/2} \tag{18.9}$$

We investigate three separate cases: case a, where $E < 2$ and the motion is oscillatory; case b, where $E = 2$ and $\dot{\theta}$ has a double zero, yielding a sticking solution; and case c, where $E > 2$, which corresponds to a nonoscillatory, nonsticking motion. Case c is periodic, however, because here the curve to which the particle is confined is closed. We call this a circulating motion, or *libration.*

Case a. $E < 2$. Let $E/2 = k^2$, $y/k = z$. Then (18.9) becomes

$$\dot{z} = [(1 - z^2)(1 - k^2z^2)]^{1/2} \tag{18.10}$$

or

$$\int \frac{dz}{[(1 - z^2)(1 - k^2z^2)]^{1/2}} = t - t_0 \tag{18.11}$$

The integral on the left of (18.11) is called an *elliptic integral* of the first kind. We shall discuss this integral briefly.

The function $\sin \theta$ may be defined in the following way. If

$$\int_0^y \frac{dy}{(1 - y^2)^{1/2}} = \theta$$

then

$$y = \sin \theta$$

The period η of the sine function is given by

$$\frac{\eta}{4} = \int_0^1 \frac{dy}{(1 - y^2)^{1/2}}$$

We proceed similarly with the integral in (18.11). Let

$$\int_0^z \frac{dz}{[(1 - z^2)(1 - k^2z^2)]^{1/2}} = u \tag{18.12}$$

Then the function sn (read *ess-en*) is defined by

$$z = \text{sn}\, u \tag{18.13}$$

In (18.13) z also depends on k, the modulus of the function, although this dependence is not usually indicated explicitly. The inverse function is denoted by F,

$$u = F(\phi, k)$$

where $z = \sin \phi$. The function F is tabulated as the incomplete elliptic integral of the first kind.*

* E. Janke and F. Emde, *Tables of Functions,* Dover, New York (1943).

The function sn is periodic, its period ζ depending on k:

$$\frac{\zeta}{4} = \int_0^1 \frac{dz}{[(1 - z^2)(1 - k^2z^2)]^{1/2}} \tag{18.14}$$

$$= F\left(\frac{\pi}{2}, k\right)$$

This function is the complete elliptic integral of the first kind. For $k < 1$ this period is real and finite.

The solution of the plane pendulum's oscillating motion is thus

$$y = k \text{ sn } (t - t_0) \tag{18.15}$$

expressed in natural units. In conventional units this is

$$y = k \text{ sn } \left(\frac{g}{l}\right)^{1/2} (t - t_0) \qquad \left(k^2 = \frac{E}{2mgl}\right)$$

The period depends on the amplitude.

For small amplitudes, which correspond to small k values, the expression for the period may be expanded in a power series in k:

$$\frac{\zeta}{4} = \int_0^1 \frac{dz}{(1 - z^2)^{1/2}} (1 + \tfrac{1}{2}k^2z^2 + \cdot \cdot \cdot) \tag{18.16}$$

$$\approx \frac{\pi}{2}\left(1 + \frac{k^2}{4}\right)$$

Now $k^2 = E/2$, and E is connected to the maximum value of θ, θ_0 say, by (18.5) with $\dot{\theta}$ put equal to zero. Hence for small θ_0,

$$k \approx \frac{\theta_0}{2} \tag{18.17}$$

and

$$\zeta \approx 2\pi\left(1 + \frac{{\theta_0}^2}{16}\right) \tag{18.18}$$

Again in conventional units, the period τ is given by

$$\tau \approx 2\pi\left(\frac{l}{g}\right)^{1/2}\left(1 + \frac{{\theta_0}^2}{16}\right)$$

The motion deviates from the isochronous only in the second order, and there with a small coefficient. Hence the motion is very nearly isochronous even for finite amplitudes.

Case b. $E = 2$. From (18.5) we see that all the energy is potential energy at the top of the circle, and here $dV/d\theta$ vanishes, so that this is a sticking motion. Equation (18.9) becomes

$$\dot{y} = 1 - y^2 \tag{18.19}$$

so that

$$y = \tanh (t - t_0) \tag{18.20}$$

In conventional units this is

$$y = \tanh \left(\frac{g}{l}\right)^{1/2} (t - t_0) \tag{18.21}$$

As $t \to \infty$, $y \to 1$, $\theta \to \pi$.

Case c. $E > 2$. The total energy is greater than the potential energy at the top of the circle, so that the kinetic energy is never zero and the particle never comes to rest. This is a circulating motion.

Let $k^2 = 2/E < 1$. Then (18.9) becomes

$$k\dot{y} = [(1 - y^2)(1 - k^2y^2)]^{1/2} \tag{18.22}$$

or

$$\int_0^y \frac{dy}{[(1 - y^2)(1 - k^2y^2)]^{1/2}} = \frac{1}{k}(t - t_0) \tag{18.23}$$

In terms of the sn function this is

$$y = \mathrm{sn}\, \frac{t - t_0}{k} \tag{18.24}$$

In conventional units

$$y = \mathrm{sn} \left(\frac{E}{2mgl}\right)^{1/2} \left(\frac{g}{l}\right)^{1/2} (t - t_0) \tag{18.25}$$

The period decreases with decreasing values of k, or with increasing values of E. The acceleration of gravity does not cancel out of (18.25) completely since E depends on g. As E increases, the potential energy gets less important relative to the kinetic energy and the dependence of the motion on g diminishes.

19. CHILD-LANGMUIR LAW

An interesting application of the conservation of energy in one-dimensional motion is afforded by the law of space-charge limitation of a one-dimensional steady electric current in vacuo.* If the particles that compose the current are all identical (usually electrons), with charge e and mass m, energy conservation for each one of them implies

$$\tfrac{1}{2}mv^2 + e\phi = \text{const} \tag{19.1}$$

where ϕ is the electrostatic potential due to all of the others. In a sufficiently dense current beam it is a good approximation to regard ϕ as a function $\phi(x)$ of the distance x along the beam, ϕ then being the electrostatic potential at x due to all of the particles

$$\nabla^2\phi = \frac{d^2\phi(x)}{dx^2} = -4\pi\rho(x) \tag{19.2}$$

and $\rho(x)$ being the charge density at a point inside the beam. We note also that the current density j is related to $\rho(x)$ by

$$j = \rho v \tag{19.3}$$

and that for a steady current beam of constant cross section A

$$j = J/A$$

where J is the total current. j is therefore independent of x. Thus, from (19.2) and (19.3)

$$\frac{d^2\phi}{dx^2} = -\frac{4\pi j}{v}$$

and from (19.1)

$$v = \left[\frac{2e}{m}(\phi_0 - \phi)\right]^{1/2}$$

where ϕ_0 is the potential at the point where $v = 0$.

Thus, writing $\phi_0 - \phi = V$, we have

$$\frac{d^2V}{dx^2} = 4\pi j\left(\frac{m}{2e}\right)^{1/2} V^{-1/2}$$

The solution of this corresponding to $V = 0$ at $x = 0$ is

$$V = Ax^{4/3} \tag{19.4}$$

* C. D. Child, *Phys. Rev.*, 32, 498 (1911). H. F. Ivey, *Advances in Electronics and Electron Phys.*, 6, 170–179 (1954). J. R. Pierce, *Theory and Design of Electron Beams*, Van Nostrand, New York (1949). I. Langmuir, *Phys. Rev.*, 33, 954 (1929). L. Jacob, *An Introduction to Electron Optics*, Methuen, London (1951).

where

$$A = (9\pi j)^{2/3} \left(\frac{m}{2e}\right)^{1/3} \tag{19.5}$$

Thus for a plane cathode and a parallel plane anode at a distance d from it, the space-charge limited current density is given by

$$\begin{aligned} j &= \frac{1}{9\pi}\left(\frac{2e}{m}\right)^{1/2} A^{3/2} \\ &= GV^{3/2} \end{aligned} \tag{19.6}$$

where V is the voltage difference, and

$$G = \frac{1}{9\pi}\left(\frac{2e}{m}\right)^{1/2} \frac{1}{d^2} \tag{19.7}$$

is called the *perveance* of the beam. For a given voltage V, the current density that it is possible to achieve therefore increases as the spacing d decreases, and for a given V and d the space-charge limited current density is greater for electrons than for positive ions by a factor which is equal to the reciprocal of the square root of the mass ratio.

The above analysis completely neglects any motion transverse to the beam arising from thermal motion of the particles, or from their mutual electrostatic repulsion, or from collisions with the residual gas.

EXERCISES

1. Obtain an expression for the potential along a space-charge limited beam of particles emitted from the point at which $\phi = 0$ with a velocity v, current density j.

2. An oscillator moves under the influence of a potential function V given by

$$V = \tfrac{1}{2}gx^2 + \epsilon x^4$$

Find the period of the motion as a function of the amplitude, and derive an approximate expression for the period of a simple pendulum as a function of the amplitude.

3. From (17.5) the period of an oscillator is

$$\tau = \oint \frac{dx}{\left[\dfrac{2}{m}(E - V)\right]^{1/2}} = (2m)^{1/2} \frac{\partial}{\partial E} \oint (E - V)^{1/2}\, dx$$

If $V(x)$ can be extended into the complex x plane, this integral may be taken around a contour surrounding the turning points, which are branch points of the integrand, and hence may often be evaluated by use of Cauchy's residue theorem. Use this or some other method to find the period when

$$(a) \qquad V = \tfrac{1}{2}kx^2 \qquad (E > 0)$$

$$(b) \qquad V = -\frac{K}{x} + \frac{l^2}{x^2} \qquad (E < 0)$$

This is a special case of (62.3).

4 TWO-PARTICLE SYSTEMS

20. INTRODUCTION

It has been stressed that much of the art of theoretical physics consists in appraising a complicated physical situation and focusing attention on the interactions which dominate it. It is relatively useless to compute an effect which is almost negligible while other more important features of the problem are left untouched. Since macroscopic systems contain some 10^n atomic particles, where n is of the order of 20 or 30, a great deal of simplification is necessary before a problem involving such a system can be handled. Fortunately, there exist a number of situations, chief among which are those described by celestial mechanics, in which the interacting systems are very small compared with their distances apart. Thus the internal structure of the sun and planets becomes of negligible importance in determining their relative motion, and it is sufficient to consider each planet as a single particle. Since the planets are small in mass compared with the sun, very little further error is introduced by neglecting their effects one on the other, and describing the motion of one planet around the sun by a two-particle theory. The results may then be used as a basis for a more accurate treatment of the problem, although the essential features of the motion already appear in this approximation.

The analogous situation of an electron in an atom may be treated in similar fashion, although here the approximation of a two-particle theory is less serious than the error introduced by applying classical mechanics to the problem. It might be thought that at least there is no error involved in treating an electron as a single particle, but that which is recorded in measuring equipment and termed "an electron" is known to have a very complex structure. Provided that electrons come no closer to each other than about 10^{-11} cm, however, this internal structure is unimportant, and the electrons may be treated as point particles.

Much information may also be obtained about atomic and subatomic systems by accelerating one system and measuring how it is scattered by another. In many cases such collisions may be adequately treated theoretically as two-particle problems.

Finally, we shall have occasion to discuss the motion of a pair of interacting particles when the effects of other matter cannot be neglected but when such effects may be accurately represented by known electromagnetic or gravitational fields.

21. REDUCED MASS

The center of mass of a two-particle system moves as a particle whose mass is the sum of the masses of the two particles and which is acted on by the sum of the external forces acting on the two particles. We restrict our discussion to those two-particle systems in which the external forces per unit mass on the two particles are the same, and in particular to the case for which both of these forces are zero. Our analysis requires modification for two charged particles in an external electric field.

A two-particle system has six degrees of freedom. Of these the center of mass represents three. The other three degrees of freedom can be treated as those belonging to a single particle acted on by appropriate forces. These forces are particularly simple if the restriction just made holds. The equations of motion of the particles are written

$$\begin{aligned} m_1\ddot{\mathbf{r}}_1 &= \mathbf{F}_1^{\text{int}} + \mathbf{F}_1^{\text{ext}} \\ m_2\ddot{\mathbf{r}}_2 &= \mathbf{F}_2^{\text{int}} + \mathbf{F}_2^{\text{ext}} \end{aligned} \tag{21.1}$$

Adding these equations gives the equation of motion of the center of mass

$$\mathbf{R} = \frac{1}{m}(m_1\mathbf{r}_1 + m_2\mathbf{r}_2)$$

Subtracting the second equation from the first after each has been divided through by the appropriate m_ρ gives

$$\ddot{\mathbf{r}}_1 - \ddot{\mathbf{r}}_2 = \frac{\mathbf{F}_1^{\text{int}}}{m_1} - \frac{\mathbf{F}_2^{\text{int}}}{m_2} + \frac{\mathbf{F}_1^{\text{ext}}}{m_1} - \frac{\mathbf{F}_2^{\text{ext}}}{m_2} \tag{21.2}$$

The last two terms cancel in the cases we are discussing. The first two terms on the right combine because $\mathbf{F}_1^{\text{int}} = -\mathbf{F}_2^{\text{int}}$, i.e., the internal force on one of the particles comes entirely from the other one. When we write $\mathbf{r} = \mathbf{r}_1 - \mathbf{r}_2$ and $\mathbf{F}$ for $\mathbf{F}_1^{\text{int}}$, (21.2) becomes

$$\ddot{\mathbf{r}} = \mathbf{F}\left(\frac{1}{m_1} + \frac{1}{m_2}\right) \tag{21.3}$$

This can be written in the form

$$\mu\ddot{\mathbf{r}} = \mathbf{F} \tag{21.4}$$

where μ is called the *reduced mass* of the system and is given by

$$\mu = \frac{m_1 m_2}{m_1 + m_2} \tag{21.5}$$

μ is less than the smaller of the two masses m_1 and m_2.

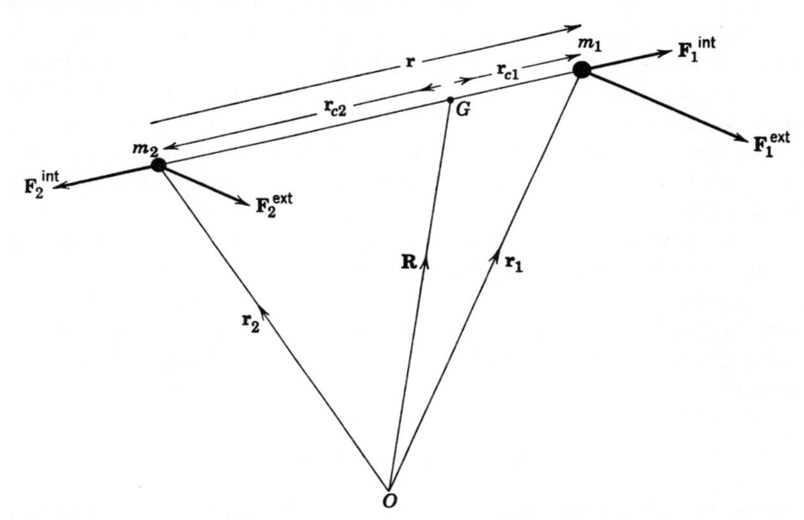

Fig. 21-1. Vectors introduced in studying the relative motion of two particles. Since it is common practice for $\mathbf{r}$ to denote the radius vector of particular interest in a problem, the meanings of $\mathbf{r}$ and $\mathbf{R}$ are different from those adopted in Sec. 15, where attention was focused on the motion of the center of mass.

The motion of the two-particle system may thus be broken down into the motions of two one-particle systems: The center of mass moves under the influence of the external forces which are assumed to be the same per unit mass on the two particles, and the relative motion is represented by the motion of a fictitious particle of mass μ acted on by the internal force $\mathbf{F}$ on the first particle.

Since the motion of the center of mass of the two-particle system does not enter into the second of these problems, in it the center of mass is considered to be at rest, and we speak of the "center-of-mass coordinate system." This name is somewhat misleading as $\mathbf{r}$ is not the radius vector of particle 1 with the center of mass as origin, but with particle 2 as origin. (See Fig. 21-1.) The radius vector $\mathbf{r}_{c1}$, of particle 1 from the center of mass, is given by

$$\mathbf{r}_{c1} = \frac{\mu}{m_1}\mathbf{r}$$

and $\mathbf{r}_{c2}$, that of particle 2 referred to the center of mass, is

$$\mathbf{r}_{c2} = -\frac{\mu}{m_2}\mathbf{r}$$

It is the use of $\mathbf{r}$ rather than $\mathbf{r}_{c1}$ as coordinate that brings with it the use of the reduced mass μ rather than the true mass m_1.

22. RELATIVE KINETIC ENERGY

From the point of view of applications, the problem of the collision of two particles is of special interest. In such problems the only relevant forces are usually the mutual forces between the two colliding particles, and so the breakdown of the two-particle problem described in the preceding section can be applied. When a collision is studied experimentally, one of the particles is initially at rest while the other one approaches it. Analytically it is simpler to treat the motion relative to the center of mass of the two particles, since this center of mass moves with constant velocity and the kinetic energy and angular momentum associated with the center of mass are constant and can therefore be disregarded in studying the relative motion. We desire to find the connection between the laboratory coordinate system and the center-of-mass coordinate system.

Let the two particles have masses m_1 and m_2 respectively, and let their radius vectors in the laboratory system be denoted by $\mathbf{r}_1$ and $\mathbf{r}_2$ respectively. Initially let $\mathbf{r}_2$ be constant, say zero, and let $\mathbf{r}_1$ vary, with $\mathbf{v}$, the velocity of particle 1, being parallel to the x axis.

$$x_2 = 0, \qquad y_2 = 0, \qquad z_2 = 0 \tag{22.1}$$

and

$$\begin{aligned} x_1 &= \alpha + vt \\ y_1 &= \beta \\ z_1 &= \gamma \end{aligned} \tag{22.2}$$

The velocity of the center of mass is easily determined, for it is constant and may be found from the initial motion before any part of the collision has taken place.

$$m\dot{\mathbf{R}} = m_1\dot{\mathbf{r}}_1{}^0 + m_2\dot{\mathbf{r}}_2{}^0 = m_1\dot{\mathbf{r}}_1{}^0 \tag{22.3}$$

so that

$$\dot{\mathbf{R}} = \frac{m_1}{m}\dot{\mathbf{r}}_1{}^0 \tag{22.4}$$

where the superscript 0 denotes the initial value.

Two things are of primary interest. The first is the initial value of the relative kinetic energy or the kinetic energy in the center-of-mass system. This is the part of the total kinetic energy produced in the laboratory which is available for any process occurring between the particles, since the kinetic energy associated with the motion of the center of mass must remain constant. The second is the angle of scattering. If the two particles composing the system do not coalesce to form a single particle, they must separate again after collision. The incident particle will in general have its velocity changed in the collision process. The angle between the initial velocity and the final velocity is the angle of scattering. It is measured in the laboratory coordinate system and calculated in the center-of-mass coordinate system. The transformation from one to the other is needed.

The total initial kinetic energy produced in the laboratory is $\frac{1}{2}m_1v^2$. The initial kinetic energy of the center of mass is $\frac{1}{2}m[(m_1/m)v]^2$. The relative kinetic energy is the difference between these two:

$$T_{\text{rel}}{}^0 = \frac{1}{2}\frac{m_1m_2}{m_1 + m_2}v^2 = \frac{1}{2}\mu v^2 \tag{22.5}$$

as might have been expected. If the two particles are equally massive, one-half of the initial kinetic energy is relative kinetic energy. In general, with m_2 fixed initially in the laboratory coordinate system, the fraction of the kinetic energy which is relative is $m_2/(m_1 + m_2)$. It is thus advantageous to have the heavier particle at rest, a conclusion which is intuitively obvious.

23. LABORATORY AND CENTER-OF-MASS SYSTEMS

The angles of scattering in the two coordinate systems differ because the x components of the velocities differ whereas the y and z components do not. Figure 23-1 illustrates the situation. The subscript C refers to the center-of-mass system; subscript L, to the laboratory system. $\mathbf{v}'$ is the velocity of the first particle after the collision has taken place. ϕ is the angle of scattering in the center-of-mass system, and θ is this angle in the laboratory system. $\mathbf{v}_C'$ and $\mathbf{v}_L'$ differ by the velocity of the center of mass.

The expression of θ in terms of ϕ in general involves the solution of a transcendental equation. The equations to be solved arise from taking transverse and longitudinal components in the velocity triangle of Fig. 23-1. They are

$$v_L' \sin\theta = v_C' \sin\phi$$
$$v_L' \cos\theta = v_C' \cos\phi + \frac{\mu}{m_2} v \tag{23.1}$$

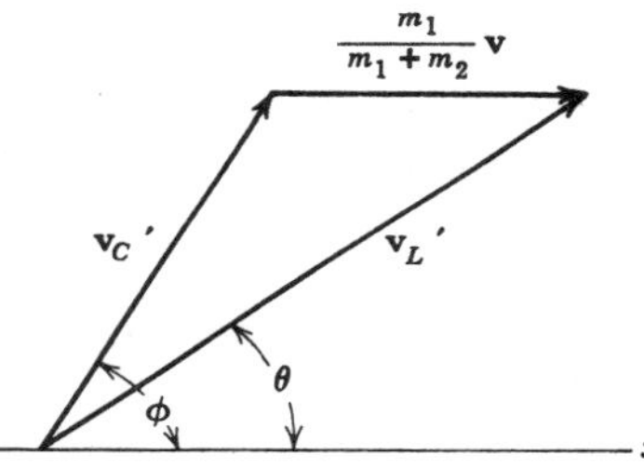

Fig. 23-1. Scattering angles in laboratory and center-of-mass coordinates.

In two special cases the result is simple. The first is that in which the incident particle has a mass negligible in comparison with the mass of the other. In this case the center of mass is at the position of the second particle, and the two coordinate systems coincide. The second is that in which the two particles are equally massive and the final kinetic energy is equal to its initial value. Then

$$v_C' = \frac{m_1}{m_1 + m_2} v = \frac{1}{2} v \tag{23.2}$$

and the velocity triangle in Fig. 23-1 is isosceles. It is then clear from elementary geometry that

$$\theta = \tfrac{1}{2}\phi \tag{23.3}$$

since $\mathbf{v}_L'$ is the diagonal of a rhombus and therefore bisects the angle between the two adjacent sides. This result is no longer valid if the relative kinetic energy is not conserved.

24. CENTRAL MOTION

One of the most frequently recurring problems in physics is that of two particles which exert forces on each other along the line joining them. In such cases the force on each particle is directed towards the center of mass of the system since that center must lie on the line joining the particles. In the center-of-mass coordinate system these forces are therefore directed toward a fixed point. This characteristic of the force leads to the name *central force*, and *central motion* is the motion of two particles about their center of mass under the influence of a central force.

In Sec. 21 it was seen that the relative motion of the two particles can be represented by the motion of a single particle of mass μ, the reduced mass of the system, acted on by the mutual force between the particles, which in the present case is directed toward or away from the origin. There are several theorems on central motion which do not depend on anything except the central character of the force.

Theorem a. Central motion takes place in a plane. This is physically obvious. The force vector and, therefore, the acceleration are parallel to the radius vector. Thus the radius vector, acceleration, and velocity lie in a plane, and the particle will never leave this plane because there is no component of the acceleration out of it. In Chapter 7 we shall see how this appears in the Lagrange treatment of the problem.

Theorem b. The orbital angular momentum is constant. The orbital angular momentum $\boldsymbol{l}$ is given by

$$\begin{aligned} \boldsymbol{l} &= \mathbf{r} \times \mathbf{p} \\ &= \mathbf{r} \times \mu\dot{\mathbf{r}} \end{aligned} \tag{24.1}$$

Thus

$$\begin{aligned} \frac{d\boldsymbol{l}}{dt} &= \dot{\mathbf{r}} \times \mu\dot{\mathbf{r}} + \mathbf{r} \times \mu\ddot{\mathbf{r}} \\ &= 0 \end{aligned} \tag{24.2}$$

We have assumed here that the particle has no intrinsic spin angular momentum $\boldsymbol{\sigma}$ or if it does that $\boldsymbol{\sigma}$ is independently conserved.

Theorem c. If the central force is independent of the time, the path of the particle, or the orbit, is symmetrical about an apse. (A point where the velocity vector is perpendicular to the radius vector is called an *apse*.) The proof of this theorem lies in the observation that the equations of motion are not changed if the sign of the time t is reversed so that the motion is reversed in direction. Since at an apse the radial velocity is zero, the effect of reversing the time when the particle is at an apse is simply to reverse the direction of the motion around the force center without affecting the radial motion. Thus the halves of the orbit on either side of the apse must be mirror images of each other in the line joining the apse to the origin. A corollary of this theorem is that there can be at most two apsidal distances, i.e., values of $|\mathbf{r}|$ corresponding to apses, since the two apses on either side of a given apse are symmetrically placed with respect to that apse, so that alternate apses are at equal distances from the origin.

EXERCISES

1. Show that the relative motion of two particles is not affected by a uniform gravitational field. Is the same true for a hydrogen atom (classical model) in a uniform electric field?

2. If θ and ϕ are the laboratory and center-of-mass scattering angles respectively, show that

$$\tan\theta = \frac{\sin\phi}{\cos\phi + \dfrac{\mu}{m_2}\dfrac{v}{v_c'}}$$

Show further that, if the relative kinetic energy of the two particles increases by an amount Q,

$$\frac{\mu}{m_2}\frac{v}{v_c'} = \frac{m_1}{m_2}\left(1 + \frac{Q}{T_{\text{rel}}^0}\right)^{-1/2}$$

3. Li^7 is bombarded with deuterons of 10 Mev energy, yielding Be^8 and a neutron. The reaction is exothermic with an energy yield of 14.5 Mev. Find the energy of the emergent neutron in the laboratory coordinate system as a function of the angle between its direction of emergence and the direction of the incident deuteron, this angle also being measured in the laboratory system. Make a polar plot of the result.

4. A particle makes an elastic collision with an equally massive particle initially at rest. Show that the final velocity vectors are mutually perpendicular.

5. A double star has its two components executing circular orbits. Express the mass ratio of the two components in terms of the radii of their orbits.

6. Two particles of masses m_1, m_2 approach each other with velocities v_1, v_2 so that if there were no interaction between them they would have missed each other by a distance d. Use the laws of conservation of energy and angular momentum to compute their closest distance of approach a if they move under their mutual gravitational attraction. Show that for $m_1 \gg m_2$ the result reduces to that derived in Exercise 13 of Chapter 2.
{*Ans.* $a = (d^2 + d_0^2)^{1/2} - d_0$, where $d_0 = [(G(m_1 + m_2)/V^2]$ and V is the initial relative velocity.}

7. Two particles connected by an elastic string of stiffness k and equilibrium length a rotate about their center of mass with angular momentum l. Show that their distance r_1 of closest approach and their maximum separation r_2 are related by

$$\frac{r_1^2 r_2^2 (r_1 + r_2 - 2a)}{r_1 + r_2} = \frac{l^2}{k\mu}$$

where μ is their reduced mass, and $r_1 > a$, $r_2 > a$.

8. A particle moves in a circle under the influence of a central attractive force which is derived from the potential $V = -\kappa/r^n$ $(\kappa > 0)$. Show that for $n < 2$ its total energy is negative, but that for $n > 2$ the particle has enough energy to escape.

5 TIME-DEPENDENT FORCES AND NONCONSERVATIVE MOTION

25. INTRODUCTION

In Sec. 11 it was noted that if a particle is moving under the influence of a force which is derivable from a time-independent potential V, the field of force is called conservative because throughout the motion $E = T + V$ is conserved. On the other hand, if the potential depends on the time, net work is done by or on the particle by the external field as shown in (11.12) and the total energy of the particle is not conserved. However, the rate of increase of the energy of the particle is equal to the rate at which work is being done on it by the external force, and in this sense mechanical energy is conserved throughout the motion.

In many situations of interest, the force on a system cannot be derived from a potential at all. Examples of this type of force include friction and the resistance offered by a medium to the motion of an object through it, forces arising from radiation pressure and radiation damping, and the forces of jet propulsion. Since such forces are not derivable from an ordinary potential function or even from a generalized potential as discussed in Sec. 32, the motion cannot be described by a Lagrangian. In some of these examples, the action of the force leads to a degradation of energy to the form of heat, so that the mechanical energy is not conserved.

In this chapter we consider an example of a system subject to a potential which depends periodically on the time (the inverted pendulum) and four examples of systems in which the force cannot be derived from a potential.

26. THE INVERTED PENDULUM

If the support of a pendulum is caused to oscillate in an appropriate manner by an externally applied force, a configuration of stable equilibrium exists in which the pendulum remains upside down. The fact that this position is stable may be demonstrated by slightly displacing the pendulum and noting how it oscillates about the inverted position.

We consider here the case in which the support is caused to oscillate vertically with an angular frequency Ω, resulting in an acceleration α of the support which is periodic with period $2\pi/\Omega$. To an observer at rest with respect to the support, the resultant downward acceleration of gravity at any time is therefore $G = g + \alpha$ if α also is measured downward, and the equation of motion of the pendulum for small oscillations about the downward vertical is (cf. Sec. 18)

$$I\ddot{\theta} + mGl\theta = 0 \tag{26.1}$$

where I is the moment of inertia about the axis. If α is sinusoidal

$$\alpha = \alpha_0 \cos \Omega t \tag{26.2}$$

the equation of motion for oscillations about the inverted position is

$$\ddot{\theta} - \frac{ml}{I}(g + \alpha_0 \cos \Omega t)\theta = 0 \tag{26.3}$$

For $\alpha_0 = 0$, the motion is clearly unstable, but a study of this equation, which is of the form of Mathieu's equation, shows that sets of values of the parameters α_0, Ω, l exist for which the motion is stable.

We write

$$a = \frac{-4mlg}{\Omega^2 I}$$

$$q = \frac{2\alpha_0 ml}{\Omega^2 I} \tag{26.4}$$

$$\xi = \frac{\Omega t}{2}$$

Thus for the inverted position of the pendulum a is negative, whereas for the normal position a is positive and equal to $(2\omega/\Omega)^2$, where ω is the natural frequency of the pendulum. From (26.2), the amplitude x_0 of the oscillations of the support is α_0/Ω^2 so that

$$q = \frac{2\omega^2 x_0}{g} \tag{26.5}$$

Using (26.4), we may write the equation of motion in the standard form of Mathieu's equation:

$$\frac{d^2\theta}{d\xi^2} + (a - 2q \cos 2\xi)\theta = 0$$

This equation is soluble in the form of an infinite series

$$\theta = Ae^{\mu\xi} \sum_{n=-\infty}^{\infty} C_{2n}e^{2in\xi} + Be^{-\mu\xi} \sum_{n=-\infty}^{\infty} C_{2n}e^{-2in\xi}$$

in which μ and the C_{2n} are functions of the parameters a, q. If μ is either real or complex, the motion builds up exponentially and is therefore unstable. If, on the other hand, $\mu = i\beta$, where β is real, the motion is stable, and the dominant term $n = 0$ in the series solution behaves like

$$A' \cos \beta\xi + B' \sin \beta\xi$$

In the present case the fundamental frequency of the resulting motion is then $\beta\Omega/2$.

Detailed discussions of the solutions of Mathieu's equation are given elsewhere.* It can be shown that there exists an infinite set of regions in the a-q diagram for which the motion is stable, the first of these being shown in Fig. 26-1. The sign of q is related only to the phase of the applied oscillating force, hence it is irrelevant to the stability of the system. The sign of a, on the other hand, is very important in determining the stability. For $q = 0$, the motion is always stable for $a > 0$ (corresponding to the normal pendulum) and always unstable for $a < 0$ (corresponding to the inverted pendulum without oscillation of the support). We note that for $q > 0.9$ and a negative the motion will be stable if $|a|$ is not too large, but if a is positive the motion is unstable. This may be demonstrated by operating an inverted pendulum in this region and then suddenly inverting it again so that it oscillates in the normal pendulum position. It is found that under these conditions the normal motion is unstable.

* N. W. McLachlan, *Theory and Application of Mathieu Functions*, Oxford, London (1947). R. Campbell, *Théorie Générale de l'Equation de Mathieu*, Masson et Cie, Paris (1955). J. Meixner and F. W. Schafke, *Mathieusche Funktionen und Sphäroidfunktionen*, Springer, Berlin (1954). J. P. den Hartog, *Mechanical Vibrations*, McGraw-Hill, New York (1956), Chapter 8. W. Paul and H. Steinwedel, *Z. Naturforsch.*, 8a, 448 (1953). R. F. Wuerker, H. Shelton, and R. V. Langmuir, *J. Appl. Phys.*, 30, 342 (1959).

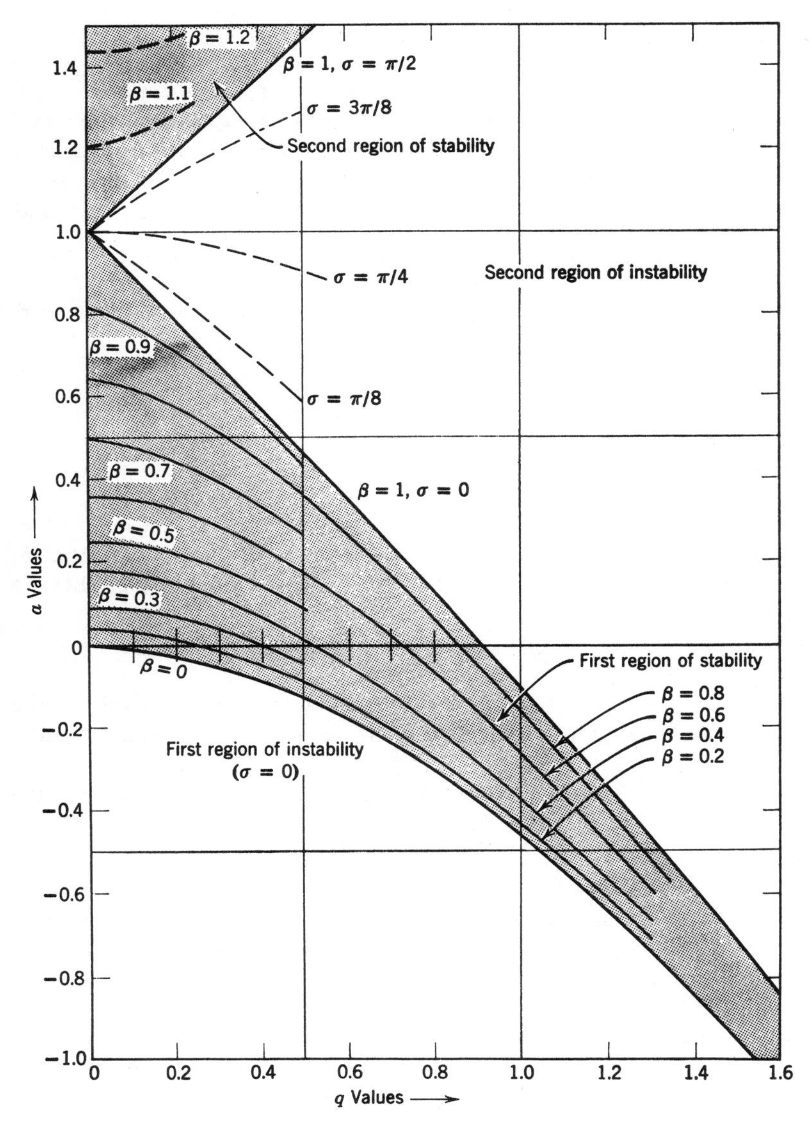

Fig. 26-1. Stability diagram for Mathieu's equation $\frac{d^2\theta}{d\xi^2} + (a - 2q\cos 2\xi)\theta = 0$.

27. ROCKET MOTION

Let us consider a rocket of instantaneous mass M moving with velocity $\mathbf{v}$, expelling propellant with a velocity $\mathbf{c}$ relative to the rocket.* At a time Δt later, when an extra amount $(-\Delta M)$ (>0) of propellant has been ejected, the velocity of the rocket is $\mathbf{v} + \Delta\mathbf{v}$ and its mass is $M + \Delta M$. The law of conservation of momentum then tells us that

$$(M + \Delta M)(\mathbf{v} + \Delta\mathbf{v}) + (\mathbf{c} + \mathbf{v})(-\Delta M) = M\mathbf{v}$$

since the velocity of the propellant relative to the ground is $\mathbf{c} + \mathbf{v}$. Hence

$$\frac{M\,\Delta\mathbf{v}}{\Delta t} = \frac{\mathbf{c}\,\Delta M}{\Delta t}$$

or

$$M\ddot{\mathbf{r}} = \dot{M}\mathbf{c} \tag{27.1}$$

where $\dot{M}$ is the rate of increase of M, and so is negative. If in addition the rocket is subjected to forces such as drag which have a resultant $\mathbf{F}$, and to the force $-M\,\nabla U$ of gravity (U = gravitational potential) the equation of motion becomes

$$M\ddot{\mathbf{r}} + M\,\nabla U = \dot{M}\mathbf{c} + \mathbf{F} \tag{27.2}$$

This is the equation of rocket motion which we consider first for the special case of free motion in one dimension ($\mathbf{F} = 0$, U = const, $\mathbf{v}$ antiparallel to $\mathbf{c}$).

In this case (27.2) reduces to (27.1) which may be integrated to give

$$\frac{v}{c} = \ln\frac{M_0}{M} \tag{27.3}$$

where M_0 is the mass of the rocket at the time when its velocity is zero. If the mass is known as a function of the time, the velocity and therefore the position may then be computed in terms of the time. In particular, if the rate of expulsion of propellant is uniform,

$$M = M_0\left(1 - \frac{t}{t_0}\right)$$

where t_0 = const, and the distance traveled in time t from rest is

$$x = c\left[t - (t - t_0)\ln\left(1 - \frac{t}{t_0}\right)\right] \tag{27.4}$$

* H. S. Seifert (ed.), *Space Technology*, Wiley, New York (1959), Chapter 3.

This is valid only up to the time $(M_F/M_0)t_0$ at which the fuel, of initial mass M_F, is exhausted. At this point the rocket has traveled a distance

$$ct_0\left[1 - \frac{M_L}{M_0}\left(1 + \ln\frac{M_0}{M_L}\right)\right]$$

where $M_L = M_0 - M_F$ is the mass of the rocket without fuel, and the velocity is

$$c \ln \frac{M_0}{M_L} \tag{27.5}$$

Unless the mass M_L of the rocket when unfueled is very small compared with the mass M_0 at take-off, the velocity reached can never be much larger than the exhaust velocity.

If the rocket is moving vertically upward in the gravitational field of the earth, assumed uniform, (27.2) becomes

$$\dot{v} = -\frac{\dot{M}c}{M} - g \tag{27.6}$$

If $\dot{M}$ is constant, so that the mass is decreasing linearly with time,

$$M = M_0(1 - at) \qquad (aM_0 = -\dot{M} > 0)$$

we have

$$v = -gt - c\ln(1 - at) = -gt + c\ln\frac{M_0}{M} \tag{27.7}$$

if $v = 0$ when $t = 0$. The height reached at time t is then

$$x = ct - \frac{1}{2}gt^2 + \frac{c}{a}(1 - at)\ln(1 - at) \qquad \left(t < t_0 = \frac{1}{a}\right) \tag{27.8}$$

$$= ct - \frac{1}{2}gt^2 - \frac{c}{a}\frac{M}{M_0}\ln\frac{M_0}{M}$$

28. ATMOSPHERIC DRAG

If a macroscopic object moves through a region of space in which it is continuously bombarded by small particles, the motion of its center of mass is modified by a drag or lift force arising from the momentum transfer between the object and the particles. The resultant force depends on whether the collisions are elastic on the one hand, or such that the particles penetrate the object and become part of it on the other. For specular reflection, each particle imparts an impulse to the object in the direction normal to the surface at the point of collision, and the resultant drag then depends critically upon the shape of this surface. At the other extreme—the case in which the particles are absorbed by the object—the shape of the latter is not so important.

If A is the cross section of a body normal to the direction of its motion, the resultant drag force arising from its motion through a fluid is approximately proportional to the square of its velocity if the fluid motion is turbulent:

$$F = \tfrac{1}{2}C_D A\rho v^2 \tag{28.1}$$

Here ρ is the density of the fluid and C_D is a coefficient which depends on the shape of the object and is a slowly varying function of v. For laminar flow, however, the force on a sphere of radius a is given by Stokes' law

$$F = 6\pi\eta va \tag{28.2}$$

where η is the viscosity of the fluid. It is therefore proportional to the first power of the velocity.

For a rocket moving vertically upward through the atmosphere, (27.6) gives, with (28.1),

$$\dot{v} = -\frac{\dot{M}c}{M} - g - \frac{1}{2}\frac{C_D A\rho v^2}{M} \tag{28.3}$$

The atmospheric density ρ at height x is given approximately by

$$\rho = \rho_0 e^{-\beta x}$$

where $\beta^{-1} \approx 3 \times 10^4$ ft.

Equation (28.3) may be solved approximately by supposing that the effect of drag is small and expanding in powers of the coefficient C_D. Thus for $\dot{M} = \text{const} = -aM_0$, we may substitute the solution (27.7), obtained by neglecting drag, into the last term of (28.3) to obtain

$$v = v_0 + C_D A v_1$$

with

$$v_1 = -\frac{1}{2}\frac{\rho_0 e^{-\beta x} v_0{}^2}{M_0(1 - at)}$$

v_0 being given by (27.7), x by (27.8). Thus v_1 is obtained as a function of t.

More powerful methods of approximation will be discussed later in this book, although the development of digital and analog computers now makes it less important to obtain algebraic solutions to such problems, since in a practical case it is simpler and more accurate to solve equations such as (28.3) by machine.

29. THE POYNTING-ROBERTSON EFFECT

In this section we consider the motion of a meteoritic dust particle in the gravitational field of the sun, but take into account the effect of radiation pressure on the motion of the particle. Since the force due to radiation pressure is proportional to the area of the particle, whereas the gravitational force is proportional to the mass and therefore to the volume, the relative importance of radiation pressure increases as the radius of the particle decreases, until for particles of radius $\sim 10^{-4}$ cm the two forces are comparable.

We may make the simplifying assumption that the particle absorbs all the radiation that is incident upon it. If $\dot{E}$ ergs/sterradian/sec are emitted from the sun with a velocity c, the flux incident on the particle at a distance r from the radius of the sun is $(\dot{E}A)/(4\pi r^2)$ ergs/sec, where A is the cross section of the particle in the plane normal to the line joining it to the center of the sun. The particle therefore accretes mass at the rate $\dot{M} = (\dot{E}A)/(4\pi r^2 c^2)$ grams/sec and acquires momentum at the rate $[(\dot{E}A)/(4\pi r^2 c)]\mathbf{r}$ per sec. Its equation of motion is therefore

$$\frac{d}{dt}(m\mathbf{v}) = -\frac{GMm}{r^3}\mathbf{r} + \frac{\dot{E}A}{4\pi r^3 c}\mathbf{r}$$

or

$$m\frac{d\mathbf{v}}{dt} = -\left(GMm - \frac{\dot{E}A}{4\pi c}\right)\frac{\mathbf{r}}{r^3} - \frac{\dot{E}A}{4\pi r^2 c^2}\mathbf{v} \tag{29.1}$$

In general, the radial force is dominant, and the effect of radiation pressure is simply to reduce the resultant attraction toward the sun. However, superimposed on this attraction is a small effective force in the direction of $-\mathbf{v}$, and this force causes the particle to spiral in towards the sun. For

$$GMm < \frac{\dot{E}A}{4\pi c} \tag{29.2}$$

this inward spiraling motion does not occur, since under these conditions the resultant radial force is in the direction away from the sun.

Spheres of density ρ and radius

$$r < \frac{3\dot{E}}{16\pi GMc\rho} \tag{29.3}$$

are therefore swept from the solar system by radiation pressure. As mentioned above, this critical radius depends on the density but using the value $\dot{E} = 3.78 \times 10^{26}$ watts we find that it is of the order of 1 μ.

The assumption made in our analysis that the dust particle absorbs all of the radiation incident on it is not valid, but the motion of the particle would not be altered if some or even all of this energy were subsequently reradiated uniformly in all directions. However, the actual radiation pattern will depend upon the spin, thermal conductivity, and shape of the particle, and a nonisotropic pattern would lead to an extra thrust on the particle not included in the above analysis.*

30. THE DAMPED OSCILLATOR

We consider here the one-dimensional oscillator which is subject to a damping force which is proportional to the velocity and in the direction opposite to the velocity. The equation of motion is therefore

$$m\ddot{x} = -kx - \lambda\dot{x} \qquad (\lambda > 0) \tag{30.1}$$

Hence, on multiplying by $\dot{x}$, we see that

$$\frac{d}{dt}\left(\tfrac{1}{2}m\dot{x}^2 + \tfrac{1}{2}kx^2\right) = -\lambda\dot{x}^2$$

expressing the rate of decrease of the kinetic plus potential energy of the particle in terms of the rate at which work is done against the particle by the dissipative force.

Even though all quantities in (30.1) are real, we may look for a solution of the form

$$x = x_0 e^{\alpha t}$$

even though the resulting equation for α

$$m\alpha^2 + \lambda\alpha + k = 0$$

implies that α may be complex:

$$\alpha = -\frac{\lambda \pm (\lambda^2 - 4km)^{1/2}}{2m} \tag{30.2}$$

* A. C. B. Lovell, *Meteor Astronomy*, Oxford, London (1954); T. R. Kaiser, *Advances in Phys.*, 2, 495 (1953).

If $\lambda^2 > 4km$, α is in fact real, and the general solution of (30.1) is a sum of two terms corresponding to the two roots (30.2):

$$(30.3) \quad x = \exp\left(-\frac{\lambda t}{2m}\right)\left[A \exp\left(\frac{(\lambda^2 - 4km)^{1/2}}{2m} t\right) + B \exp\left(-\frac{(\lambda^2 - 4km)^{1/2}}{2m}\right)\right]$$

Since both values of α are negative for $\lambda > 0$, the motion is damped, and since $\lambda^2 > 4km$, there is no oscillatory motion. For the special case $\lambda^2 = 4km$, the general solution in terms of the arbitrary constants A, B is

$$x = (A + Bt) \exp\left(-\frac{\lambda t}{2m}\right)$$

The motion is then said to be *critically damped.* For $\lambda^2 < 4km$, the solution continues to have the form (30.3), with A, B complex numbers such that the expression for x is real. Alternatively, we may now express the solution as

$$x = \exp\left(-\frac{\lambda t}{2m}\right)\left(C \cos \frac{(4km - \lambda^2)^{1/2}}{2m} t + D \sin \frac{(4km - \lambda^2)^{1/2}}{2m} t\right)$$

where C, D are real constants. The oscillator therefore moves with angular frequency

$$\omega = \left(\frac{k}{m} - \frac{\lambda^2}{4m^2}\right)^{1/2} = \frac{2\pi}{T}$$

but its amplitude decreases by the factor

$$\exp\left(-\frac{\lambda T}{2m}\right) = \exp \frac{2\pi}{\left(\dfrac{4mk}{\lambda^2} - 1\right)^{1/2}}$$

during each period T. It will therefore decrease by a factor e during Q periods, where

$$Q = \frac{1}{2\pi}\left(\frac{4mk}{\lambda^2} - 1\right)^{1/2}$$

For $4mk \gg \lambda^2$, the frequency is decreased only slightly by the damping, and

$$Q \approx \frac{m\omega}{\pi\lambda}$$

Further discussion of dissipative systems is given in Sec. 46.

EXERCISES

1. A dust particle of radius 10 μ, density 5 grams/cm^3 is left at a height of 80 km from the burning up of a meteor. Find the time for it to reach the ground if convection currents in the atmosphere are neglected.

2. Find the rate at which the earth is spiraling into the sun under the influence of the Poynting-Robertson effect.
(*Ans.* Approximately 5×10^{-4} cm per year!)

3. Write down and solve approximately the equation of motion of a rocket moving vertically upward, taking account of the variation of gravity.

4. An external force $F_0 \cos \Omega t$ is applied to the particle discussed in Sec. 30. Find the resulting steady state motion. Show the dependence of the amplitude and phase difference between the force and the velocity of the particle on the frequency of the applied force.

5. Discuss the case of a one-dimensional oscillator which is subject to a damping force equal to $\lambda\dot{x}^2$. Find the effect of this force on the motion, neglecting powers of λ higher than the first.

6. Lunar tides on the earth cause a dissipation of kinetic energy of rotation by friction. Discuss the cause of the concomitant decrease of angular momentum.

7. Examine the analogy between simple, electric circuits and the types of motion discussed in Secs. 26 and 30 and Exercises 4 and 5.

6 LAGRANGE'S EQUATIONS OF MOTION

31. DERIVATION OF LAGRANGE'S EQUATIONS

In Chapter 2 the laws of motion were formulated in the language of vectors in a Euclidean three-dimensional space. The fact that these laws can be taken as statements in a $3N = f'$ dimensional Euclidean space helps in the visualization of some properties of the system but does not contribute to the solution of a given problem, since the only change in the mathematics entailed by this point of view is a relabeling of the coordinates with a single index running from 1 to f' instead of with two, perhaps i and ρ.

A problem in dynamics is usually given in the form of the question: What motion results from the action of specified forces acting on a system with specified constraints? The problem is thus to find a solution of the equation of motion

$$m_\rho \mathbf{a}_\rho = \mathbf{F}_\rho$$

starting from a given initial state, where $\mathbf{F}_\rho$ is a given function of the coordinates and the time.

Elimination of the constraints by the method of Sec. 2 reduces the number of coordinates to the minimum, but these coordinates are not generally cartesian in nature. Writing down the components of the acceleration in generalized coordinates is tedious. We seek a form for the equations of motion which is valid in all coordinate systems and which involves only functions easily determined in generalized coordinates, namely functions of the coordinates and velocities.

A system of N particles is described by $3N$ cartesian coordinates $x_{\rho,i}$. This system is subject to k constraints, $k < 3N$. The $3N$

cartesian coordinates can then be expressed as functions of the $f = 3N - k$ generalized coordinates q_m. From Sec. 3 we obtain expressions for the cartesian velocity components.

$$\dot{x}_{\rho,i} = \frac{\partial x_{\rho,i}}{\partial q_m} \dot{q}_m + \frac{\partial x_{\rho,i}}{\partial t} \qquad (m = 1, 2, \cdots, f) \tag{31.1}$$

If now for purposes of differentiation we consider q_m and $\dot{q}_m$ as independent variables (which is certainly permissible since the coordinate differentials dq_m which enter the $\dot{q}_m$ do not depend on the q_m), we obtain

$$\frac{\partial \dot{x}_{\rho,i}}{\partial \dot{q}_m} = \frac{\partial x_{\rho,i}}{\partial q_m} \tag{31.2}$$

The equations of motion in cartesian coordinates are

$$F_{\rho,i} = m_\rho \ddot{x}_{\rho,i} \tag{31.3}$$

We now proceed to calculate the amount of work done by these forces in an arbitrary small displacement $dx_{\rho,i}$ which is consistent with the constraints. This is given by

$$W = \sum_{\rho,j} F_{\rho,j}\, dx_{\rho,j} \tag{31.4}$$

When expression (31.3) is used for the $F_{\rho,i}$, this becomes

$$\begin{aligned}
W &= \sum_\rho m_\rho \ddot{x}_{\rho,i}\, dx_{\rho,i} \\
&= \sum_\rho m_\rho \ddot{x}_{\rho,i} \left(\frac{\partial x_{\rho,i}}{\partial q_m}\, dq_m + \frac{\partial x_{\rho,i}}{\partial t}\, dt \right) \\
&= \sum_\rho m_\rho \left[\frac{d}{dt} \left(\dot{x}_{\rho,i} \frac{\partial \dot{x}_{\rho,i}}{\partial \dot{q}_m} \right) - \dot{x}_{\rho,i} \frac{d}{dt} \frac{\partial x_{\rho,i}}{\partial q_m} \right] dq_m \\
&\qquad\qquad + \sum_\rho m_\rho \ddot{x}_{\rho,i} \frac{\partial x_{\rho,i}}{\partial t}\, dt \\
&= \sum_\rho \left[\frac{d}{dt} \frac{\partial}{\partial \dot{q}_m} \left(\frac{1}{2}\, m_\rho \dot{x}_{\rho,i} \dot{x}_{\rho,i} \right) - \frac{\partial}{\partial q_m} \left(\frac{1}{2}\, m_\rho \dot{x}_{\rho,i} \dot{x}_{\rho,i} \right) \right] dq_m \\
&\qquad\qquad + \sum_\rho m_\rho \ddot{x}_{\rho,i} \frac{\partial x_{\rho,i}}{\partial t}\, dt
\end{aligned} \tag{31.5}$$

where use has been made of (31.2).

Equation (31.5) gives us an expression for the work done in an arbitrary small displacement, a quantity independent of any particular choice of coordinates, in terms of derivatives of a function with respect to the q_m, $\dot{q}_m$, and t. The function occurring is the kinetic energy of the system T. Since T is a function of the $\dot{x}_{\rho,i}$, it can easily be expressed in terms of the $\dot{q}_m$ and the indicated differentiations carried out. Thus

$$W = \left(\frac{d}{dt}\frac{\partial T}{\partial \dot{q}_m} - \frac{\partial T}{\partial q_m}\right) dq_m + \sum_{\rho,i} m_\rho \ddot{x}_{\rho,i} \frac{\partial x_{\rho,i}}{\partial t}\, dt \tag{31.6}$$

Now (31.4) may be written

$$\begin{aligned} W &= \sum_{\rho,i} F_{\rho,i} \left(\frac{\partial x_{\rho,i}}{\partial q_m}\, dq_m + \frac{\partial x_{\rho,i}}{\partial t}\, dt\right) \\ &= Q_m\, dq_m + Q_t\, dt \end{aligned} \tag{31.7}$$

where

$$Q_m = \sum_{\rho,i} F_{\rho,i} \frac{\partial x_{\rho,i}}{\partial q_m} \qquad Q_t = \sum_{\rho,i} F_{\rho,i} \frac{\partial x_{\rho,i}}{\partial t} \tag{31.8}$$

may be called the generalized force components. The two expressions (31.6) and (31.7) must be equal for any displacement dq_m, dt, and so the coefficients of dq_m must be equal:

$$\frac{d}{dt}\frac{\partial T}{\partial \dot{q}_m} - \frac{\partial T}{\partial q_m} = Q_m \tag{31.9}$$

These are the equations of motion in Lagrange's form. The equality of the coefficients of dt is trivial.

It should be noted that the only forces $F_{\rho,i}$ which contribute to the generalized forces Q_m defined by (31.7) are what we call the applied forces. The forces exerted by the constraints do not contribute to the Q_m since they do not contribute to W. This arises from the fact that the displacements were chosen so that they did not violate the constraints. For example, in discussing the motion of a bead along a smooth wire, we should consider in (31.4) only those coordinate changes which are consistent with the condition that the bead remain on the wire, i.e., displacements that are normal to the constraining force. This force of constraint therefore does no work during the displacement. This is an important simplification because the forces exerted by the constraints are generally unknown and their discovery is necessary to solve the problem if the form (31.3) for the equations of motion is used.

If none of the constraints on the system involves the time and if a moving coordinate system is not introduced, the kinetic energy is a homogeneous quadratic function of the generalized velocity components. Equation (31.1) then becomes

$$\dot{x}_{\rho,i} = \frac{\partial x_{\rho,i}}{\partial q_m} \dot{q}_m \tag{31.10}$$

and thus

$$\begin{aligned} T &= \frac{1}{2} \sum_{\rho,i} m_\rho (\dot{x}_{\rho,i})^2 \\ &= \tfrac{1}{2} g_{mn} \dot{q}_m \dot{q}_n \end{aligned} \tag{31.11}$$

where

$$g_{mn} = \sum_{\rho,i} m_\rho \frac{\partial x_{\rho,i}}{\partial q_m} \frac{\partial x_{\rho,i}}{\partial q_n} \tag{31.12}$$

If there are moving constraints or if a moving coordinate system has been introduced, the kinetic energy T will involve terms linear in and terms independent of the generalized velocity components.

If the applied forces acting on the system are derivable from a potential function, the generalized force components are derivable from the same potential function expressed in terms of the generalized coordinates. If

$$F_{\rho,j} = -\frac{\partial V}{\partial x_{\rho,j}} \tag{31.13}$$

then from (31.7)

$$\begin{aligned} Q_m &= \sum_{\rho,j} F_{\rho,j} \frac{\partial x_{\rho,j}}{\partial q_m} \\ &= -\sum_{\rho,j} \frac{\partial V}{\partial x_{\rho,j}} \frac{\partial x_{\rho,j}}{\partial q_m} \\ &= -\frac{\partial V}{\partial q_m} \end{aligned} \tag{31.14}$$

Thus, if the system is in a lamellar force field, (31.9) can be written

$$\frac{d}{dt} \frac{\partial T}{\partial \dot{q}_m} - \frac{\partial T}{\partial q_m} + \frac{\partial V}{\partial q_m} = \frac{d}{dt} \frac{\partial T}{\partial \dot{q}_m} - \frac{\partial}{\partial q_m} (T - V) = 0 \tag{31.15}$$

32. THE LAGRANGIAN FUNCTION

The equations of motion for a system in a lamellar field of force (31.15) suggest that the function $T - V$ plays an important role. In generalized coordinates T depends on both the coordinates and the velocity components, whereas V depends only on the coordinates and perhaps the time explicitly. We may therefore introduce a term $(d/dt)(\partial V/\partial \dot{q}_m)$ into (31.15) without changing it.

We define a function L by the equation

$$L = T - V \tag{32.1}$$

and call it the *Lagrangian function* or, briefly, the *Lagrangian*. The equations of motion (31.15) then can be written as

$$\frac{d}{dt}\frac{\partial L}{\partial \dot{q}_m} - \frac{\partial L}{\partial q_m} = 0 \tag{32.2}$$

Knowledge of the Lagrangian alone is sufficient to predict the motion of a system in a lamellar field from a given initial state.

A Lagrangian exists if the system is in a lamellar field. This is not a necessary condition. A Lagrangian also exists if those forces Q_m which are not derivable from a potential, Q_m' say, can be expressed in the form

$$Q_m' = \frac{d}{dt}\frac{\partial M}{\partial \dot{q}_m} - \frac{\partial M}{\partial q_m} \tag{32.3}$$

where M is a function of the coordinates and the velocity components. If this is the case, the Lagrangian is given by

$$L = T - V - M \tag{32.4}$$

where V is the ordinary potential and M might be called a generalized potential. Such M functions are needed to describe the motion of a charged particle in an electromagnetic field.

The Lagrange equations of motion are a set of f ordinary differential equations of the second order. A complete solution will contain $2f$ arbitrary constants. These constants are usually taken to specify the state of the system at some initial time. Instead of giving the initial state of the system one might give the initial configuration and a later configuration. These conditions may not be self-consistent, because the second configuration may not result from the first one under the action of the given forces no matter how the initial velocity components are chosen.

One of the most useful devices for solving the Lagrange equations of motion is to discover the *first integrals* of the motion. A first integral of a set of differential equations is a function of the unknowns which contains derivatives of one order lower than the order of the differential equations themselves and which remains constant by virtue of the differential equations. Examples of such integrals are the energy and the momentum of isolated systems. The advantage of having an integral of the motion is that it reduces the order of the system of equations to be solved. Often a problem can be completely solved by using integrals of the motion without writing down the equations of motion. This was done in Chapter 3 by use of the energy integral. We proceed to investigate some of these integrals.

33. THE JACOBIAN INTEGRAL

In Sec. 12 the conservation of energy was discussed. If a system has a potential energy which does not depend on the time explicitly, the total mechanical energy of the system is constant. A theorem which reduces to the conservation of mechanical energy for most systems of interest is the following: *In a system described by a Lagrangian which does not contain the time explicitly, the quantity*

$$E = \frac{\partial L}{\partial \dot{q}_m} \dot{q}_m - L \tag{33.1}$$

is constant. The proof follows from (32.2).

$$\begin{aligned} \frac{dE}{dt} &= \frac{d}{dt}\left(\frac{\partial L}{\partial \dot{q}_m}\right) \dot{q}_m + \frac{\partial L}{\partial \dot{q}_m} \ddot{q}_m - \frac{\partial L}{\partial \dot{q}_m} \ddot{q}_m - \frac{\partial L}{\partial q_m} \dot{q}_m - \frac{\partial L}{\partial t} \\ &= \left(\frac{d}{dt}\frac{\partial L}{\partial \dot{q}_m} - \frac{\partial L}{\partial q_m}\right) \dot{q}_m - \frac{\partial L}{\partial t} \\ &= -\frac{\partial L}{\partial t} \end{aligned} \tag{33.2}$$

The quantity E is called the *Jacobian integral* of the motion.

In most systems of interest E represents the total mechanical energy of the system. Being of the form (32.1), the Lagrangian contains terms at most quadratic in the velocity components. Let us write L as

$$L = L_2 + L_1 + L_0 \tag{33.3}$$

where L_2 is that part of L quadratic in the velocities, L_1 is the part linear in the velocities, and L_0 is the part independent of the velocities.

Then, by Euler's theorem on homogeneous functions,

$$\frac{\partial L}{\partial \dot{q}_m} \dot{q}_m = 2L_2 + L_1 \tag{33.4}$$

so that

$$E = L_2 - L_0 \tag{33.5}$$

Now if the system is described in a stationary coordinate system and if the potential is not of the generalized M type, the part L_1 of the Lagrangian is zero, the part L_2 is the kinetic energy, and the part L_0 is the negative of the potential energy. A system with this type of Lagrangian is called a *natural* one. Thus in a natural system

$$E = T + V \tag{33.6}$$

and represents the total mechanical energy. *The mechanical energy of a natural system whose Lagrangian does not contain the time explicitly is constant.*

34. MOMENTUM INTEGRALS

The Jacobian integral exists when the Lagrangian is independent of the time. When the Lagrangian is independent of a coordinate, another kind of integral exists which we call a *momentum integral.* The missing coordinate is called *ignorable* or *cyclic.*

Suppose that the Lagrangian does not contain q_m explicitly. Then $\partial L/\partial q_m$ is zero, and the corresponding Lagrange equation reduces to

$$\frac{d}{dt}\frac{\partial L}{\partial \dot{q}_m} = 0 \tag{34.1}$$

which gives upon integration

$$\frac{\partial L}{\partial \dot{q}_m} = \beta_m \tag{34.2}$$

where β_m is a constant. In general, the quantity $\partial L/\partial \dot{q}_m$ is called the m component of the momentum and is denoted by p_m:

$$\frac{\partial L}{\partial \dot{q}_m} = p_m \tag{34.3}$$

Thus, if a coordinate q_m is ignorable, the corresponding momentum p_m is a constant. If the system is a natural one and if the coordinates are cartesian, the momenta defined by (34.3) are the same as those originally defined by (9.1), namely

$$p_{\rho,i} = m_\rho \dot{x}_{\rho,i} \tag{34.4}$$

Momentum integrals may exist even though no coordinate is ignorable, since a change of coordinates may make a Lagrangian which contained only $f - 1$ coordinates contain all f of the new coordinates. Similarly, on a fortunate change of coordinates, the new Langrangian may contain fewer coordinates than the old.

An example of an ignorable coordinate is given by a two-dimensional harmonic oscillator described in plane polar coordinates. Let the mass of the oscillator be m and the spring constant be k. Then in plane polar coordinates

$$\begin{aligned} T &= \tfrac{1}{2}m(\dot{r}^2 + r^2\dot{\theta}^2) \\ V &= \tfrac{1}{2}kr^2 \end{aligned} \tag{34.5}$$

so that the Lagrangian is

$$L = T - V = \tfrac{1}{2}m(\dot{r}^2 + r^2\dot{\theta}^2) - \tfrac{1}{2}kr^2 \tag{34.6}$$

θ is an ignorable coordinate. Hence $p_\theta = \beta_\theta =$ const.

$$mr^2\dot{\theta} = \beta_\theta \tag{34.7}$$

β_θ is seen to be the angular momentum about the origin, and to remain constant throughout the motion. Another constant of the motion is provided by the Jacobian integral (33.6),

$$E = T + V = \tfrac{1}{2}m(\dot{r}^2 + r^2\dot{\theta}^2) + \tfrac{1}{2}kr^2 \tag{34.8}$$

Equations (34.7) and (34.8) can be solved for r and θ as functions of the time, the solution involving only the evaluation of two integrals.

If the system described above were given in terms of cartesian coordinates, there would be no ignorable coordinate, though both the energy and the angular momentum would still be conserved. In these coordinates

$$\begin{aligned} T &= \tfrac{1}{2}m(\dot{x}^2 + \dot{y}^2) \\ V &= \tfrac{1}{2}k(x^2 + y^2) \end{aligned} \tag{34.9}$$

so that the Lagrangian is

$$L = \tfrac{1}{2}m(\dot{x}^2 + \dot{y}^2) - \tfrac{1}{2}k(x^2 + y^2) \tag{34.10}$$

It is not hard to show that the quantity

$$m(x\dot{y} - y\dot{x})$$

is constant as a consequence of the equations of motion. The existence of this integral is concealed by the unfortunate choice of coordinates.

In Sec. 13 the conservation of angular momentum was established for an isolated system of particles under the condition that the forces between any two particles of the system acted along the line joining the particles. We may now prove a more general theorem: the total angular momentum of an isolated system is conserved.

To prove this theorem we choose as one of the coordinates θ of the system the angular position of one particular particle of the system about an axis. The angular positions of all the other particles of the system about this axis are then measured relative to the above particle, so that a displacement $\delta\theta$ causes the entire system to revolve about the axis through the angle $\delta\theta$. The Lagrangian of the system will contain $\dot{\theta}$, and $\partial L/\partial\dot{\theta}$ will be the component of the total angular momentum about one axis since $\partial L/\partial\theta$ or Q_θ is of the nature of a torque. Now, because the system is isolated, its angular position θ about an arbitrary axis cannot affect its motion, and so θ cannot appear in the Lagrangian and Q_θ must be zero. Hence θ is an ignorable coordinate, and the angular momentum is

$$p_\theta = \frac{\partial L}{\partial \dot{\theta}} = \text{const} \tag{34.11}$$

This is true for any axis; therefore the total angular momentum is constant.

The distinction between this theorem and that of Sec. 13 lies in the way the angular momentum is defined. If "spin" angular momentum is present, the quantity $\partial L/\partial\dot{\theta}$ may differ from $\mathbf{l} \cdot \mathbf{n}$, where $\mathbf{l}$ is the "orbital" angular momentum defined in Sec. 15, and $\mathbf{n}$ is a unit vector in the direction of the above axis. Nonrelativistically, this difference is a constant if forces within the system are central in character. If noncentral forces are acting or relativistic effects are considered, only the total angular momentum is conserved (see Sec. 78).

35. CHARGED PARTICLE IN AN ELECTROMAGNETIC FIELD

In Sec. 32 we mentioned that a Lagrangian exists for systems in which the forces are not derivable from an ordinary potential but are expressible in the form

$$Q_m' = \frac{d}{dt}\frac{\partial M}{\partial \dot{q}_m} - \frac{\partial M}{\partial q_m} \tag{35.1}$$

where M is a function of the coordinates, velocities, and perhaps the time. Of particular interest are those cases where the forces Q_m' are those on a charged particle in an electromagnetic field, since these forces depend on the velocity and cannot, therefore, be derived from an ordinary potential.

The electric and magnetic fields *in vacuo* can be expressed in the form

$$\mathbf{B} = \nabla \times \mathbf{A}$$

$$\mathbf{E} = -\nabla\phi - \frac{1}{c}\frac{\partial \mathbf{A}}{\partial t} \tag{35.2}$$

where **A** is the vector potential and ϕ the scalar potential. We use gaussian units so that c is the velocity of light *in vacuo*.

The force on a particle with charge e is given by the Lorentz formula

$$\mathbf{F} = e\left(\mathbf{E} + \frac{\mathbf{v}}{c} \times \mathbf{B}\right) \tag{35.3}$$

To show how this can be derived from an M function we write down the equations of motion

$$\begin{aligned} \frac{d}{dt}(m\mathbf{v}) &= \mathbf{F} = e\left(\mathbf{E} + \frac{\mathbf{v}}{c} \times \mathbf{B}\right) \\ &= e\left[-\nabla\phi - \frac{1}{c}\frac{\partial \mathbf{A}}{\partial t} + \frac{\mathbf{v}}{c} \times (\nabla \times \mathbf{A})\right] \end{aligned} \tag{35.4}$$

The part of **E** coming from the scalar potential ϕ is already in the required form, since it is derivable from an ordinary potential. Using a result of Exercise 9, Chapter 1,

$$\mathbf{v} \times (\nabla \times \mathbf{A}) = \nabla(\mathbf{v} \cdot \mathbf{A}) - (\mathbf{v} \cdot \nabla)\mathbf{A}$$

we may rewrite (35.4) thus:

$$\begin{aligned} \frac{d}{dt}(m\mathbf{v}) &= e\left[-\nabla\phi - \frac{1}{c}\left(\frac{\partial \mathbf{A}}{\partial t} + \mathbf{v} \cdot \nabla\mathbf{A}\right) + \nabla\frac{\mathbf{v} \cdot \mathbf{A}}{c}\right] \\ &= e\left[-\nabla\left(\phi - \frac{\mathbf{v} \cdot \mathbf{A}}{c}\right) - \frac{1}{c}\frac{d\mathbf{A}}{dt}\right] \end{aligned} \tag{35.5}$$

The form of (35.5) suggests combining the total time derivatives on the left, to give

$$\frac{d}{dt}\left(m\mathbf{v} + \frac{e}{c}\mathbf{A}\right) = -e\nabla\left(\phi - \frac{\mathbf{v} \cdot \mathbf{A}}{c}\right) \tag{35.6}$$

With a total time derivative on the left and a partial derivative with respect to the coordinates on the right, this has the general form of a set of Lagrange equations

$$\frac{d}{dt}\left(\frac{\partial L}{\partial \dot{q}_j}\right) = \frac{\partial L}{\partial q_j} \tag{35.7}$$

and, with q_j ($j = 1, 2, 3$) denoting the cartesian coordinates, (35.6) is identical with (35.7) if

$$\frac{\partial L}{\partial v_i} = mv_i + \frac{e}{c} A_i$$

$$\frac{\partial L}{\partial x_i} = \frac{\partial}{\partial x_i}\left[-e\left(\phi - \frac{v_j A_j}{c}\right)\right]$$

or, in vector notation,

$$\frac{\partial L}{\partial \mathbf{v}} = m\mathbf{v} + \frac{e}{c}\mathbf{A}$$

$$\frac{\partial L}{\partial x} = \frac{\partial}{\partial x}\left[-e\left(\phi - \frac{\mathbf{v}\cdot\mathbf{A}}{c}\right)\right]$$

i.e., if

$$L = \tfrac{1}{2}mv^2 - e\phi + \frac{e\mathbf{v}\cdot\mathbf{A}}{c} \tag{35.8}$$

The Lorentz force equation (35.4) may therefore be derived from this Lagrangian, the effects of the vector potential **A** on the motion of the particle being described very simply by the extra M function [cf. (32.4)]

$$M = -\frac{e\mathbf{v}\cdot\mathbf{A}}{c}$$

The subtraction of the M from the Lagrangian changes the momentum as defined by (34.3):

$$\mathbf{p} = \frac{\partial L}{\partial \mathbf{v}} = m\mathbf{v} + \frac{e}{c}\mathbf{A} \tag{35.9}$$

Thus the momentum of a charged particle moving in an electromagnetic field is not defined as its mass times its velocity. If such a particle moves under the influence of vector and scalar potentials which are independent of a particular coordinate, then the momentum of the particle in that direction is a constant of the motion, momentum being defined by (35.9).

GENERAL REFERENCES

M. Abraham and R. Becker, *The Classical Theory of Electricity and Magnetism.* Blackie, London (1932).

W. K. H. Panofsky and Melba Phillips, *Classical Electricity and Magnetism.* Addison-Wesley, Cambridge (1955).

J. C. Slater and N. H. Frank, *Electromagnetism.* McGraw-Hill, New York (1947).

J. A. Stratton, *Electromagnetic Theory.* McGraw-Hill, New York (1941).

EXERCISES

1. Write down the expressions for the kinetic energy of the following systems, using the minimum number of coordinates: (*a*) a free particle; (*b*) a particle constrained to remain on a sphere; (*c*) a particle constrained to remain on a circular cylinder; (*d*) a particle constrained to remain on a paraboloid of revolution.

2. A system is described by a Lagrangian $L(q_m, \dot{q}_m, t)$ involving f coordinates. If k additional constraints

$$\phi_s(q_1 \cdots q_f, t) = 0 \qquad (s = 1 \cdots k)$$

are imposed on the system, show that the Lagrange equations may be written

$$\frac{d}{dt}\frac{\partial L}{\partial \dot{q}_m} - \frac{\partial L}{\partial q_m} = \lambda_s(t)\frac{\partial \phi_s}{\partial q_m}$$

where the $\lambda_s(t)$ are unknown functions of t. The equations of motion together with the k equations of constraint provide $f + k$ equations for the $f + k$ unknowns, the q_m and the λ_s. (The λ's are known as *Lagrange multipliers.*)

3. Write down the Lagrangian for a particle confined to a horizontal plane in cartesian coordinates. Introduce the additional constraint $x^2 + y^2 = a^2$ by means of a Lagrange multiplier λ, and show that λ is proportional to the centripetal force exerted by the constraint upon the particle.

4. Show that in general the reaction on a constraint ϕ introduced with a Lagrange multiplier λ is given by

$$\lambda\left(g_{ij}\frac{\partial \phi}{\partial q_i}\frac{\partial \phi}{\partial q_j}\right)^{1/2}$$

5. An isolated system is described first in a fixed cartesian coordinate system and then in a uniformly moving one. Write down the Lagrangian for the system in both coordinate systems, and show that the equations of motion are identical. (The potential energy of an isolated system depends only on the relative positions of the particles and so only on the difference of coordinates.)

6. An isolated system is described first in fixed cartesian coordinates and then in a cartesian coordinate system which is rotating uniformly about the

z axis with angular velocity ω. Write down the Lagrangian in both coordinate systems. If the motion of each particle of the system is constrained to be either parallel or perpendicular to the plane containing that particle and the z axis, show that the effect of the rotation is to modify the potential energy of the system. (Either of the above constraints is sufficient to eliminate the Coriolis force, which cannot be derived from a potential.)

7. Compare the values of the function $(\partial L/\partial \dot{q}_m)\dot{q}_m - L$ in the stationary and moving coordinate systems of Exercises 5 and 6.

8. Show that, if the electromagnetic potentials ϕ and $\mathbf{A}$ are independent of the time, the Jacobian integral for the system of a charged particle in the electromagnetic field is the energy and is independent of $\mathbf{A}$. (This reflects the fact that a magnetic field can do no work on a charged particle.)

9. It is remarked in Sec. 32 that the solution of the equations of motion cannot always be specified by giving the configuration of the system at two distinct times. Show that in the case of the harmonic oscillator this can be done except when the two times are separated by an integral number of periods. Show that, if the two times are separated by almost a period, the motion is very sensitive to the conditions given.

10. Show that the equations of motion derivable from a Lagrangian are unchanged if to the Lagrangian there is added the total time derivative of an arbitrary function of q_m, t.

11. If F is a known function of the generalized velocities $\dot{q}_j$, discuss the applicability of the equations

$$\frac{d}{dt}\left(\frac{\partial L}{\partial \dot{q}_j}\right) - \frac{\partial L}{\partial q_j} = -\frac{\partial F}{\partial \dot{q}_j}$$

to the motion of a dissipative system, with particular reference to (29.1), (30.1), and (28.3). F is called the *Rayleigh dissipation function.*

7 APPLICATIONS OF LAGRANGE'S EQUATIONS

36. ORBITS UNDER A CENTRAL FORCE

Some of the properties of the motion of a particle under the influence of a force directed toward a fixed point were discussed in Sec. 24. Such a central force is not necessarily derivable from a potential, the only condition on it being that in spherical polar coordinates with the origin at the center of force the generalized forces Q_θ, Q_ϕ in the θ, ϕ directions should be zero. The kinetic energy expressed in terms of such coordinates is

$$T = \tfrac{1}{2}\mu(\dot{r}^2 + r^2\dot{\theta}^2 + r^2 \sin^2 \theta \dot{\phi}^2) \tag{36.1}$$

and the equations of motion are

$$\begin{aligned} \frac{d}{dt}\left(\frac{\partial T}{\partial \dot{r}}\right) - \frac{\partial T}{\partial r} &= Q_r \\ \frac{d}{dt}\left(\frac{\partial T}{\partial \dot{\theta}}\right) - \frac{\partial T}{\partial \theta} &= 0 \\ \frac{d}{dt}\left(\frac{\partial T}{\partial \dot{\phi}}\right) - \frac{\partial T}{\partial \phi} &= 0 \end{aligned} \tag{36.2}$$

Since T does not depend on ϕ, the last of equations (36.2) may be integrated to yield

$$\frac{\partial T}{\partial \dot{\phi}} = \mu r^2 \sin^2 \theta \dot{\phi} = m \tag{36.3}$$

where m is a constant. This is an angular momentum integral, and it gives the component of the angular momentum about the polar axis.

(In quantum mechanics this angular momentum component has to have the value of an integral multiple of $\hbar = h/2\pi$, where h is Planck's constant. This integer is denoted by m and is called the magnetic quantum number. In the classical case m can assume any value.)

Since the direction of the polar axis is arbitrary, all components of the angular momentum are constant and, therefore, so is the square of the length of this vector. This square is given by

$$
\begin{aligned}
l^2 &= (\mathbf{r} \times \mathbf{p})^2 \\
&= \mathbf{r}^2\mathbf{p}^2 - (\mathbf{r} \cdot \mathbf{p})^2 \\
&= 2\mu r^2 T - (\mu \dot{r} r)^2 \\
&= p_\theta{}^2 + m^2 \operatorname{cosec}^2 \theta
\end{aligned}
\tag{36.4}
$$

That this is a constant may also be verified directly:

$$\frac{d}{dt} l^2 = 2(p_\theta \dot{p}_\theta - m^2 \dot{\theta} \cos \theta \operatorname{cosec}^3 \theta)$$

since from (36.1), (36.2), and (36.3) it follows that

$$
\begin{aligned}
p_\theta &= \frac{\partial T}{\partial \dot{\theta}} = \mu r^2 \dot{\theta} = m \operatorname{cosec}^2 \theta (\dot{\theta}/\dot{\phi}) \\
\dot{p}_\theta &= \frac{\partial T}{\partial \theta} = \mu r^2 \sin \theta \cos \theta \dot{\phi}^2 \\
&= m \cot \theta \dot{\phi}
\end{aligned}
$$

Thus $p_\theta \dot{p}_\theta = m^2 \cos \theta \operatorname{cosec}^3 \theta \dot{\theta}$ and l^2 is constant. Note that $|m| \leq l$.

Sometimes it is more convenient to reorient the coordinate system so that l^2 has a simple form. If we choose the polar axis to lie in the plane of the orbit, i.e., if we choose the direction of the polar axis so that it intersects the orbit at some point, m vanishes, because at the point of intersection r and $\dot{\phi}$ must be finite and $\sin \theta$ vanishes. When we use this fact, the expression (36.1) for the kinetic energy becomes simplified because the term involving $\dot{\phi}$ disappears. Thus

$$T = \tfrac{1}{2}\mu(\dot{r}^2 + r^2\dot{\theta}^2) \tag{36.5}$$

This is the expression for the kinetic energy of a particle moving in a plane and described in plane polar coordinates.

The total orbital angular momentum now appears as the momentum

p_θ, and it is a constant because T does not depend on θ:

$$\frac{\partial T}{\partial \dot{\theta}} = \mu r^2 \dot{\theta} = l \tag{36.6}$$

To obtain an equation for the orbit we eliminate the time between (36.6) and the first of equations (36.2) with (36.5) for T. From the former,

$$\frac{d}{dt} = \dot{\theta}\frac{d}{d\theta} = \frac{l}{\mu r^2}\frac{d}{d\theta} \tag{36.7}$$

The latter equation is

$$\mu \ddot{r} - \mu r \dot{\theta}^2 = Q_r \tag{36.8}$$

Thus the equation of the orbit is

$$\frac{1}{r^2}\frac{d}{d\theta}\left(\frac{1}{r^2}\frac{dr}{d\theta}\right) - \frac{1}{r^3} = \frac{\mu Q_r}{l^2} \tag{36.9}$$

This becomes simpler on the introduction of $u = 1/r$ as the independent variable:

$$\frac{d^2u}{d\theta^2} + u = -\frac{\mu Q_r}{l^2 u^2} \tag{36.10}$$

In many cases occurring in nature, especially central gravitational and electrostatic forces, the central force on a particle may be derived from a potential function V. It then follows that V can depend only on r and, possibly, t, for if it were a function of θ or ϕ it would yield a force on the particle that would not be directed toward the origin.

In systems where the force Q_r is derivable from a potential, more information can be obtained about the orbit without actually solving (36.10). If the potential energy is independent of the time, the total energy $E = T + V$ is an integral of the motion. We can discuss the orbits in terms of E.

Using the angular momentum integral (36.6) in the radial equation of motion (36.8), we obtain

$$\mu \ddot{r} - \frac{l^2}{\mu r^3} + \frac{\partial V}{\partial r} = 0 \tag{36.11}$$

This is precisely the radial equation of motion that would be obtained from the one-dimensional Lagrangian

$$L = \frac{1}{2}\mu \dot{r}^2 - \frac{1}{2}\frac{l^2}{\mu r^2} - V \tag{36.12}$$

Thus the radial motion in a conservative central force problem is the same as the motion of a conservative one-dimensional system with a potential function V^* given by

$$V^* = V + \frac{1}{2}\frac{l^2}{\mu r^2} \tag{36.13}$$

V^* depends on the angular momentum of the original system. The addition to the potential energy, $l^2/2\mu r^2$, is called the centrifugal potential. The negative of its gradient gives the centrifugal force. (Note that this result cannot be obtained by eliminating ϕ from the Lagrangian $T - V$ directly by means of the angular momentum integral. The reason is that the operator $\partial/\partial r$ occurring in the radial equation of motion requires $\dot{\theta}$ to be held constant, and not l. Thus $\partial l/\partial r \neq 0$, even though $dl/dt = 0$.)

When $V(r)$ is known, $V^*(r)$ can be found for any value of l. The results of Sec. 17 may then be used to get a qualitative picture of the orbits. If the r motion is oscillatory, the particle is in a bound orbit about the origin. If the r motion is nonoscillatory, the orbit is not bounded and the particle goes to infinity. A sticking motion in the r direction always gives an orbit which approaches a circle asymptotically. As in Sec. 17, which of these orbits actually occur depends on the form of the potential and the value of the energy.

37. KEPLER MOTION

The most important instance of central motion is that under the influence of an inverse square law of force. This law comprises both gravitational and electrostatic forces; the electrostatic forces may be either attractive or repulsive, whereas the gravitational forces are always attractive.

An inverse square law of force is derivable from a potential

$$Q_r = \frac{\kappa}{r^2} = -\frac{\partial V}{\partial r} \tag{37.1}$$

so that we obtain the potential

$$V = \frac{\kappa}{r} \tag{37.2}$$

the constant having been chosen so that $V(\infty)$ vanishes. If the force is attractive, κ is negative; if the force is repulsive, κ is positive.

The equation for the orbit (36.10) becomes

$$\frac{d^2u}{d\theta^2} + u = -\frac{\kappa\mu}{l^2} \tag{37.3}$$

the right-hand side becoming a constant for this form of V. This equation can be integrated to give

$$u = \frac{1}{r} = -\frac{\kappa\mu}{l^2}[1 - e\cos(\theta - \theta_0)] \tag{37.4}$$

where e and θ_0 are the constants of integration. θ_0 represents the orientation of the curve. It is important in finding the motion resulting from given initial conditions, but for our purpose it is unimportant and will be taken to be zero. The shape of the orbit depends on the constant e. Equation (37.4) is the equation of a conic section. A conic section is the locus of a point whose distance from a given point, the focus, bears a constant ratio to its distance from a given line, the directrix. This ratio is e, the eccentricity of the conic. Figure 37-1 shows the geometrical situation. The focus is taken as the origin, and the directrix is a distance s to the left of the focus. Then we have

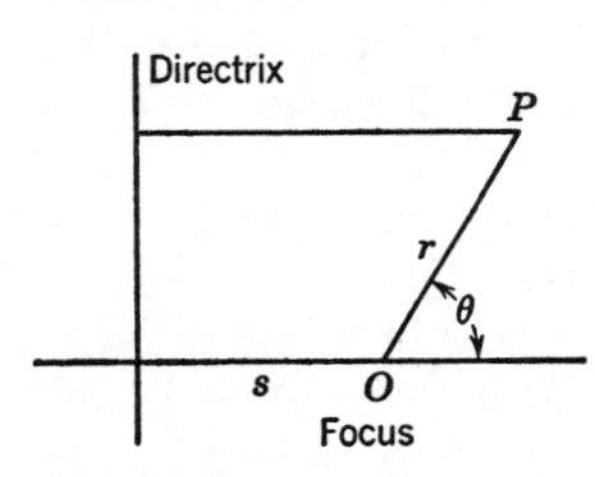

Fig. 37-1. Geometry of conic sections.

$$\begin{aligned} r &= e(s + r\cos\theta) \\ u &= \frac{1}{r} = \frac{1}{es}(1 - e\cos\theta) \end{aligned} \tag{37.5}$$

the second equation being similar in form to (37.4).

The character of the orbit depends on the value of e. If $e < 1$, $1/r$ never vanishes and so r remains finite. This curve is an ellipse. If $e = 1$, $1/r$ vanishes when $\theta = 0$ only; therefore, at $\theta = 0$, r becomes infinite. This curve is a parabola. If $e > 1$, $1/r$ vanishes for two values of θ, hence r becomes infinite for two values of θ. This curve is a hyperbola. From (37.4) it is clear that, since u is essentially non-negative, if κ is positive e must be greater than unity and the orbit must be a hyperbola. Furthermore, in this case $e\cos\theta > 1$, so that θ is confined to a range of values which is less than π in extent. Thus the hyperbola has the origin as an exterior focus. If κ is negative, e may be greater than unity, so that again the orbit is a hyperbola; but in this case the restriction on θ is that $e\cos\theta < 1$, which confines θ to a range of values which is greater than π in extent. Thus the hyperbola has the origin as an interior focus. When κ is negative, e may be unity or less than unity, giving parabolic and elliptic orbits respectively.

Most of the above information can be summed up in the physically obvious statement that when the force is repulsive the orbit is convex to the origin, whereas when the force is attractive the orbit is concave to the origin.

The Lagrangian for Kepler motion is

$$L = T - V = \tfrac{1}{2}\mu(\dot{r}^2 + r^2\dot{\theta}^2) - \frac{\kappa}{r}$$

Since L is independent of the time, the total energy

$$E = T + V = \frac{1}{2}\mu\dot{r}^2 + \frac{1}{2}\frac{l^2}{\mu r^2} + \frac{\kappa}{r} \tag{37.6}$$

is a constant of the motion.

For $e < 1$, the maximum and minimum values of r allowed by (37.5) are called the *apsidal distances*,

$$r_1 = \frac{es}{1 - e}, \qquad r_2 = \frac{es}{1 + e}$$

respectively, their average

$$a = \frac{es}{1 - e^2} \tag{37.7}$$

being the major semiaxis of the ellipse. At an apsidal distance, $\dot{r} = 0$, and E may be most conveniently related to the parameters specifying the ellipse by evaluating (37.6) at one of these points. Setting $r = r_1$ or r_2 and $\dot{r} = 0$ in (37.6), and using the equation (37.4) for the ellipse, we obtain

$$e = \left(1 + \frac{2El^2}{\kappa^2\mu}\right)^{1/2} \tag{37.8}$$

and

$$a = \frac{\kappa}{2E} = -\frac{l^2}{\kappa\mu}(1 - e^2)^{-1} \tag{37.9}$$

Equation (37.8) is valid for elliptic or hyperbolic motion, although in the latter case it is necessary to set $r = r_2$, the minimum apsidal distance. Thus for $e < 1$, it follows that $E < 0$, a result which corresponds to the fact that a particle moving in an ellipse does not have enough energy to escape to infinity. If $E < 0$, it follows from (37.8) that $\kappa < 0$, i.e., such elliptical motion can occur only under the influence of an attractive potential. Further, from (37.9), one notes that the energy depends only on κ and the major axis of the ellipse, i.e., all elliptical orbits with the same major axis about the same attractive center correspond to the same energy per unit mass in the case of gravitational attraction and the same energy per unit charge in the case of electrostatic attraction.

The magnitude of the angular momentum l is

$$l = |r \times \mu\dot{r}| = 2\mu\dot{A} \tag{37.10}$$

where $\dot{A}$ is the rate at which the radius vector sweeps over the area of the ellipse. Thus $\dot{A}$ = const, a result known for the motion of the planets as *Kepler's second law.*

The area A of the ellipse is

$$\begin{aligned} A &= \pi a^2(1 - e^2)^{1/2} \tag{37.11} \\ &= \pi a^2\left(-\frac{2El^2}{\kappa^2\mu}\right)^{1/2} \quad \text{by (37.8)} \\ &= \pi a^{3/2}\left(-\frac{l^2}{\kappa\mu}\right)^{1/2} \quad \text{by (37.9)} \end{aligned}$$

The period of the motion is then

$$\tau = \frac{A}{\dot{A}} = 2\pi\left(-\frac{\mu}{\kappa}\right)^{1/2} a^{3/2} \tag{37.12}$$

In the case of a planet of mass m in the solar system

$$\frac{\mu}{\kappa} = \frac{1}{G(M + m)}$$

(G = gravitational constant, M = mass of sun) so that the factor $(\mu/\kappa)^{1/2}$ in (37.12) has almost the same value for each planet. Thus $\tau^2 \approx a^3$, a result known as *Kepler's third law.*

The equation of the orbit and the angular momentum integral enable us to find the time dependence of the motion. Using (37.4), with $\theta_0 = 0$, we may write the angular momentum integral in the form

$dt/d\theta = \mu/(lu^2)$ so that

$$(37.13)\quad \frac{\kappa^2\mu}{l^3}(t - t_0) = \int_0^\theta \frac{d\theta}{(1 - e\cos\theta)^2}$$

$$= \frac{e\sin\theta}{(1 - e\cos\theta)(1 - e^2)} + \frac{2}{(1 - e^2)^{3/2}}$$

$$\times \tan^{-1}\left[\left(\frac{1 + e}{1 - e}\right)^{1/2} \tan\frac{\theta}{2}\right] \qquad (-\pi < \theta < \pi)$$

The period τ is twice the limit of $t - t_0$ as $\theta \to \pi$, which yields (37.12).

The potential V^* of the equivalent one-dimensional system is, from (36.13),

$$(37.14)\qquad V^* = \frac{\kappa}{r} + \frac{1}{2}\frac{l^2}{\mu r^2}$$

This potential is illustrated in Fig. 37-2 for both signs of κ. If $\kappa \geq 0$, there is no potential minimum and the orbits, being hyperbolas, are

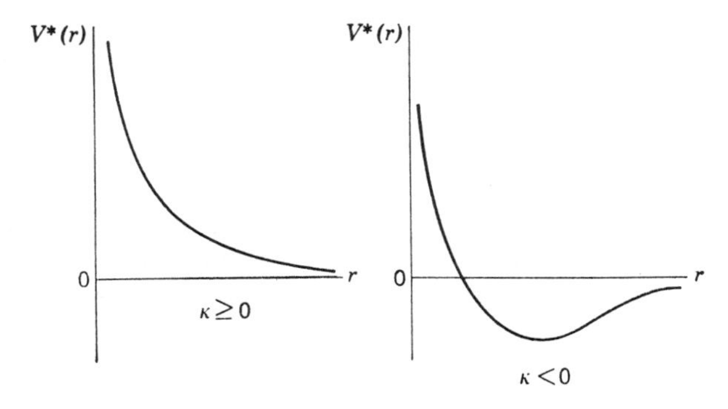

Fig. 37-2. Effective Kepler potentials.

unbounded. If $\kappa < 0$, there are bound orbits for negative values of E. The bound orbits are ellipses, the unbounded ones hyperbolas for $E > 0$, parabolas for $E = 0$.

The position and depth of the potential well in the right-hand side of Fig. 34-2 depend on the value of l. The value of r for which V^* is a minimum is

$$(37.15)\qquad r_{\min} = -\frac{l^2}{\mu\kappa} \qquad (K < 0)$$

and the corresponding value of V^* is

$$V_{\min}^* = -\frac{\kappa^2 \mu}{2l^2}$$

The smaller l is, the deeper the potential well and the closer it is to the origin. The product $r_{\min} V_{\min}^* = \kappa/2$ is independent of both the angular momentum and the reduced mass of the system, depending only on the original potential V.

By choosing l and E sufficiently small, the motion can be confined within arbitrarily narrow limits. One would expect this argument to break down on the atomic scale because of the existence of a smallest unit of angular momentum. Let us take $l = \hbar$; let μ be the reduced mass of the electron and proton; and let $\kappa = -e^2$, where e is the charge on an electron. Then, from (37.15),

$$r_{\min} = \frac{\hbar^2}{\mu e^2}$$

This value of $r_{\min}$ is the radius of the first Bohr orbit in the hydrogen atom in the classical picture of the atom. The argument that the motion can be confined arbitrarily closely does indeed break down here, but the details of the failure are beyond the scope of this volume.

The motion of a satellite or a ballistic missile under the influence of the earth's gravitational field is also described approximately by the equation of motion (37.3) but it is more convenient to express the equation of the orbit in terms of quantities measured at burn-out.* Thus if γ is the angle to the local upward vertical made by the velocity vector at burn-out, and r_0 is the distance of the burn-out point from the center of the earth, the orbit during free flight is described by the equation

$$\frac{r_0}{r} = \frac{1 - \cos\theta}{\alpha \sin^2\gamma} + \frac{\sin(\gamma - \theta)}{\sin\gamma} \tag{37.16}$$

the oblateness and rotation of the earth being neglected. Here the orbit is described by r and θ coordinates as shown in Fig. 37-3, and α is twice the ratio of the kinetic to potential energy at launch. Equation (37.15) is of the form (37.5) which describes an ellipse, and for $\theta = 0$ it gives $r = r_0$. The equation also implies that

$$\left.\frac{dr}{r\,d\theta}\right|_{\theta=0} = \cot\gamma$$

* A. D. Wheelon and L. Blitzer, *Am. J. Phys.*, 25, 21 (1957).

as required. We leave to the reader to verify that

$$\alpha = \frac{r_0 v_0^2}{GM}$$

where v_0 is the initial velocity.

The following theorem holds for motion in a Kepler field between any two points A and B: A projectile fired from A with the minimum

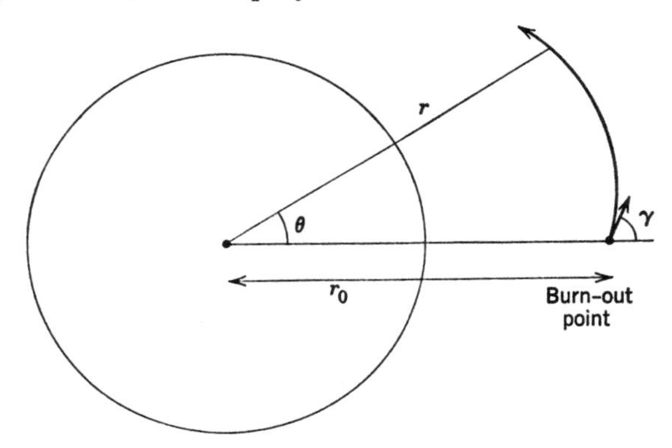

Fig. 37-3. Parameters specifying an elliptic orbit in the field of the earth.

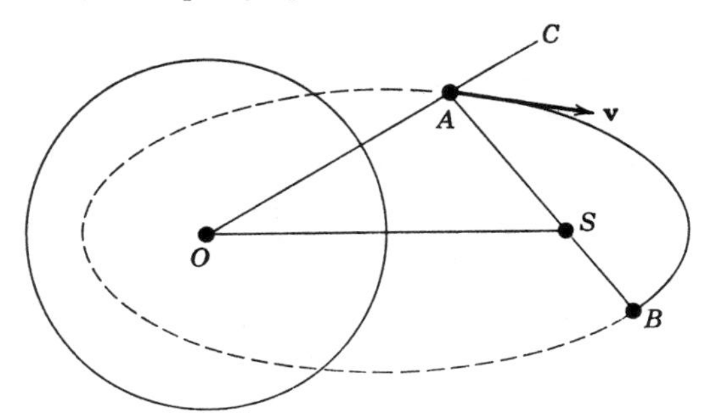

Fig. 37-4. Geometrical construction for minimum energy orbit.

kinetic energy required to reach B must be projected so that its initial velocity vector bisects the angle between AB and the upward drawn vertical at A. A special case of this theorem occurs when A and B are close together on the horizontal surface of the earth. In this case the initial velocity vector lies at an angle of 45° to the horizontal.

The result may be established from (37.16) which describes the orbit, but it is simpler to use a geometrical construction as shown in Fig. 37-4. Let O be the center of the Kepler potential (e.g., the center of the earth) and let A and B be the two points in question, OAC being the upward drawn vertical at A. If S is the other focus of that ellipse which requires a minimum energy to pass through A and B, we have

$$OA + AS = 2a = OB + BS$$

or

$$AS + BS = 4a - (OA + OB) \tag{37.17}$$

From (37.9)

$$a = \frac{GM}{-2(T + V)}$$

so that for minimum kinetic energy T at A it is required that a should also be as small as possible. From (37.17) since OA and OB are fixed, this implies that $AS + BS$ should be a minimum, i.e., that the other focus S should lie on the line AB. Since it is a property of the ellipse that the tangent at any point of the ellipse bisects the angle between the lines joining that point to the focuses, the result follows. A more general optimization theorem for powered flight is given in Sec. 80.

38. RUTHERFORD SCATTERING

In Secs. 22 and 23 some preliminaries to a solution of scattering problems were discussed. We now wish to treat the scattering of particles caused by an inverse square law force. As in Sec. 22, one particle is initially at rest while the other has a component of velocity toward the first. This system is treated in the center-of-mass system, where it reduces to the motion of a particle of mass μ under the influence of a fixed force center, the scattering center.

In atomic and nuclear physics, where scattering experiments are made, the problem is not to find the orbit of a particular particle acted on by the scattering center. The details of the orbit are not observed, but only the initial and final velocities. Even these are observed only statistically in most cases; usually the distribution in angle of the scattered particles is observed when a great many particles are scattered. The problem is to predict this distribution.

The solution of collision problems is given in terms of a *cross section.* If there is a flux of N particles per unit area toward a force center and if n of these particles undergo the process in question, the cross section for that process σ is defined as

$$\sigma = \frac{n}{N} \tag{38.1}$$

The value of n must be large enough to preclude the possibility of large statistical fluctuations in the value of σ. The dimensions of σ are those of an area. In nuclear problems σ is expressed in units of 10^{-24} cm^2, a unit which is called a *barn*.

The process of concern to us here is the scattering of the incident particle. This scattering requires two angles for its specification, one angle θ being the angle through which the particle is scattered, the other angle ϕ giving the orientation of the plane in which the initial and final velocity vectors lie. Thus we shall be interested in the cross section for scattering through an angle between θ and $\theta + d\theta$ in a plane whose azimuth lies between ϕ and $\phi + d\phi$. The differentials of these two angles combine naturally into the differential of solid angle $d\omega$:

$$d\omega = \sin\theta\, d\theta\, d\phi \tag{38.2}$$

The *differential cross section* $\sigma(\theta, \phi)$ for scattering into the element of solid angle $d\omega$ about the direction specified by the angles θ and ϕ is the cross section per unit solid angle, and therefore the cross section for this scattering is $\sigma(\theta, \phi)\, d\omega$.

If the force is a central one, the cross section is independent of ϕ, and we may write the cross section for scattering through the angle θ in any plane as

$$\int_{\phi=0}^{2\pi} \sigma(\theta) \sin\theta\, d\theta\, d\phi = 2\pi\sigma(\theta) \sin\theta\, d\theta \tag{38.3}$$

The *total scattering cross section* σ is the cross section for scattering through any angle at all,

$$\int \sigma(\theta, \phi)\, d\omega \tag{38.4}$$

In the present case this integral diverges; this means merely that the force between charged bodies of such long range that the path of an incident particle is deflected even at very large distances from the force center. This does not occur in nature because there every charge is more or less closely surrounded by a charge of opposite sign which cuts down the range of the force.

We now wish to find $\sigma(\theta)$ when the force between the particles is an inverse square force.

The orbit of the particle will be a hyperbola. The direction of one asymptote of the hyperbola is given by the direction of the initial velocity $\mathbf{v}_0$. The direction of the other asymptote is determined by the initial speed v_0 and the collision or impact parameter γ. γ is the distance from the force center S that the particle would pass if there were no force acting. The angle of scattering is the angle between the asymptotes. If ψ is the angle so marked in Fig. 38-1, the angle of scattering θ is given by

(38.5) $$\theta = \pi - 2\psi$$

If there is a flux of one particle per unit area in the direction of $\mathbf{v}_0$, the cross section for the particle having a collision parameter between γ and $\gamma + d\gamma$ is the area of the ring of radius γ and width $d\gamma$.

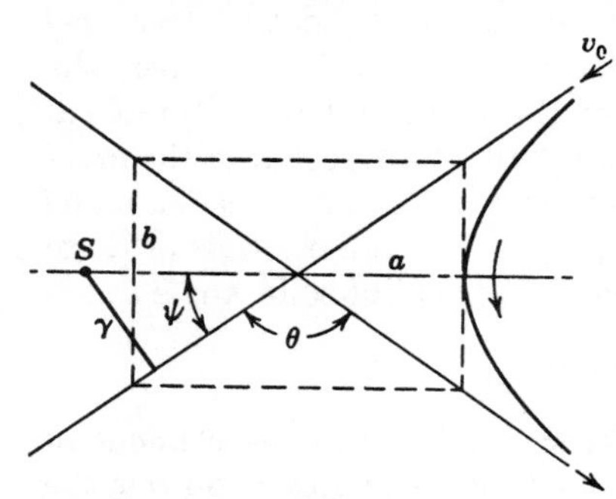

Fig. 38-1. Rutherford scattering.

(38.6) $$\sigma(\gamma, d\gamma) = 2\pi\gamma\, d\gamma$$

Now if θ is known as a function of γ, the cross section $2\pi\sigma(\theta) \sin \theta\, d\theta$ for scattering through an angle between θ and $\theta + d\theta$ is given by

(38.7) $$2\pi\sigma(\theta) \sin \theta\, d\theta = -2\pi\gamma\, d\gamma$$

where the negative sign occurs because θ decreases as γ increases. Our problem thus reduces to finding θ as a function of γ.

The equation of the orbit is, from (37.4),

(38.8) $$u = \frac{\kappa\mu}{l^2}(e \cos \theta - 1) \qquad (e > 1)$$

The asymptotic directions are those directions for which r is infinite or for which u vanishes. Thus

(38.9) $$e \cos \psi = 1, \qquad e = \sec \psi$$

It now remains to find γ as a function of θ. Since the angular momentum is given by

(38.10) $$l = \mu v_0 \gamma$$

and the total energy E is

(38.11) $$E = \frac{1}{2}\mu {v_0}^2 = \frac{\kappa^2\mu}{2l^2}(e^2 - 1) \quad \text{by (37.8)}$$

$$= \frac{\kappa^2\mu}{2l^2}\cot^2\frac{\theta}{2} \quad \text{by (38.5), (38.9)}$$

it follows that

(38.12) $$\gamma = \frac{l}{\mu v_0} = \frac{\kappa}{\mu {v_0}^2}\cot\frac{\theta}{2}$$

by substituting for l in terms of v_0, κ, θ from (38.11).

Thus, from (38.7),

$$2\pi\sigma(\theta)\sin\theta\,d\theta = -2\pi\gamma\,d\gamma$$

$$= \frac{\pi\kappa^2}{\mu^2 v_0^4}\cot\frac{\theta}{2}\operatorname{cosec}^2\frac{\theta}{2}\,d\theta$$

or

$$\sigma(\theta) = \frac{\kappa^2}{4\mu^2 v_0^4}\operatorname{cosec}^4\frac{\theta}{2} \tag{38.13}$$

This is the famous *Rutherford scattering formula.*

Formula (38.13) was the basis for a series of experiments by Geiger and by Geiger and Marsden.* In those experiments the incident particle was an alpha particle from naturally radioactive material, and the scattering center was a gold or silver nucleus. The mass ratio between the particles is so great that the laboratory and the center-of-mass coordinates system can be taken as coincident. The object of the experiment was not, of course, to check (38.13) but to see if the scattering produced by an atom is that resulting from a central force given by Coulomb's law.

The $\operatorname{cosec}^4(\theta/2)$ dependence was verified over a range of angles which made this quantity vary by a factor of 250,000, and the velocity dependent factor was varied by 10. The prediction of the Rutherford formula was fulfilled; thus strong evidence for the nuclear model of the atom was provided.

39. THE SPHERICAL PENDULUM

The spherical pendulum is a generalization of the plane pendulum in which the particle comprising the system is constrained to move on a sphere rather than on a circle. The force is not a central one but is constant in the vertical direction. We choose the mass of the particle and the radius of the sphere to be unity, and we measure time in such units that the acceleration of gravity is also unity (cf. Sec. 8).

Spherical coordinates are used to describe the particle. Because of the constraint, $r = 1$ and $\dot{r} = 0$. The expression for the kinetic energy is therefore

$$T = \tfrac{1}{2}(\dot{\theta}^2 + \sin^2\theta\,\dot{\phi}^2) \tag{39.1}$$

and that for the potential energy is

$$V = 1 - \cos\theta \tag{39.2}$$

* E. Rutherford, J. Chadwick, and C. D. Ellis, *Radiation from Radioactive Substances,* Cambridge Univ. Press, Cambridge (1930).

where the polar axis is taken in the downward direction and the constant is chosen so that the potential vanishes at the bottom of the sphere. The Lagrangian is thus

$$L = \tfrac{1}{2}(\dot{\theta}^2 + \sin^2\theta\dot{\phi}^2) - (1 - \cos\theta) \tag{39.3}$$

This is a conservative, natural system.

The coordinate ϕ is ignorable. Hence

$$\frac{\partial L}{\partial\dot{\phi}} = \sin^2\theta\dot{\phi} = \beta \tag{39.4}$$

The total energy is constant:

$$E = T + V = \tfrac{1}{2}(\dot{\theta}^2 + \sin^2\theta\dot{\phi}^2) + (1 - \cos\theta) \tag{39.5}$$

$\dot{\phi}$ can be eliminated by use of (39.4) to yield

$$E = \tfrac{1}{2}(\dot{\theta}^2 + \beta^2\operatorname{cosec}^2\theta) + (1 - \cos\theta) \tag{39.6}$$

This is a differential equation for θ.

$$\dot{\theta}^2 = 2(E - 1) + 2\cos\theta - \beta^2\operatorname{cosec}^2\theta \tag{39.7}$$

If θ is not zero, we may make the substitution

$$z = \cos\theta \tag{39.8}$$

upon which (39.7) becomes

$$\dot{z}^2 = -2z^3 - 2(E - 1)z^2 + 2z - \beta^2 + 2(E - 1) \tag{39.9}$$

The turning points of the θ motion are given by the zeros of (39.7) or (39.9). Equation (39.9) gives a cubic equation in z. We have

$$\begin{array}{llll} z = -\infty, & z = -1, & z = +1, & z = +\infty \\ \dot{z}^2 = +\infty, & \dot{z}^2 = -\beta^2, & \dot{z}^2 = -\beta^2, & \dot{z}^2 = -\infty \end{array}$$

There is thus one root between $z = -\infty$ and $z = -1$, which is of no interest since it means that θ is imaginary, and there are an even number of roots between $z = -1$ and $z = +1$, i.e., either none or two. Since the initial value of z does not make $\dot{z}^2$ negative, there must be one root, and so there are two real roots in this region. They may not be distinct.

These two real roots in the physical region give two horizontal circles on the sphere between which the motion takes place. If the root is double, the two circles coincide and the orbit is a circle. A spherical pendulum moving in a circular orbit is called a *conical* pendulum.

The equation for $\dot{z}$ (39.9) could be solved in terms of elliptic functions. We do not carry out the exact solution but look for an approximate one by means of a perturbation method. (A discussion of perturbation methods in general is given in Sec. 74.)

The idea of a perturbation method is this: We seek a system similar to the one of interest but enough simpler so that it can be solved exactly. The difference between the actual system and the simpler system is then treated as a small quantity, and the solution of the actual system is expressed as a power series in this quantity. Usually only the first two or three terms are retained. The simple system is called the zero-order approximation, and the solution linear in the difference terms is called the first-order, that quadratic in the difference terms the second-order, etc. We shall make only a first-order calculation here.

We take the zero-order system to be the conical pendulum. Its angle with the polar axis is a constant, θ_0. Then in general

$$\theta = \theta_0 + \lambda\theta_1 \tag{39.10}$$

where λ is a so-called *parameter of smallness* in powers of which the expansion is to be made. The θ equation of motion is

$$\ddot{\theta} - \sin\theta\cos\theta\dot{\phi}^2 + \sin\theta = 0 \tag{39.11}$$

and the angular momentum integral is

$$\beta = \sin^2\theta\dot{\phi} \tag{39.12}$$

Introduction of (39.12) into (39.11) gives

$$\ddot{\theta} - \frac{\cos\theta}{\sin^3\theta}\beta^2 + \sin\theta = 0 \tag{39.13}$$

We now insert the value (39.10) in (39.13), obtaining

$$\lambda\ddot{\theta}_1 - \frac{\cos(\theta_0 + \lambda\theta_1)}{\sin^3(\theta_0 + \lambda\theta_1)}\beta^2 + \sin(\theta_0 + \lambda\theta_1) = 0 \tag{39.14}$$

Expanding this in powers of λ and retaining only linear terms, we get

$$\lambda\ddot{\theta}_1 - \left[\frac{\cos\theta_0}{\sin^3\theta_0} - \left(\frac{1}{\sin^2\theta_0} + \frac{3\cos^2\theta_0}{\sin^4\theta_0}\right)\lambda\theta_1\right]\beta^2 + \sin\theta_0 + \lambda\theta_1\cos\theta_0 = 0 \tag{39.15}$$

Coefficients of the various powers of λ must vanish separately because this is to be an identity in λ. Thus

$$-\frac{\cos\theta_0}{\sin^3\theta_0}\beta^2 + \sin\theta_0 = 0 \tag{39.16}$$

and

$$\ddot{\theta}_1 + \left[\left(\frac{1}{\sin^2\theta_0} + \frac{3\cos^2\theta_0}{\sin^4\theta_0}\right)\beta^2 + \cos\theta_0\right]\theta_1 = 0 \tag{39.17}$$

Equation (39.16) can be solved for β^2 and the value inserted in (39.17). This yields

$$\ddot{\theta}_1 + \left(\frac{1 + 3\cos^2\theta_0}{\cos\theta_0}\right)\theta_1 = 0 \tag{39.18}$$

θ_1 executes harmonic oscillations:

$$\theta_1 = \cos\omega t, \qquad \omega = \left(\frac{1 + 3\cos^2\theta_0}{\cos\theta_0}\right)^{1/2} \tag{39.19}$$

Thus the solution for θ is

$$\theta = \theta_0 + \lambda\cos\omega t \tag{39.20}$$

The pendulum oscillates about the value $\theta = \theta_0$. The corresponding ϕ motion may now be found.

From (39.12) we see that

$$\begin{aligned}\dot{\phi} &= \frac{\beta}{\sin^2(\theta_0 + \lambda\theta_1)} \\ &= \frac{\beta}{\sin^2\theta_0 + 2\lambda\theta_1\sin\theta_0\cos\theta_0} \\ &= \frac{\beta}{\sin^2\theta_0}(1 - 2\lambda\theta_1\cot\theta_0)\end{aligned} \tag{39.21}$$

Inserting the value (39.19) for θ_1 and integrating, we obtain

$$\phi = \frac{\beta}{\sin^2\theta_0}\left(t - \frac{2\lambda}{\omega}\cot\theta_0\sin\omega t\right) \tag{39.22}$$

On the average $\dot{\phi}$ has its unperturbed value, but it oscillates about this value with the angular frequency ω.

The orbit of the particle will be approximately an ellipse, but the major axis of the ellipse will precess in the direction of rotation.

Using (39.20) and (39.22), we see that at $t = 0$,

$$\theta = \theta_0 + \lambda, \qquad \phi = 0$$

and at $t = \pi/\omega$,

$$\theta = \theta_0 - \lambda, \qquad \phi = \frac{\pi}{(1 + 3\cos^2\theta_0)^{1/2}} > \frac{\pi}{2}$$

Thus the azimuth ϕ increases by more than $\pi/2$ in the time that the particle goes from the outer limit to the inner limit of its motion. The ϕ motion *gains* on the θ motion, causing the advance of the major axis. The amount of advance per revolution can be found easily if θ_0 is small. Then, at $t = \pi/\omega$,

$$\text{(39.23)} \qquad \phi = \frac{\pi}{(1 + 3\cos^2\theta_0)^{1/2}} = \frac{\pi}{(4 - 3\sin^2\theta_0)^{1/2}}$$

$$\approx \frac{\pi}{2}\left(1 + \frac{3}{8}\sin^2\theta_0\right)$$

The relative gain in ϕ is on the average $\frac{3}{8}\sin^2\theta_0$, so that in one revolution the gain is $(3\pi/4)\sin^2\theta_0$. If we return to metric units in which the radius of the sphere is a, the last value can be written

$$\frac{3\pi}{4}\sin^2\theta_0 = \frac{3\pi}{4}\frac{\bar{d}^2}{a^2}$$

$$\approx \frac{3\pi}{4}\frac{d_{max}d_{min}}{a^2}$$

where d_{max} and d_{min} are the major and minor semiaxes of the ellipse respectively.

40. LARMOR'S THEOREM

A theorem concerning the motion of a system in a uniform magnetic field is of importance. Consider a system with a potential energy which is symmetrical about the z axis of a coordinate system, so that

$$\text{(40.1)} \qquad V = V(\rho, z)$$

where ρ is the distance of a point from the z axis. If the system is placed in a magnetic field parallel to the z axis, an M function must be subtracted from the Lagrangian of the system. The vector potential of a uniform magnetic field is given by

$$\text{(40.2)} \qquad A_x = -\tfrac{1}{2}B_z y, \qquad A_y = \tfrac{1}{2}B_z x, \qquad A_z = 0$$

or in cylindrical coordinates

$$A_\rho = A_z = 0, \qquad A_\theta = \tfrac{1}{2}B_z\rho \tag{40.3}$$

Thus in this case

$$M = -\frac{e}{c}\mathbf{v}\cdot\mathbf{A} = -\frac{eB_z}{2c}\rho^2\dot{\theta} \tag{40.4}$$

From now on we drop the subscript from the B, since only the one component is present.

The Lagrangian of the system is now

$$\begin{aligned} L &= T - V - M \\ &= \frac{m}{2}(\dot{\rho}^2 + \rho^2\dot{\theta}^2 + \dot{z}^2) - V(\rho, z) + \frac{eB}{2c}\rho^2\dot{\theta} \end{aligned} \tag{40.5}$$

θ is an ignorable coordinate.

$$\frac{\partial L}{\partial \dot{\theta}} = \beta = m\rho^2\dot{\theta} + \frac{eB}{2c}\rho^2 \tag{40.6}$$

Let

$$\omega = \frac{eB}{2mc} \tag{40.7}$$

This ω is called the *Larmor frequency*. Then

$$\beta = m\rho^2(\dot{\theta} + \omega) \tag{40.8}$$

We introduce a coordinate system rotating about the negative z axis with angular velocity ω,

$$\bar{\rho} = \rho, \qquad \bar{z} = z, \qquad \bar{\theta} = \theta + \omega t \tag{40.9}$$

and in this coordinate system

$$\beta = m\bar{\rho}^2\dot{\bar{\theta}} \tag{40.10}$$

Thus in the rotating coordinate system, angular momentum about the z axis is conserved.

The Lagrangian may be expressed in terms of the rotating coordinate system:

$$\begin{aligned} \bar{L} &= \frac{m}{2}[\dot{\bar{\rho}}^2 + \dot{\bar{z}}^2 + \bar{\rho}^2(\dot{\bar{\theta}} - \omega)^2] - V(\bar{\rho}, \bar{z}) + \frac{eB}{2c}\bar{\rho}^2(\dot{\bar{\theta}} - \omega) \\ &= \frac{m}{2}[\dot{\bar{\rho}}^2 + \dot{\bar{z}}^2 + \bar{\rho}^2(\dot{\bar{\theta}}^2 - \omega^2)] - V(\bar{\rho}, \bar{z}) \end{aligned} \tag{40.11}$$

Thus ω enters only in the term $\dot{\theta}^2 - \omega^2$, and, if ω is small compared to $\dot{\theta}$, then ω can be neglected. A small magnetic field can thus be entirely eliminated from the problem by viewing the system from a rotating coordinate system.

The meaning of the word *small* depends on the mechanical system. Let us examine the situation in an atomic problem, the lowest state of the hydrogen atom according to the most elementary Bohr picture. The electron revolves in a circular orbit around the proton, the angular momentum being just $\hbar$. The equations of motion are

$$mr\dot{\theta}^2 = \frac{e^2}{r^2} \tag{40.12}$$

and

$$mr^2\dot{\theta} = \hbar \tag{40.13}$$

Solving these equations for the radius of the orbit,

$$r = \frac{\hbar^2}{me^2} \tag{40.14}$$

Now a magnetic field which we should call *large* would cause the electron, traveling at the speed of the electron in the hydrogen atom, to travel in a circle whose radius is of the same order of magnitude as the r in (40.14). Thus the magnetic force on the electron must be equal in order of magnitude to the electrostatic force on the electron in the hydrogen atom:

$$\frac{eBv}{c} = \frac{eBr\dot{\theta}}{c} = \frac{e^2}{r^2} \tag{40.15}$$

or

$$\begin{aligned} B &= \frac{ec}{r^3\dot{\theta}} = \frac{ecm}{r\hbar} \\ &= \frac{e^3m^2c}{\hbar^3} \end{aligned} \tag{40.16}$$

To evaluate this we may break it up into familiar combinations of the fundamental constants

$$\begin{aligned} B &= e\left(\frac{e^2}{\hbar c}\right)^3\left(\frac{mc^2}{e^2}\right)^2 \\ &= \frac{e\alpha^3}{r_0^{\,2}} \approx 5 \times 10^9 \text{ gauss} \end{aligned} \tag{40.17}$$

where α is the fine structure constant $e^2/\hbar c$ and r_0 is the classical electron radius. The numerical result shows that the magnetic fields obtainable in the laboratory, so far about 10^6 gauss at most, are very small fields in the sense of Larmor's theorem.

In larger scale systems where the charged particles describe larger orbits, the range of magnetic fields which can be considered as small is very much more restricted.

41. THE CYLINDRICAL MAGNETRON

As an illustration of the use of Larmor's theorem we consider the cylindrical magnetron, a device in which electrons move in an axially symmetric electric field and in a uniform magnetic field parallel to the axis. Let the electrostatic potential energy be $V(\rho)$ with $V(0) = 0$. The Lagrangian for the system is given by (40.5).

We introduce a coordinate system rotating around the negative z axis with the angular velocity ω given by Larmor's theorem:

$$\dot{\theta}^* = \dot{\theta} + \omega, \qquad \omega = \frac{eB}{2mc} \tag{41.1}$$

In this system the z component of the angular momentum is constant:

$$m\rho^2\dot{\theta}^* = \beta \tag{41.2}$$

z is also an ignorable coordinate, so that $\dot{z}$ is constant. We can make this constant vanish by choosing a coordinate system moving in the z direction with the appropriate speed.

In the stationary system there is the Jacobian integral

$$E = \tfrac{1}{2}m(\dot{\rho}^2 + \rho^2\dot{\theta}^2) + V(\rho) \tag{41.3}$$

If we take as initial conditions that the kinetic energy is zero at $\rho = 0$ and that $\beta = 0$, which are appropriate if the electron is emitted from a cathode of small diameter, then $E = 0$.

For a given strength of magnetic field the maximum value of ρ which the electron can attain is given by

$$\tfrac{1}{2}m(\dot{\rho}^2 + \rho^2\dot{\theta}^2) + V(\rho) = 0 \tag{41.4}$$

with $\dot{\rho}$ put equal to zero. If we denote this value of ρ by ρ^*, (41.4) becomes

$$\tfrac{1}{2}m\rho^{*2}\omega^2 + V(\rho^*) = 0 \tag{41.5}$$

Equation (41.5) may be looked at another way. ρ^* may be the distance from the axis to the plate of the magnetron, and $V(\rho^*)$ may be an externally fixed potential energy. Then (41.5) is an equation for the magnetic field which will just allow electrons to reach the plate.

The magnetic field in this case is not small, since $\dot{\theta} = -\omega$. In order to obtain Larmor frequencies of 10^{10} radians/sec a magnetic field of only the order of 1000 gauss is necessary.

EXERCISES

1. A particle of mass m is acted on by a force

$$F_x = -k_1 x, \qquad F_y = -k_2 y, \qquad F_z = 0$$

If initially $\dot{z} = 0$, show that the motion is plane. Show that the force is noncentral unless $k_1 = k_2$. Find the orbit of the plane motion, and verify the fact that the rate of change of angular momentum is the torque. Find the ratios of k_1 to k_2 which make the motion periodic.

2. A particle of mass m with initial velocity v_0 strikes a smooth sphere of radius a and mass M which is at rest. Find the cross section for scattering of the particle through an angle between θ and $\theta + d\theta$ in the center-of-mass system. Find the average value of the logarithm of E_1/E_2, where E_1 is the initial energy and E_2 is the final energy in the laboratory system. Show that an approximate expression for this is given by $2m/M$.

3. Find the differential scattering cross section of a scattering center described by a potential energy such that

$$V = 0 \quad (r > a), \qquad V = -E_0 \quad (r \leq a)$$

for particles of reduced mass μ. (The particle receives an impulse when crossing the sphere $r = a$. Otherwise it experiences no forces.)

4. Charged particles (electrons) are emitted from a source at the origin with a continuous energy spectrum. If the source is placed in a homogeneous magnetic field B in the z direction, and if only those particles which start out at an angle between θ and $\theta + d\theta$ with the field are considered, show that particles of a given energy are focused at a point on the z axis. Evaluate the derivative $\partial E/\partial z$, assuming θ fixed, and $\partial E/\partial\theta$ at fixed z. (This principle is used in the construction of some beta-ray spectrometers. The resolving power of such an instrument is given by E/dE, where

$$dE = \frac{\partial E}{\partial \theta}\, d\theta + \frac{\partial E}{\partial z}\, dz$$

and where $d\theta$ is the angular range selected and dz is the effective extension along the z axis of the source and detector.)

5. Particles of charge e and mass m pass between segments of concentric cylinders of radius a, $a + \delta$ respectively, the angle of the segments being ψ. If there is a potential difference V between the cylinders, find the velocity which will permit a particle to travel in a circular orbit concentric with the cylinders. Show that all particles with this velocity starting out in the

pencil of directions between $-d\alpha$ and $+d\alpha$ with the circular orbit will be brought to a focus at a point if ψ is properly chosen. Neglect the fringing field and treat δ as small. (This system is used as a velocity selector for ions in connection with accelerators of the van de Graaff type.)

6. Particles of charge e and mass m are emitted in various directions with various speeds in a uniform magnetic field B in the x direction and a uniform electric field E in the y direction, $B > E$ (in gaussian units). Show that the paths of the particles for which $\dot{x} = 0$ are cycloids progressing in the z direction.

7. Show that, if a particle of mass m and charge e is moving in a uniform electric field, that field may be eliminated from the problem by viewing the system from a coordinate system which is under an acceleration $\mathbf{a} = e\mathbf{E}/m$ relative to the system in which the electric field is $\mathbf{E}$.

8 SMALL OSCILLATIONS

42. OSCILLATIONS OF A NATURAL SYSTEM

In this chapter we shall study the motion of a system when its configuration is close to a configuration of static equilibrium. A system of particles is in static equilibrium when all the particles of the system are at rest and the total force on each particle is permanently zero. Many systems which are not in static equilibrium may be reduced to equivalent systems which are. In Sec. 36 the motion of a particle in a plane under the action of a central force derivable from a potential was reduced to an equivalent one-dimensional problem with a potential including the centrifugal potential. This equivalent system may have a configuration of static equilibrium for a particular value of r, namely the r for which $\partial V^*/\partial r$ vanishes. This "static" equilibrium really represents a motion of the particle about the force center in a circular orbit. The restriction to cases of static equilibrium is thus not as stringent as might at first be imagined.

The system of particles is described by f generalized coordinates q_j. We first consider a natural system so that the kinetic energy is a homogeneous quadratic function of the generalized velocity components and the potential is independent of the time.

$$T = \tfrac{1}{2} g_{ij} \dot{q}_i \dot{q}_j \qquad V = V(q_1, \cdot \cdot \cdot , q_f) \tag{42.1}$$

The equilibrium values of the coordinates $q_{j,0}$ must be such that

$$\left(\frac{\partial V}{\partial q_j}\right)_0 = 0 \tag{42.2}$$

where the subscript indicates that the derivative is to be evaluated at $q_j = q_{j,0}$.

If we restrict our attention to a small region in configuration space about the position of equilibrium, we may expand the potential in a power series and may neglect all but the first nonvanishing, nonconstant term:

$$V = V_0 + \left(\frac{\partial V}{\partial q_i}\right)_0 (q_i - q_{i,0}) + \frac{1}{2}\left(\frac{\partial^2 V}{\partial q_i \partial q_j}\right)_0 (q_i - q_{i,0})(q_j - q_{j,0}) + \cdots$$

The first term on the right is a constant and may be chosen to be zero. The second term vanishes because of (42.2). The third term is the first one of interest. We introduce new coordinates with $q_{j,0}$ as origin, so that $q_j - q_{j,0}$ becomes simply q_j. Then we can write

$$V = \frac{1}{2}\left(\frac{\partial^2 V}{\partial q_i \partial q_j}\right)_0 q_i q_j = \tfrac{1}{2} V_{ij} q_i q_j \tag{42.3}$$

where the quantities V_{ij} are constants and $V_{ij} = V_{ji}$. The V_{ij} thus form a symmetric matrix.

The kinetic energy is already a positive definite quadratic form in the velocity components, and $g_{ij} = g_{ji}$. If we again restrict our attention to a small region of configuration space about the equilibrium configuration, the g_{ij} may be considered as constants, having the values they have at equilibrium. We put $(g_{ij})_0 = G_{ij}$. Thus G_{ij} is a symmetric matrix with constant elements and positive eigenvalues. Then

$$T = \tfrac{1}{2} G_{ij} \dot{q}_i \dot{q}_j \tag{42.4}$$

Near the equilibrium configuration of the system the Lagrangian may therefore be written as

$$L = \tfrac{1}{2} G_{ij} \dot{q}_i \dot{q}_j - \tfrac{1}{2} V_{ij} q_i q_j \tag{42.5}$$

Since the matrices G and V are constant, the equations of motion derived from the Lagrangian (42.5) are

$$G_{ij} \ddot{q}_j + V_{ij} q_j = 0 \tag{42.6}$$

or, in matrix notation,

$$G\ddot{q} + Vq = 0 \tag{42.7}$$

This is a set of simultaneous differential equations with constant coefficients representing the motion of a set of coupled oscillators. Multiplication by the matrix G^{-1}, the reciprocal of G, which always exists, transforms the equation to

$$\ddot{q} + Aq = 0 \tag{42.8}$$

where

$$A = G^{-1}V \tag{42.9}$$

The coupling between the oscillators then occurs through the off-diagonal terms of the matrix A, for in general A will not be diagonal.

We look for harmonic solutions of (42.8) of the form

$$\ddot{q} + \omega^2 q = 0 \tag{42.10}$$

Note that in this equation q is a vector with f components. It may be imagined as a vector in f-dimensional space or as a definite linear combination of the components $q_1, q_2, \cdots, q_f$ corresponding to a set of simultaneous values for the various degrees of freedom. From (42.8) and (42.10) it follows that a simple harmonic motion of the vector q with angular frequency ω is possible if

$$Aq = \omega^2 q \tag{42.11}$$

i.e., if q is an eigenvector of the operator A defined by (42.9) belonging to the eigenvalue ω^2. From Appendix II these eigenvalues are the roots of the equation

$$\det (A_{kj} - \omega^2 \delta_{kj}) = 0$$

or, on multiplying by $\det G_{ik}$,

$$\det (V_{ij} - \omega^2 G_{ij}) = 0 \tag{42.12}$$

The eigenfrequencies may therefore be determined directly from the matrices V, G which appear in the Lagrangian (42.5). The eigenvectors may be obtained directly from (42.11) or from the equivalent equation

$$(V - \omega^2 G)q = 0 \tag{42.13}$$

Since G and V are real symmetric matrices, it follows that the eigenvalues ω^2 of A are real. To see this we take the conjugate complex of (42.13)

$$Vq^* - \omega^{*2} G q^* = 0 \tag{42.14}$$

so that

$$q^*Vq = \omega^2 q^*Gq \quad \text{from (42.13)}$$

$$qVq^* = \omega^{*2} qGq^* \quad \text{from (42.14)}$$

Since V and G are symmetric, it then follows that $\omega^2 = \omega^{*2}$, i.e., that ω^2 is real since $q^*Gq \neq 0$, G being positive definite.

Each oscillation with a definite frequency is called an *eigenvibration* or *normal mode* of the system. During such a vibration the corresponding vector q changes only in length and not in direction, i.e., the mode consists of the simultaneous oscillation of several, possibly all, degrees of freedom, the ratios of the amplitudes of the displacements of the various degrees of freedom staying constant. If A is degenerate, there is an arbitrary choice of the normal modes that correspond to the same eigenfrequencies. New coordinates x_i along the directions of the eigenvectors q_i of A may be introduced to describe the system. The equations of motion for the x_i then become

$$\ddot{x}_i + \omega_{i|}{}^2 x_i = 0 \tag{42.15}$$

These coordinates are called the *normal coordinates* of the system. If these eigenvectors are chosen as a set of basic vectors, the matrix A becomes transformed to a diagonal matrix (see Appendix II)

$$A_{ij} \to \bar{A}_{ij} = (SAS^{-1})_{ij} = \omega_{i|}{}^2 \delta_{ij} \tag{42.16}$$

where

$$x = Sq \tag{42.17}$$

Equation (42.15) is then seen to follow directly from (42.8). This diagonalization procedure shows how appropriate linear combinations of the coordinates behave as uncoupled oscillators, each with its own characteristic frequency.

Equation (42.15) may be formally derived from the Lagrangian

$$L = \frac{1}{2} \sum_i (\dot{x}_i{}^2 - \omega_{i|}{}^2 x_i{}^2) \tag{42.18}$$

although this is not necessarily the same Lagrangian as (42.5) (cf. Exercises 1, 2).

If the determinantal equation is too involved to solve directly, the variational expression (Appendix II, Eq. 51) for the eigenvalues may be used. It becomes

$$\omega_k|^2 = \min_{y_{l|}} \max_q \frac{V_{ij}q_iq_j}{G_{ij}q_iq_j} \tag{42.19}$$

$$G_{ij}y_{l|i}q_j = 0 \qquad (l = 1, 2, \cdots, k-1)$$

43. SYSTEMS WITH FEW DEGREES OF FREEDOM

(a) The Double Pendulum. As an example of the method derived in the last section, consider two pendulums of equal mass and length suspended from a nonrigid support in such a way that the displacement of one pendulum influences the potential energy of the other, and vice versa. Confining ourselves to small oscillations, we have

$$\begin{aligned} T &= \tfrac{1}{2}(\dot{\theta}_1^2 + \dot{\theta}_2^2) \\ V &= \tfrac{1}{2}(\theta_1^2 + \theta_2^2 - 2\epsilon\theta_1\theta_2) \end{aligned} \tag{43.1}$$

where θ_i is the angle of the ith pendulum from the vertical, where ϵ is the coupling constant between the pendulums, and where natural units are used with $m = l = g = 1$. The sign and magnitude of ϵ will depend on the actual structure of the pendulum support.

The two matrices G_{ij} and V_{ij} may be identified:

$$G_{ij} = \begin{pmatrix} 1 & 0 \\ 0 & 1 \end{pmatrix}, \qquad V_{ij} = \begin{pmatrix} 1 & -\epsilon \\ -\epsilon & 1 \end{pmatrix} \tag{43.2}$$

The determinantal equation for the eigenfrequencies is thus

$$\begin{vmatrix} 1 - \omega^2 & -\epsilon \\ -\epsilon & 1 - \omega^2 \end{vmatrix} = 0 \tag{43.3}$$

or

$$1 - \omega^2 = \pm\epsilon \tag{43.4}$$

Finally, if ϵ is small, we obtain

$$\omega = 1 \mp \tfrac{1}{2}\epsilon \tag{43.5}$$

In conventional units

$$\omega = \left(\frac{g}{l}\right)^{\frac{1}{2}} (1 \mp \tfrac{1}{2}\epsilon) \tag{43.6}$$

The normal modes of oscillation are now obtained from (42.13) with the values of ω^2 from (43.4). These equations are

$$\begin{aligned} \pm\epsilon\theta_1 - \epsilon\theta_2 &= 0 \\ -\epsilon\theta_1 \pm \epsilon\theta_2 &= 0 \end{aligned} \tag{43.7}$$

Since they are compatible, either equation may be solved for the ratio θ_1/θ_2:

$$\frac{\theta_1}{\theta_2} = \pm 1 \tag{43.8}$$

Assuming ϵ to be positive, if the upper sign is chosen, the two pendulums oscillate together with an angular frequency less than that of one of the pendulums singly; if the lower sign holds, they oscillate in opposite directions at a frequency larger than the uncoupled frequency. The general oscillation is a superposition of these two modes of oscillation. The oscillation of either pendulum will in general show the phenomenon of beats.

(b) The Triple Pendulum, a Degenerate System. The double pendulum has two degrees of freedom and therefore two eigenfrequencies which are distinct. If a third pendulum is added coupled symmetrically to the others, the system becomes degenerate. If all pendulums are alike, we have, using natural units,

$$\begin{aligned} T &= \tfrac{1}{2}(\dot{\theta}_1^{\,2} + \dot{\theta}_2^{\,2} + \dot{\theta}_3^{\,2}) \\ V &= \tfrac{1}{2}[\theta_1^{\,2} + \theta_2^{\,2} + \theta_3^{\,2} - 2\epsilon(\theta_1\theta_2 + \theta_2\theta_3 + \theta_3\theta_1)] \end{aligned} \tag{43.9}$$

The matrices G_{ij} and V_{ij} are identified to be

$$G_{ij} = \begin{pmatrix} 1 & 0 & 0 \\ 0 & 1 & 0 \\ 0 & 0 & 1 \end{pmatrix}, \qquad V_{ij} = \begin{pmatrix} 1 & -\epsilon & -\epsilon \\ -\epsilon & 1 & -\epsilon \\ -\epsilon & -\epsilon & 1 \end{pmatrix} \tag{43.10}$$

so that the determinantal equation is

$$\begin{vmatrix} 1 - \omega^2 & -\epsilon & -\epsilon \\ -\epsilon & 1 - \omega^2 & -\epsilon \\ -\epsilon & -\epsilon & 1 - \omega^2 \end{vmatrix} = 0 \tag{43.11}$$

Let $1 - \omega^2 = \mu$. Then (43.11) becomes

$$\mu^3 - 3\epsilon^2\mu - 2\epsilon^3 = 0 \tag{43.12}$$

The three roots of this cubic are

$$\begin{aligned} \mu &= -\epsilon,\ -\epsilon,\ +2\epsilon \\ \omega^2 &= 1 + \epsilon,\ 1 + \epsilon,\ 1 - 2\epsilon \end{aligned} \tag{43.13}$$

The system is degenerate, two frequencies being equal.

To find the normal modes corresponding to these frequencies, we may solve any pair of the equations

$$\begin{aligned} (1 - \omega^2)\theta_1 - \epsilon\theta_2 - \epsilon\theta_3 &= 0 \\ -\epsilon\theta_1 + (1 - \omega^2)\theta_2 - \epsilon\theta_3 &= 0 \\ -\epsilon\theta_1 - \epsilon\theta_2 + (1 - \omega^2)\theta_3 &= 0 \end{aligned} \tag{43.14}$$

for, say, the ratios θ_1/θ_3, θ_2/θ_3. Using the single root of (43.11), we obtain

$$\frac{\theta_1}{\theta_3} = \frac{\theta_2}{\theta_3} = 1 \tag{43.15}$$

This shows us that one eigenvector of the matrix A_{ij} (which here is V_{ij} since G_{ij} is the unit matrix) has all three components equal:

$$x_{1|} = \frac{1}{\sqrt{3}}(1, 1, 1) \tag{43.16}$$

If the double root of (43.11) is used, all three of equations (43.14) are equivalent:

$$\epsilon(\theta_1 + \theta_2 + \theta_3) = 0 \tag{43.17}$$

We take as the second eigenvector one satisfying (43.17) and having its second component zero:

$$x_{2|} = \frac{1}{\sqrt{2}}(1, 0, -1) \tag{43.18}$$

The third eigenvector is chosen to satisfy (43.17) and to be orthogonal to $x_{2|}$. Thus, if its components are (a, b, c), we must have

$$\begin{aligned} (a^2 + b^2 + c^2)^{1/2} &= 1 \quad \text{[normalization]} \\ a - c &= 0 \quad [\text{orthogonality to } x_{2|}] \\ a + b + c &= 0 \quad [(43.17)] \end{aligned} \tag{43.19}$$

The result is

$$\begin{aligned} x_{3|} &= (a, b, c) \\ &= \frac{1}{\sqrt{6}}(1, -2, 1) \end{aligned} \tag{43.20}$$

Any combination of the vectors $x_{2|}$ and $x_{3|}$ is again an eigenvector.

The normal coordinates of the system are

$$x_1 = \frac{1}{\sqrt{3}}(\theta_1 + \theta_2 + \theta_3)$$

(43.21)
$$x_2 = \frac{1}{\sqrt{2}}(\theta_1 - \theta_3)$$

$$x_3 = \frac{1}{\sqrt{6}}(\theta_1 - 2\theta_2 + \theta_3)$$

The reader will readily verify that in these coordinates the Lagrangian of the system is

(43.22)
$$L = \tfrac{1}{2}[(\dot{x}_1)^2 + (\dot{x}_2)^2 + (\dot{x}_3)^2] - \tfrac{1}{2}[(1 - 2\epsilon)(x_1)^2 + (1 + \epsilon)(x_2)^2 + (1 + \epsilon)(x_3)^2]$$

The choice (43.18) for $x_{2|}$, hence the resulting expression (43.20) for $x_{3|}$, is arbitrary. In the case of a twofold degeneracy such as this, there is a one-parameter set of possible eigenvibrations. Any linear combination of $x_{2|}$ and $x_{3|}$ is again an eigenvibration. We can choose as eigenvibrations any two linear combinations of $x_{2|}$ and $x_{3|}$ (four parameters) subject to the restrictions that both be normalized and that the two be orthogonal (three conditions). There is thus one "degree of freedom" left in the choice of the eigenvibrations. In general, the choice of eigenvibrations with an n-fold degeneracy involves $n(n - 1)/2$ parameters.

(c) The CO_2-like Molecule. We consider a system consisting of a particle of mass μ situated midway between two particles of mass unity. To specify the configuration of the system we introduce a set of cartesian coordinate axes with the z axis along the line joining the particles. The coordinates x_1, y_1 measure the displacement of the first of the two particles of unit mass away from the z axis, and z_1 measures the displacement along the z axis away from the position of equilibrium. x_2, y_2, z_2 do the same for the other particle of unit mass, and x_3, y_3, z_3 describe the particle of mass μ.

We place the origin of the coordinate system at the center of mass, in this case at the equilibrium position of the central (carbon) atom. In this system the linear momentum is zero

(43.23)
$$\dot{x}_1 + \dot{x}_2 + \mu\dot{x}_3 = 0$$

and similarly for y and z, so that

(43.24)
$$x_3 = -\frac{x_1 + x_2}{\mu}, \text{ etc.}$$

We shall consider the case in which the angular momentum about the center of mass is also zero. Since during a vibration the central atom is displaced only an infinitesimal distance from the center, only the other two atoms contribute a finite amount to the angular momentum and we may write

$$l_x = a(\dot{y}_1 - \dot{y}_2) = 0$$
$$l_y = a(\dot{x}_2 - \dot{x}_1) = 0 \tag{43.25}$$

where a is the equilibrium separation of the particles. Thus

$$\dot{y}_1 = \dot{y}_2, \qquad \dot{x}_1 = \dot{x}_2$$
$$y_1 = y_2, \qquad x_1 = x_2 \tag{43.26}$$

From (43.24), then,

$$x_3 = -\frac{2x_1}{\mu}, \qquad y_3 = -\frac{2y_1}{\mu} \tag{43.27}$$

The kinetic energy of the system is

$$T = \tfrac{1}{2}(\dot{x}_1^2 + \dot{y}_1^2 + \dot{z}_1^2) + \tfrac{1}{2}(\dot{x}_2^2 + \dot{y}_2^2 + \dot{z}_2^2) + \tfrac{1}{2}\mu(\dot{x}_3^2 + \dot{y}_3^2 + \dot{z}_3^2) \tag{43.28}$$

and the potential energy is

$$V = \tfrac{1}{2}k[(z_1 - z_3)^2 + (z_2 - z_3)^2] + \tfrac{1}{2}\kappa[(x_1 - x_3)^2 + (y_1 - y_3)^2 + (x_2 - x_3)^2 + (y_2 - y_3)^2] + \tfrac{1}{2}k'(z_2 - z_1)^2 \tag{43.29}$$

Here the force constant κ has been assumed to be the same in all directions perpendicular to the axis of the system. k is the force constant in the z direction between the carbon and one oxygen, and k' is that of the force between the two oxygens. Insertion of (43.26) and (43.27) into (43.28) and (43.29) yields

$$T = \frac{1}{2}\left(2 + \frac{4}{\mu}\right)(\dot{x}_1^2 + \dot{y}_1^2) + \frac{1}{2}\left(1 + \frac{1}{\mu}\right)(\dot{z}_1^2 + \dot{z}_2^2) + \frac{1}{\mu}\dot{z}_1\dot{z}_2 \tag{43.30}$$

and

$$V = \frac{1}{2}k\left[\left(1 + \frac{2}{\mu} + \frac{2}{\mu^2}\right)(z_1^2 + z_2^2) + \left(\frac{4}{\mu^2} + \frac{4}{\mu}\right)z_1z_2\right] + \frac{1}{2}\kappa\left[2\left(1 + \frac{2}{\mu}\right)^2(x_1^2 + y_1^2)\right] + \frac{1}{2}k'(z_1 - z_2)^2 \tag{43.31}$$

Only four coordinates remain in T and V. The matrices G and V may be identified. They are

$$G_{ij} = \begin{pmatrix} 2+\frac{4}{\mu} & 0 & 0 & 0 \\ 0 & 2+\frac{4}{\mu} & 0 & 0 \\ 0 & 0 & 1+\frac{1}{\mu} & \frac{1}{\mu} \\ 0 & 0 & \frac{1}{\mu} & 1+\frac{1}{\mu} \end{pmatrix} \tag{43.32}$$

and

(43.33) $V_{ij} =$

$$\begin{pmatrix} 2\kappa\left(1+\frac{2}{\mu}\right)^2 & 0 & 0 & 0 \\ 0 & 2\kappa\left(1+\frac{2}{\mu}\right)^2 & 0 & 0 \\ 0 & 0 & k\left(1+\frac{2}{\mu}+\frac{2}{\mu^2}\right)+k' & k\left(\frac{2}{\mu}+\frac{2}{\mu^2}\right)-k' \\ 0 & 0 & k\left(\frac{2}{\mu}+\frac{2}{\mu^2}\right)-k' & k\left(1+\frac{2}{\mu}+\frac{2}{\mu^2}\right)+k' \end{pmatrix}$$

Equation (42.12) for the eigenfrequencies has the form

(43.34)

$$\begin{vmatrix} \alpha-\alpha'\omega^2 & 0 & 0 & 0 \\ 0 & \alpha-\alpha'\omega^2 & 0 & 0 \\ 0 & 0 & k(1+\beta)+k'-\omega^2(1+\beta') & k\beta-k'-\omega^2\beta' \\ 0 & 0 & k\beta-k'-\omega^2\beta' & k(1+\beta)+k'-\omega^2(1+\beta') \end{vmatrix} = 0$$

where

$$\alpha = 2\kappa\left(1+\frac{2}{\mu}\right)^2, \qquad \alpha' = 2+\frac{4}{\mu}$$
$$\beta = \frac{2}{\mu}+\frac{2}{\mu^2}, \qquad \beta' = \frac{1}{\mu} \tag{43.35}$$

The determinant is the product of three factors, two of them linear in ω^2 and one quadratic.

Two of the eigenfrequencies are identical and are given by

$$\omega_1|^2 = \omega_2|^2 = \frac{\alpha}{\alpha'} = \kappa\left(1+\frac{2}{\mu}\right) \tag{43.36}$$

The corresponding eigenvibrations involve x_1, y_1, and therefore, by (43.26) and (43.27), x_2, y_2 and x_3, y_3 also. Since a degeneracy exists, the choice of the eigenvibrations is arbitrary to a certain extent. A possible choice would be to take vibrations lying in two mutually perpendicular planes, the xz plane and the yz plane. Thus the first eigenvector can have components

$$e_{1|} = \frac{1}{(2 + 4/\mu^2)^{1/2}} \left(1, 0, 0;\ 1, 0, 0;\ -\frac{2}{\mu}, 0, 0\right) \tag{43.37}$$

where the first three components are the coordinates of one of the particles of unit mass, the second three those of the other, and the third three those of the central particle. Similarly,

$$e_{2|} = \frac{1}{(2 + 4/\mu^2)^{1/2}} \left(0, 1, 0;\ 0, 1, 0;\ 0, -\frac{2}{\mu}, 0\right) \tag{43.38}$$

Neither $e_{1|}$ nor $e_{2|}$ involves any angular momentum of the system. A linear combination of $e_{1|}$ and $e_{2|}$ with proper phase can, however, involve angular momentum. Using the complex time factor $e^{i\omega t}$, the unit vectors

$$\frac{1}{\sqrt{2}} (e_{1|} + ie_{2|}) \qquad \text{and} \qquad \frac{1}{\sqrt{2}} (e_{1|} - ie_{2|})$$

represent rotations of the system about the z axis in opposite directions. Thus linear combinations of degenerate eigenvibrations may have the nature of rotations. This fact is of importance in the theory of molecular spectra. It leads to a coupling between the rotational and vibrational states of motion of a molecule.

The other two eigenfrequencies of our system are determined by the equation

(43.39)
$$\begin{vmatrix} k(1 + \beta) + k' - \omega^2(1 + \beta') & k\beta - k' - \omega^2\beta' \\ k\beta - k' - \omega^2\beta' & k(1 + \beta) + k' - \omega^2(1 + \beta') \end{vmatrix} = 0$$

the roots of which are

$$\omega^2 = k + 2k' \qquad \text{and} \qquad k\,\frac{1 + 2\beta}{1 + 2\beta'} = \left(1 + \frac{2}{\mu}\right) k \tag{43.40}$$

The corresponding eigenvectors are

$$e_{3|} = \frac{1}{\sqrt{2}}(0, 0, 1;\ 0, 0, -1;\ 0, 0, 0) \tag{43.41}$$

and

$$e_{4|} = \frac{1}{(2 + 4/\mu^2)^{1/2}}\left(0, 0, 1;\ 0, 0, 1;\ 0, 0, -\frac{2}{\mu}\right) \tag{43.42}$$

These two frequencies being nondegenerate, the corresponding eigenvectors are uniquely determined.

Figure 43-1 shows the eigenvibrations of the system.

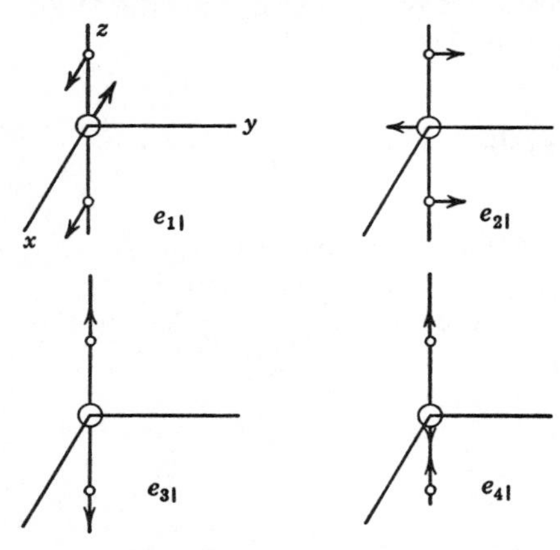

Fig. 43-1. Eigenvibrations of CO_2.

For a discussion of the application of group theory to similar analysis for more complicated molecules, the reader is referred to Appendix III.

44. THE STRETCHED STRING, DISCRETE MASSES

As an example of a system with many degrees of freedom we take a light string loaded at equal intervals by equal masses. Consider a length $l = (n + 1)a$ of a string for which coordinates $x_0 \cdots x_{n+1}$ label the equilibrium positions of the ends of the segments. Equal masses μ are located at positions $x_1 \cdots x_n$ $(x_\rho - x_{\rho-1} = a)$.

We shall consider both transverse and longitudinal motion of the

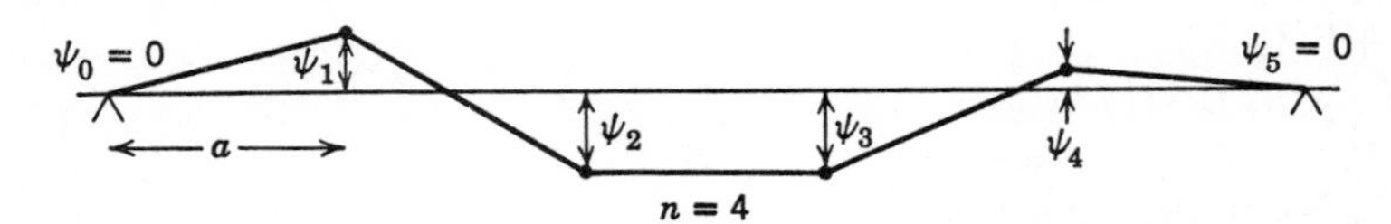

Fig. 44-1. Massless string with uniformly spaced masses (transverse oscillations).

masses, since much of the analysis is applicable to either situation. First of all, if the displacements are transverse (Fig. 44-1) the potential energy comes from the tension of the string. Let the transverse displacements of the masses be $\psi_0 \cdots \psi_{n+1}$ with $\psi_0 = \psi_{n+1} = 0$. The potential energy associated with one segment is then the tension, K, times the extension of the segment; the total potential energy is the sum over the segments. Thus

$$V = \sum_{\rho=1}^{n+1} K\{[a^2 + (\psi_\rho - \psi_{\rho-1})^2]^{1/2} - a\} \tag{44.1}$$

$$\approx \frac{1}{2} \sum_{\rho=1}^{n+1} Ka \left(\frac{\psi_{\rho-1} - \psi_\rho}{a}\right)^2$$

where the term within $[\ \ ]^{1/2}$ has been expanded in terms of $(\psi_\rho - \psi_{\rho-1})/a$.

If the displacements are longitudinal (Fig. 44-2), the potential energy comes from the elasticity of the string. Let the force required to stretch a length a of the string an amount ψ be $F = K'\psi/a$. Then

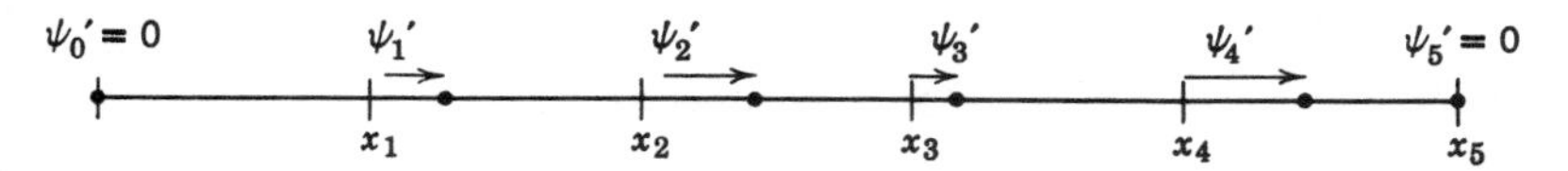

Fig. 44-2. Massless string with equal masses spaced uniformly at equilibrium (longitudinal oscillations).

if the longitudinal displacements are $\psi_0' \cdots \psi_{n+1}'$, the potential energy per segment is $\frac{1}{2}(K'/a)(\Delta\psi')^2$, and for the entire string is given by

$$V = \frac{1}{2} \sum_{\rho=1}^{n+1} \frac{K'}{a} (\psi_\rho' - \psi_{\rho-1}')^2 \tag{44.2}$$

which is formally identical with (44.1) though the symbols have distinct meanings. We drop the primes and consider either situation. For either transverse or longitudinal oscillations the kinetic energy is

$$T = \frac{1}{2} \sum_{\rho=0}^{n+1} \mu\dot{\psi}_\rho^2 \tag{44.3}$$

The matrices G and V may be identified. G is a multiple of the unit matrix. V is a matrix with diagonal elements and with elements lying on either side of the principal diagonal. Thus (42.12) for the eigenfrequencies becomes

$$
\begin{vmatrix}
\frac{2K}{a} - \mu\omega^2 & -\frac{K}{a} & 0 \cdots \\
-\frac{K}{a} & \frac{2K}{a} - \mu\omega^2 & -\frac{K}{a} \cdots \\
0 & -\frac{K}{a} & \frac{2K}{a} - \mu\omega^2 \cdots \\
\cdot & \cdot & \cdot \\
\cdot & \cdot & \cdot \\
\cdot & \cdot & \cdot
\end{vmatrix} = 0 \tag{44.4}
$$

If n is large a direct solution of this equation becomes difficult.

Except for the first and last row, each row is obtained from the one above it by stepping it over one space. Thus each ψ_ρ is related to $\psi_{\rho-1}$ in the same way, except for $\rho = 0, n + 1$. We know that the time dependence will be harmonic with angular frequency ω. We assume

$$\psi_{\rho+1} = e^{i\phi}\psi_\rho \tag{44.5}$$

where finally either real or imaginary parts of all expressions are to be taken. Then

$$\psi_\rho = e^{i\rho\phi}\psi_0 \tag{44.6}$$

Insertion of this form of ψ_ρ into the equation (42.12) for the eigenvectors of the problem leads to

$$-\frac{K}{a}e^{-i\phi} + \left(\frac{2K}{a} - \mu\omega^2\right) - \frac{K}{a}e^{i\phi} = 0$$

or

$$\omega^2 = \frac{2K}{\mu a}(1 - \cos\phi) \tag{44.7}$$

The values of ω^2 thus depend on ϕ, the phase difference of the motion of adjacent masses.

The values permitted to ϕ depend on the conditions placed on the ends of the string. For fixed end points, we take

$$\text{Im}\,\psi_0 = \text{Im}\,\psi_{n+1} = 0 \tag{44.8}$$

The equation for ψ_0 may be satisfied by taking

$$\psi_0 = A\sin\omega t$$

so that

$$\psi_\rho = Ae^{i\rho\phi} \sin \omega t, \qquad \text{Im } \psi_\rho = A \sin \rho\phi \sin \omega t \tag{44.9}$$

To satisfy (44.8) for ψ_{n+1} we must have

$$\sin (n + 1)\phi = 0, \qquad (n + 1)\phi = m\pi$$

where m is an integer, so that

$$\phi = \frac{\pi m}{n + 1} \tag{44.10}$$

The eigenfrequencies are given by

$$\omega_{m|}{}^2 = \frac{2K}{\mu a}\left(1 - \cos \frac{\pi m}{n + 1}\right) \tag{44.11}$$

The n values of m ($m = 1, 2, \cdots, n$) give all possible values of the frequency, although any integer m is permissible. The eigenvectors have components

$$e_{m|\rho} = \left(\frac{2}{n + 1}\right)^{1/2} \sin \frac{\pi m \rho}{n + 1} \tag{44.12}$$

The particles all lie on a sine curve whose wavelength depends on the value of m and thus on the frequency $\omega_{m|}$.

The solution just obtained is a *standing wave*. By changing the boundary conditions we can get *traveling wave* solutions. Let us choose $\psi_0 = \psi_n$ and $\psi_1 = \psi_{n+1}$. These are known as *periodic boundary conditions* because we can imagine the string to be duplicated indefinitely on either side, and with these boundary conditions the motion of each duplicate would be the same.

The periodic boundary conditions are satisfied by ψ_ρ given by (44.9) provided

$$e^{in\phi} = 1 \tag{44.13}$$

so that

$$\phi = \frac{2\pi m}{n} \qquad (m = 1, 2, \cdots) \tag{44.14}$$

The eigenfrequencies are now, by analogy with (44.11),

$$\omega_{m|}{}^2 = \frac{2K}{\mu a}\left(1 - \cos \frac{2\pi m}{n}\right) \qquad (m = 1, 2, \cdots, n - 1) \tag{44.15}$$

This differs from (44.11) in two respects. The range of m is reduced by one because our new boundary condition gives ψ_n in terms of ψ_0 whereas before ψ_n was free. More important, in this case the string must contain an integer number of whole wavelengths of the oscillation whereas before an integer number of half wavelengths would do. Thus m multiplies 2π rather than π in the argument of the cosine.

This traveling wave solution is not exactly in the form of an eigenvibration as we have used the term, because when real variables are employed the time dependence is not described by a factor common to all ψ_ρ. Instead we have

$$\psi_0 = Ae^{-i\omega t} = |A|e^{i\theta}e^{-i\omega t}$$

as ψ_0 must share the periodic motion, and thus

$$\begin{aligned} \psi_\rho &= A \exp\left[i\left(\frac{2\pi m\rho}{n} - \omega t + \theta\right)\right] \\ \operatorname{Im}\psi_\rho &= A \sin\left(\frac{2\pi m\rho}{n} - \omega t + \theta\right) \end{aligned} \tag{44.16}$$

Remembering that ρ measures distances along the string in units of a, we may write this as

$$\begin{aligned} \operatorname{Im}\psi_\rho &= A \sin(k\rho a - \omega t + \theta) \\ k &= \frac{2\pi m}{na} \end{aligned} \tag{44.17}$$

which is the usual form for a traveling wave. The phase velocity v_p is thus

$$v_p = \frac{\omega}{k} = \frac{\omega na}{2\pi m} \tag{44.18}$$

which, for $m \ll n$, becomes independent of m

$$v_p = \left(\frac{Ka}{\mu}\right)^{\frac{1}{2}} \qquad (m \ll n) \tag{44.19}$$

obtained from expanding (44.15) and (44.18).

The group velocity of a wave is defined by

$$v_g = \frac{\partial\omega}{\partial k} \tag{44.20}$$

From (44.15) and (44.17) we have

$$\omega_k|^2 = \frac{2K}{\mu a}(1 - \cos ka)$$

Thus

$$v_g = \frac{K \sin ka}{\mu\omega_k} \tag{44.21}$$

This becomes zero for $ka = \pi$ or $m = n$. The corresponding frequency, the maximum, is therefore known as the *cutoff frequency*. At this frequency adjacent masses move oppositely.

Values of $m > n$ can occur. We see what they mean if we consider ρ as a continuous variable. If $m' = m + nN$, then from (44.16) we see that

$$\begin{aligned} \operatorname{Im} \psi_\rho &= A \sin\left(\frac{2\pi m'\rho}{n} - \omega t + \theta\right) \\ &= A \sin\left(\frac{2\pi m\rho}{n} + 2\pi\rho N - \omega t + \theta\right) \end{aligned}$$

However, ψ_ρ has meaning only for integer ρ, so that the $2\pi\rho N$ in the argument contributes nothing. If we want to draw a sine curve through the points $(\rho a, \psi_\rho)$, we can make it have as many oscillations as we wish between the points ρa. The range $0 < m < n$ yields the longest wavelengths possible.

45. REDUCTION OF THE NUMBER OF DEGREES OF FREEDOM

Consider a system, described by a quadratic Lagrangian, whose eigenfrequencies and eigenvibrations are known. We inquire about the effect on this system of a constraint which is independent of the time and which does not alter the equilibrium configuration of the system. Let us describe the configuration of the system by the normal coordinates x_i. The Lagrangian is then given by (42.18).

The new constraint may also be described in terms of the normal coordinates:

$$\phi(x_1, x_2, \cdots, x_f) = 0 \tag{45.1}$$

We expand ϕ in a Taylor series and retain only the linear terms.

$$\phi = \phi_0 + \left(\frac{\partial\phi}{\partial x_i}\right)_0 x_i + \cdots \tag{45.2}$$

The constant term must vanish because the equilibrium configuration, $x_i = 0$ for all i, must satisfy the constraint. Hence the equation of constraint may be written

$$a_i x_i = 0 \tag{45.3}$$

where

$$a_i = \left(\frac{\partial \phi}{\partial x_i}\right)_0$$

The effect of the constraint (45.3) is most easily found when the variational definition of the eigenfrequencies and eigenvectors is used. According to (42.19), the unconstrained eigenfrequency ω_k is determined from

$$\omega_k|^2 = \min_{y_m|} \max_x \frac{\Sigma \omega_i^2 x_i^2}{\Sigma x_i^2} \tag{45.4}$$

$$y_{m|i} x_i = 0 \qquad (m = 1, 2, \cdots, k-1) \tag{45.5}$$

where the expression (42.18) has been used for the Lagrangian. The conditions for determining the kth constrained eigenfrequency are the same except for the addition of the constraint (45.3) to the set of equations (45.5).

If the vector a is parallel to an eigenvibration, say the mth, then in the normal coordinate system of the unconstrained problem the vector a has only one nonzero component, the mth. In this case (45.3) cannot affect the definition of any of the other eigenvectors because these eigenvibrations are mutually orthogonal, and therefore none of them has an m component. Equation (45.3) is automatically satisfied for all but the mth eigenvibration. The normal coordinate x_m is frozen, but none of the others is affected.

If the vector a is not parallel to an eigenvibration, then (45.3) is independent of all of (45.5) and may affect any of the eigenvibrations. The effect of (45.3) is to restrict the variations in x allowed in finding the maximum in (45.4). In general, then, this maximum will be lower than it was without the constraint. Denoting the kth constrained eigenfrequency by $\bar{\omega}_k$, we see that

$$\omega_k^2 \geq \bar{\omega}_k^2 \tag{45.6}$$

We can say more than this. ω_{k+1}^2 is found by adding an equation of the type (45.3) to the set of equations (45.5) used to determine ω_k^2, and then varying the vector a to obtain the minimum. This minimum occurs when a is the kth eigenvibration. However, we are not free to vary a to obtain the minimum, and so we must have

$$\bar{\omega}_k|^2 \geq \omega_{k+1}|^2 \tag{45.7}$$

Combining (45.6) and (45.7), we see that

$$\omega_k|^2 \geq \bar{\omega}_k|^2 \geq \omega_{k+1}|^2 \tag{45.8}$$

The constrained eigenfrequencies lie between the unconstrained ones. If two or more of the unconstrained eigenfrequencies are equal, i.e., if the unconstrained system is degenerate, there will be at most one fewer constrained eigenfrequencies of that frequency. Thus a single constraint can reduce the multiplicity of a degenerate eigenfrequency by at most one. It does not necessarily reduce it at all. In general, only those eigenfrequencies are affected which have a component in the direction of the constraining vector a.

46. LAPLACE TRANSFORMS AND DISSIPATIVE SYSTEMS

In the study of oscillations in linear coupled systems, the method of Laplace transforms is particularly useful. Although applicable to systems such as those discussed already in this chapter, the method finds its widest application in the many problems of engineering importance in which the individual elements of a coupled system are subject to linear dissipative forces and driven by an externally applied periodic force. Without discussing Laplace transforms in detail, we shall summarize the chief relevant results of this analysis and apply these to the problem of linear dissipative coupled oscillators.

If $f(t)$ is a function of the time t which is single-valued for $t \geq 0$, the Laplace transform of $f(t)$ is defined by

$$\mathfrak{L}f(t) = F(s) = \int_0^\infty f(t)e^{-st}\,dt \tag{46.1}$$

where the real part of the complex variable s is large enough [Re $(s) > \sigma$] to make this integral absolutely convergent. The inverse transformation is then given by

$$f(t) = \mathfrak{L}^{-1}F(s) = \frac{1}{2\pi i}\int_{c-i\infty}^{c+i\infty} F(s)e^{st}\,ds \tag{46.2}$$

with $t \geq 0$, where we require $c > \sigma$, since $F(s)$ is defined only for Re $(s) > \sigma$. It then follows on integration by parts that

$$\mathfrak{L}\left[\frac{df(t)}{dt}\right] = sF(s) - f(0) \tag{46.3}$$

where, if there is a discontinuity in $f(t)$ at $t = 0$, $f(0)$ is to be defined as the limit of $f(t)$ as $t \to 0$ from positive values. Similarly,

$$\mathfrak{L}\left[\frac{d^2f(t)}{dt^2}\right] = s^2F(s) - sf(0) - f'(0) \tag{46.4}$$

or in general

$$\mathcal{L}[f^{(n)}(t)] = s^n F(s) - \sum_{m=1}^{n} f^{(m-1)}(0) s^{n-m} \tag{46.5}$$

where

$$f^{(n)}(t) = \frac{d^n f(t)}{dt^n}$$

Thus, on taking the Laplace transform of any linear differential equation of the form

$$a_0 f(t) + \sum_{n=1}^{N} a_n \frac{d^n f(t)}{dt^n} = 0 \tag{46.6}$$

we have

$$F(s) = \frac{\sum_{n=1}^{N} a_n \sum_{m=1}^{n} f^{(m-1)}(0) s^{n-m}}{\sum_{n=0}^{N} a_n s^n} \tag{46.7}$$

giving $F(s)$ explicitly in terms of s and the initial values of $f(t)$ and its first $N - 1$ derivatives. The function $f(t)$ is then given for $t > 0$ by (46.2), which becomes

$$f(t) = \frac{1}{2\pi i} \sum_{n=1}^{N} a_n \sum_{m=1}^{n} f^{m-1}(0) \int_{c-i\infty}^{c+i\infty} \frac{s^{n-m} e^{st}}{\sum_{0}^{N} a_n s^n} ds \tag{46.8}$$

The integral is evaluated by determining the poles of the integrand, i.e., the solutions of the characteristic equation

$$\sum_{n=0}^{N} a_n s^n = 0 \tag{46.9}$$

For each such root $s = s_k$ $(k = 1 \cdots N)$ there will be a term in $f(t)$ proportional to $e^{s_k t}$, as may be seen by direct inspection of (46.6). The coefficients of these terms are given by (46.8) when the initial conditions are specified.

An equation of the form

$$a_0 f(t) + \sum_{n=1}^{N} a_n \frac{d^n f(t)}{dt^n} = e(t) \tag{46.10}$$

between two dependent variables $f(t)$, $e(t)$ does not of course provide enough information to give a solution, but using the above method we may write

$$F(s) = \frac{\sum_{n=1}^{N} a_n \sum_{m=1}^{n} f^{(m-1)}(0)s^{n-m}}{\sum_{n=0}^{N} a_n s^n} + \frac{E(s)}{\sum_{n=0}^{N} a_n s^n} \tag{46.11}$$

where $F(s)$, $E(s)$ are the Laplace transforms of $f(t)$, $e(t)$ respectively. If initially $f(t)$ and its first $N - 1$ derivatives vanish, we have

$$F(s) = G(s)E(s) \tag{46.12}$$

where in this particular case

$$G(s) = \frac{1}{\sum_{n=0}^{\infty} a_n s^n}$$

$G(s)$ is called the *open-loop transfer function* from $e(t)$ to $f(t)$. It describes the response $f(t)$ of the system to the external stimulus $e(t)$.

We may also consider the case in which the external driving force or other stimulus in (46.10) is replaced by the *difference* between $e(t)$ and $f(t)$. This could be realized in an electrical network by feeding a signal from the response back into the stimulus in such a way that it is subtracted from $e(t)$. The system then receives a stimulus only when the output differs from the input. The transfer function $H(s)$ so obtained between $e(t)$ and $f(t)$ is called the *closed-loop transfer function*

$$F(s) = H(s)E(s) \tag{46.13}$$

where we leave to the reader to verify that

$$H(s) = \frac{G(s)}{1 + G(s)} \tag{46.14}$$

The wide application of these concepts to the synthesis of electrical and mechanical networks reaches far beyond the scope of this book. However, transfer functions are also often used in the solution of problems in theoretical physics, although they are rarely referred to as such.

The extension of this analysis to a system of coupled linear oscillators which are subject to damping and driving forces is direct. The equations of motion (42.6) now assume the form

$$G_{ij}\ddot{q}_j + \Lambda_{ij}\dot{q}_j + V_{ij}q_j = e_i(t) \tag{46.15}$$

where the matrix Λ_{ij} describes the damping of the system and the $e_i(t)$ represent the external forces. The Laplace transformation of this equation is

$$AQ = B$$

where the elements of the operator A and the vector B are known functions of s if the $e_i(t)$ are prescribed

$$\begin{aligned} A_{ij} &= s^2 G_{ij} + s\Lambda_{ij} + V_{ij} \\ B_i &= E_i(s) + (sG_{ij} + \Lambda_{ij})q_j(0) + G_{ij}q_j'(0) \end{aligned} \tag{46.16}$$

and the $E_i(s)$ and the components $Q_j(s)$ of Q are the Laplace transforms of $e_i(t)$, $q_j(t)$ respectively. The solution is therefore given by

$$Q = A^{-1}B$$

where

$$A_{jk}^{-1} = \frac{\text{cofactor of } A_{kj} \text{ in det } A}{\det A}$$

provided that the reciprocal matrix exists. If the reciprocal matrix does not exist, we have

$$\det A = \det (s^2 G + s\Lambda + V) = 0 \tag{46.17}$$

a direct generalization of (42.12), to which it reduces in the special case $\Lambda = 0$, $s = i\omega$. The solutions of (46.17) correspond to the eigenfrequencies of Sec. 42, but in the present case the "normal modes" are damped.

GENERAL REFERENCES

G. Hertzberg, *Infrared and Raman Spectra of Polyatomic Molecules.* Van Nostrand, New York (1945).

J. E. Rosenthal and G. M. Murphy, *Rev. Mod. Phys.*, 8, 317 (1936).

M. F. Gardner and J. L. Barnes, *Transients in Linear Systems*, Vol. 1. Wiley, New York (1942).

John G. Truxall, *Automatic Feedback Control Synthesis.* McGraw-Hill, New York (1955).

EXERCISES

1. Show how the Lagrangian (42.5) leads to the equations of motion (42.13) by virtue of the transformations (42.16), (42.17).

2. Show that for $G_{ij} = \delta_{ij}$ the Lagrangians (42.5), (42.18) are identical.

In each of the following exercises, find the eigenfrequencies and normal modes of small oscillations, sketch graphs of the normal modes, and understand physically why certain eigenfrequencies do not depend on parameters which appear in the exercises or are less than or greater than other eigenfrequencies.

3. A string of length $4a$ is held tight with a tension K, both ends being held fixed, and three particles, each of mass m, are placed at distances a from each other and from the ends. Investigate small transverse oscillations of the system.

$$e_{1|} = \frac{1}{2}(1, -\sqrt{2}, 1), \qquad \omega_{1|} = \left(\frac{K}{ma}\right)^{1/2}(2+\sqrt{2})^{1/2}$$

$$e_{2|} = \frac{1}{\sqrt{2}}(1, 0, -1), \qquad \omega_{2|} = \left(\frac{K}{ma}\right)^{1/2}2^{1/2}$$

$$e_{3|} = \frac{1}{2}(1, \sqrt{2}, 1), \qquad \omega_{3|} = \left(\frac{K}{ma}\right)^{1/2}(2-\sqrt{2})^{1/2}$$

4. Two particles, each of unit mass, are connected by a spring of strength k to each other and by springs of unit strength to two fixed supports, the whole system lying in a straight line. Investigate small longitudinal oscillations of the system.

$$e_{1|} = \frac{1}{\sqrt{2}}(1, -1), \qquad \omega_{1|} = (1+2k)^{1/2}$$

$$e_{2|} = \frac{1}{\sqrt{2}}(1, 1), \qquad \omega_{2|} = 1$$

5. Two unit masses are coupled by springs of unit strength to a mass m. Investigate small longitudinal oscillations without imposing any conditions on the momentum of the system.

$$e_{1|} = \frac{1}{\left(2+\dfrac{4}{m^2}\right)^{1/2}}\left(1, 1, -\frac{2}{m}\right), \qquad \omega_{1|} = \left(1+\frac{2}{m}\right)^{1/2}$$

$$e_{2|} = \frac{1}{\sqrt{2}}(1, -1, 0), \qquad \omega_{2|} = 1$$

$$e_{3|} = \frac{1}{\sqrt{3}}(1, 1, 1), \qquad \omega_{3|} = 0$$

Compare these results with those for the CO_2-like molecule [Sec. 43(c)].

6. A uniform horizontal rectangular plate rests on four similar springs at the corners. Investigate small oscillations for which the plate remains rigid.

7. A uniform bar of length $3l/2$ is suspended by a string of length l. Investigate small oscillations in one plane.

9 RIGID BODIES

47. DISPLACEMENTS OF A RIGID BODY

In Sec. 14 a rigid body was defined as a system of at least three non-collinear particles which move so that the distance between any two of them stays constant throughout the motion. This is an idealization of the situation that occurs in practice, in which the separations of the various atoms in a solid oscillate about equilibrium values that are determined by the interatomic forces. However, it is an idealization that is unimportant when discussing the motion of the solid as a whole.

The number of coordinates needed to describe the configuration of a rigid body is six. To see this, consider three non-collinear particles of the body. These particles have three coordinates apiece, but there are three constraints, so that there is a total of six free coordinates. Any other particle of the body also has three coordinates, but its distances from the first three particles are fixed, so none of the additional coordinates are free. The six coordinates may be thought of as the three coordinates of any point in the body and the three coordinates needed to give the orientation of the body about that point. Thus the N particles in a rigid body are subjected to $3N - 6$ constraints, leaving six degrees of freedom for the body as a whole.

We may consider a set of cartesian coordinate axes to be a part of a rigid body. During a displacement of the body, this set of axes will retain its cartesian character. If we consider a second set of cartesian coordinate axes, fixed in space, that coincides with the body system before the displacement, we can describe the displacement by giving the connection between the space coordinates of a point and the body coordinates of that same point, in other words, by giving the transformation from the space system to the body system. This transformation will be a linear one because of the cartesian nature of the coordinates.

Let us denote the coordinates of a particle of the body referred to the body coordinate system by $\bar{x}_i$, and the coordinates of the same particle referred to the space coordinate system by x_i. Before the displacement the two coordinate systems coincide. Afterwards we must have

$$x_i = S^{\mathsf{T}}{}_{ij}\bar{x}_j + a_i \tag{47.1}$$

where the $S^{\mathsf{T}}{}_{ij}$ and a_i are constants specifying the displacement. The a_i are the space coordinates of the origin of the body system. The $S^{\mathsf{T}}{}_{ij}$ specify the orientation of the body coordinate axes relative to the space coordinate axes. S^{T} is introduced here instead of S itself in order that our notation will agree with that of Equation (6.2). The a_i are independent, but the $S^{\mathsf{T}}{}_{ij}$ are subject to conditions imposed by the rigid character of the displacement. The distance of any particle of the body from the origin of the body coordinate system must be the same when measured in either coordinate system. Thus

$$\begin{aligned} \bar{x}_j\bar{x}_j &= (x_i - a_i)(x_i - a_i) \\ &= S^{\mathsf{T}}{}_{ik}\bar{x}_k S^{\mathsf{T}}{}_{il}\bar{x}_l \end{aligned} \tag{47.2}$$

and so

$$\begin{aligned} S^{\mathsf{T}}{}_{ik}S^{\mathsf{T}}{}_{il} &= S_{ki}S^{\mathsf{T}}{}_{il} = \delta_{kl} \\ SS^{\mathsf{T}} &= I \end{aligned} \tag{47.3}$$

This is the condition for the matrix S to be orthogonal. There are nine elements in S, and (47.3) provides six conditions, since for any matrix A the product AA^{T} is symmetric. There are therefore three free parameters in S.

The orthogonal matrix S with three rows and three columns possesses one eigenvalue of unity. That it has one real eigenvalue follows from the fact that the eigenvalue equation is a cubic in the unknown with real coefficients. That the real eigenvalue is ± 1 follows from the fact that an orthogonal matrix does not change the length of vectors. It is $+1$ because a right-handed set of axes remains right-handed during a displacement. The eigenvector corresponding to the eigenvalue unity remains unchanged during the part of the displacement described by S. This part of the displacement is called a *rotation*, and the eigenvector is the *axis of the rotation.*

In the above discussion we applied a rotation S about an axis through the common origin of the body and space coordinate systems and then a translation a to the body. Let us now apply the rotation S about an axis through the point whose radius vector is w and then apply a translation b to the body. We obtain

$$x_i - w_i = S^{\mathsf{T}}{}_{ij}(\bar{x}_j - w_j) + b_i \tag{47.4}$$

or

$$x_i = S^{\mathsf{T}}{}_{ij}\bar{x}_j - (S^{\mathsf{T}}{}_{ij} - \delta_{ij})w_j + b_i \tag{47.5}$$

This is the same displacement as the one described by (47.1) if

$$b_i - a_i = (S^{\mathsf{T}}{}_{ij} - \delta_{ij})w_j \tag{47.6}$$

or

$$b - a = (S^{\mathsf{T}} - I)w$$

For a given w, this determines b uniquely. However, this does not determine w in terms of a, b, and S because the determinant of the coefficients of the w_i vanishes. Let n be a unit vector in the direction of the axis of the rotation. Then

$$Sn = S^{\mathsf{T}}n = n \tag{47.7}$$

so that

$$\begin{aligned} (b - a, n) &= (S^{\mathsf{T}}w, n) - (w, n) \\ &= (w, Sn) - (w, n) \\ &= 0 \end{aligned} \tag{47.8}$$

Thus $b - a$ is perpendicular to the axis of rotation, and so the difference in the effect of the rotation S about two parallel axes is a translation in the plane perpendicular to these axes.

In (47.6), w can be chosen so that the components of the translation b in this plane are zero. The condition is that

$$Sb = S^{\mathsf{T}}b = b \tag{47.9}$$

Eliminating b between (47.9) and (47.6) yields

$$(S + S^{\mathsf{T}} - 2I)w = (S - I)a \tag{47.10}$$

Again, in (47.10) the determinant of the coefficients of the w_i vanishes, but so does that of the a, so that (47.10) is not a meaningless equation for arbitrary a. Equation (47.10) does not determine the vector w uniquely, for if w is a solution so is $w + \alpha n$, where n is the unit vector of (47.7). If we require that w be orthogonal to n, namely that

$$(w, n) = 0 \tag{47.11}$$

then (47.10) and (47.11) determine w completely. Geometrically this means that only the new axis of rotation is important, no one point on the axis being distinguished.

The result of this analysis is that the most general displacement of a rigid body is a rotation around an axis plus a translation parallel to the axis. This result is due to Chasles.

The general displacement of a rigid body involves both rotation and translation. The most interesting mechanical applications, however, involve rotation alone. This comes about because in problems of interest either one point of the body is constrained to remain fixed, or the motion is separable into a motion of the center of mass and a rotation about the center of mass. In the latter situation the translation is that of a particle, the center of mass, and the rigid body character enters only into the motion relative to the center of mass. Thus from now on we neglect the translation and confine our attention to displacements given by

$$x_i = S^{\mathsf{T}}{}_{ij}\bar{x}_j \tag{47.12}$$

$$x = S^{\mathsf{T}}\bar{x}, \qquad \bar{x} = Sx$$

We call systems whose displacements are described by (47.12) *rigid rotators.*

The rotations of a rigid body about a fixed point constitute a *group;* i.e., the result of two successive rotations is a rotation; to every rotation there corresponds an inverse rotation; there is a rotation which leaves the body unchanged; rotations obey the associative law. These follow from the properties of the orthogonal matrices which represent the rotations: the product of two orthogonal matrices is orthogonal; every orthogonal matrix has an inverse since its determinant is ± 1; the unit matrix is orthogonal; matrices in general obey the associative law. The rotations themselves constitute the abstract three-dimensional rotation group. The 3×3 orthogonal matrices form a representation of the group. This representation may be described most conveniently in terms of the Euler angles described in the next section.

An alternative representation of the rotation group is provided by the group of quaternions of unit norm. This representation and its expression in terms of the homogeneous parameters of Euler and the Cayley-Klein parameters, as well as its relation to the orthogonal matrices S_{ij} and to the *Pauli spin matrices*, are described in Appendix IV.

48. EULER'S ANGLES

One way of describing the most general rotation of a rigid body about a point is in terms of the following three rotations, taken in the order given:

1. Rotation through an angle ϕ about the 3 axis.
2. Rotation through an angle θ about the new 1 axis (OK).
3. Rotation through an angle ψ about the new 3 axis ($\bar{3}$).

These rotations are shown in Fig. 48-1. Once the convention is decided upon, these three *Euler angles* ϕ, θ, ψ specify unambiguously the configuration of the body by specifying the directions of the three orthogonal axes, $\bar{1}$, $\bar{2}$, $\bar{3}$ which are fixed in the body.

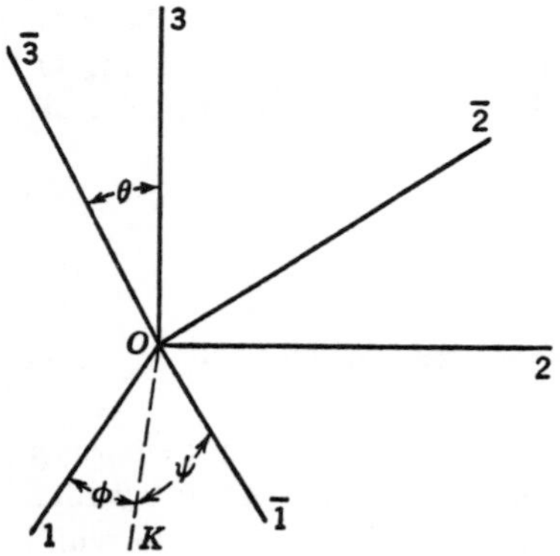

Fig. 48-1. Euler's angles.

The orthogonal matrix S of (47.12), which gives the transformation from the fixed space coordinate system to the body coordinate system, may be expressed directly in terms of the Euler angles. Let $\mathbf{z}$ be a vector fixed in the body and described in the body coordinate system by the components $\bar{z}_i$. Then

$$\bar{z}_i = S_{ij}z_j, \qquad z_j = S^{\mathsf{T}}{}_{ji}\bar{z}_i \tag{48.1}$$

The displacement takes place in three steps, each step being a rotation about a coordinate axis. Thus

$$S = S_\psi S_\theta S_\phi \tag{48.2}$$

Each of these three matrices is a simple generalization to three dimensions of the matrix given in Appendix II.

$$S_\phi = \begin{pmatrix} \cos\phi & \sin\phi & 0 \\ -\sin\phi & \cos\phi & 0 \\ 0 & 0 & 1 \end{pmatrix}, \qquad S_\theta = \begin{pmatrix} 1 & 0 & 0 \\ 0 & \cos\theta & \sin\theta \\ 0 & -\sin\theta & \cos\theta \end{pmatrix}$$

(48.3)

$$S_\psi = \begin{pmatrix} \cos\psi & \sin\psi & 0 \\ -\sin\psi & \cos\psi & 0 \\ 0 & 0 & 1 \end{pmatrix}$$

so that

$$(48.4)\quad S = \begin{pmatrix} \cos\psi\cos\phi - \sin\psi\cos\theta\sin\phi & \cos\psi\sin\phi + \sin\psi\cos\theta\cos\phi & \sin\psi\sin\theta \\ -\sin\psi\cos\phi - \cos\psi\cos\theta\sin\phi & -\sin\psi\sin\phi + \cos\psi\cos\theta\cos\phi & \cos\psi\sin\theta \\ \sin\theta\sin\phi & -\sin\theta\cos\phi & \cos\theta \end{pmatrix}$$

It is possible to find the axis of rotation and the angle of rotation, χ, directly from S. The axis of rotation is the direction of the eigenvector whose eigenvalue is unity. It is thus determined by the equations

$$(S_{ij} - \delta_{ij})x_j = 0 \tag{48.5}$$

If this eigenvector were introduced as a coordinate axis, the matrix S would have a form similar to one of those in (48.3), and we would have

$$\operatorname{tr} S = 1 + 2\cos\chi \tag{48.6}$$

The value of the trace of a matrix is independent of the coordinate system, so that in any case

$$\cos\chi = \frac{\operatorname{tr} S - 1}{2} \tag{48.7}$$

49. KINEMATICS OF ROTATION

The motion of a rigid body with one point fixed is described by giving the matrix S as a function of the time. The rate of change of the components of a vector fixed in the body is found by differentiating the first of equations (48.1)

$$\dot{\bar{z}}_i = 0 = \dot{S}_{ij}z_j + S_{ij}\dot{z}_j \tag{49.1}$$

or, in matrix notation,

$$\dot{S}\mathbf{z} + S\dot{\mathbf{z}} = 0$$

where $\mathbf{z}$ has the components z_i, the components relative to axes fixed in space of a vector which is fixed in the body. Multiplying by S^{T} we have

$$\dot{\mathbf{z}} = -S^{\mathsf{T}}\dot{S}\mathbf{z} = -\Omega\mathbf{z} \tag{49.2}$$

where

$$\Omega = S^{\mathsf{T}}\dot{S} \tag{49.3}$$

Differentiating $S^{\mathsf{T}}S = 1$ with respect to time, we have

$$S^{\mathsf{T}}\dot{S} = -\dot{S}^{\mathsf{T}}S = -(S^{\mathsf{T}}\dot{S})^{\mathsf{T}}$$

so that $\Omega = -\Omega$, or Ω is antisymmetrical. We write it thus

$$\Omega = \begin{pmatrix} 0 & \omega_3 & -\omega_2 \\ -\omega_3 & 0 & \omega_1 \\ \omega_2 & -\omega_1 & 0 \end{pmatrix} \tag{49.4}$$

or

$$\Omega_{ij} = \epsilon_{ijk}\omega_k$$

so that

$$\Omega\boldsymbol{\omega} = 0 \tag{49.5}$$

Equation (49.2) becomes

$$\begin{aligned} \dot{z}_i &= -\epsilon_{ijk}\omega_k z_j \\ &= (\boldsymbol{\omega} \times \mathbf{z})_i \end{aligned} \tag{49.6}$$

Thus the ω_i introduced by (49.4) are the components of the angular velocity vector in the space coordinate system [cf. (14.5)]. The components of this same vector relative to axes fixed in the body may be found by transforming the vector $\dot{\mathbf{z}}$ to the body coordinate system. Thus, from (48.1) and (49.2)

$$S\dot{\mathbf{z}} = -S\Omega\mathbf{z} = -\bar{\Omega}\bar{\mathbf{z}}$$

where

$$\bar{\Omega} = S\Omega S^{\mathsf{T}} = \dot{S}S^{\mathsf{T}} \tag{49.7}$$

We leave it to the reader to verify that $\bar{\Omega}$ is also antisymmetric. We may therefore write

$$\bar{\Omega}_{ij} = \epsilon_{ijk}\bar{\omega}_k \tag{49.8}$$

so that, from (49.6),

$$S_{ij}\dot{z}_j = (\bar{\boldsymbol{\omega}} \times \bar{\mathbf{z}})_i \tag{49.9}$$

which is the same result as (49.6) expressed in terms of the components in the body coordinate system.

The angular velocity components may be expressed in terms of S, S^{T} from (6.15) and (49.4) as follows:

$$\begin{aligned} \omega_j &= \tfrac{1}{2}\epsilon_{jkl}\Omega_{kl} = \tfrac{1}{2}\epsilon_{jkl}S^{\mathsf{T}}{}_{km}\dot{S}_{ml} \\ \bar{\omega}_j &= \tfrac{1}{2}\epsilon_{jkl}\bar{\Omega}_{kl} = \tfrac{1}{2}\epsilon_{jkl}\dot{S}_{km}S^{\mathsf{T}}{}_{ml} \end{aligned} \tag{49.10}$$

It is possible to derive relations between the angular velocity components and the rate of change of the Eulerian angles by studying

Fig. 48-1. Since in that figure the direction cosines of the $\bar{3}$ axis and the OK axis relative to the space axes are respectively ($\sin\theta\sin\phi$, $-\sin\theta\cos\phi$, $\cos\theta$) and ($\cos\phi$, $\sin\phi$, 0), the components of the angular velocity of the body along the space axes are immediately seen to be

$$\begin{aligned} \omega_1 &= \cos\phi\dot{\theta} + \sin\theta\sin\phi\dot{\psi} \\ \omega_2 &= \sin\phi\dot{\theta} - \sin\theta\cos\phi\dot{\psi} \qquad (49.11) \\ \omega_3 &= \dot{\phi} + \cos\theta\dot{\psi} \end{aligned}$$

Equivalent expressions for $\bar{\omega}_i$, the components of the angular velocity vector referred to the body axes $\bar{1}$, $\bar{2}$, $\bar{3}$, may also be obtained in a similar manner. They are

$$\begin{aligned} \bar{\omega}_1 &= \cos\psi\dot{\theta} + \sin\theta\sin\psi\dot{\phi} \\ \bar{\omega}_2 &= -\sin\psi\dot{\theta} + \sin\theta\cos\psi\dot{\phi} \qquad (49.12) \\ \bar{\omega}_3 &= \dot{\psi} + \cos\theta\dot{\phi} \end{aligned}$$

As an example of (49.7) we may consider

$$\bar{\Omega}_\psi \equiv \dot{S}_\psi S^{\mathsf{T}}{}_\psi$$

where S_ψ is given by (48.3). It then follows that

$$\bar{\Omega}_\psi = \dot{\psi}\begin{pmatrix} 0 & 1 & 0 \\ -1 & 0 & 0 \\ 0 & 0 & 0 \end{pmatrix} \qquad (49.13)$$

corresponding to rotation about the $\bar{3}$ axis with an angular velocity $\dot{\psi}$. Similarly,

$$\bar{\Omega}_\theta = \dot{\theta}\begin{pmatrix} 0 & 0 & 0 \\ 0 & 0 & 1 \\ 0 & -1 & 0 \end{pmatrix} \quad \text{and} \quad \bar{\Omega}_\phi = \dot{\phi}\begin{pmatrix} 0 & 1 & 0 \\ -1 & 0 & 0 \\ 0 & 0 & 0 \end{pmatrix} \qquad (49.14)$$

Using these equations we are able to derive (49.12) analytically. The angular velocity vector has components referred to axes fixed in the body given by

$$\begin{aligned} \bar{\Omega} &= \dot{S}S^{\mathsf{T}} \qquad (49.15) \\ &= (\dot{S}_\psi S_\theta S_\phi + S_\psi \dot{S}_\theta S_\phi + S_\psi S_\theta \dot{S}_\phi)(S^{\mathsf{T}}{}_\phi S^{\mathsf{T}}{}_\theta S^{\mathsf{T}}{}_\psi) \\ &= \bar{\Omega}_\psi + S_\psi \bar{\Omega}_\theta S^{\mathsf{T}}{}_\psi + S_\psi S_\theta \bar{\Omega}_\phi S^{\mathsf{T}}{}_\theta S^{\mathsf{T}}{}_\psi \\ &= \begin{pmatrix} 0 & \dot{\psi} + \cos\theta\dot{\phi} & \sin\psi\dot{\theta} - \sin\theta\cos\psi\dot{\phi} \\ - & 0 & \cos\psi\dot{\theta} + \sin\theta\sin\psi\dot{\phi} \\ - & - & 0 \end{pmatrix} \end{aligned}$$

where the dashes represent elements that may be obtained from the condition that $\bar{\Omega}$ is antisymmetric. This result agrees with (49.12).

The introduction of the angular velocity has been much more involved than that of the linear velocity. The reason for this is not difficult to find. The linear displacement of a body, a translation, is representable by a vector. The velocity is then the derivative of this vector with respect to the time. Angular displacements are representable by vectors, but the angular velocity vector may not always be obtained from an angular displacement vector by simple differentiation. Suppose that there were functions Λ_i of the Euler angles such that the angular velocities were their time derivatives:

$$\Lambda_i = \Lambda_i(\phi, \theta, \psi) \tag{49.16}$$

Then

$$\begin{aligned} \dot{\Lambda}_i &= \frac{\partial \Lambda_i}{\partial \phi} \dot{\phi} + \frac{\partial \Lambda_i}{\partial \theta} \dot{\theta} + \frac{\partial \Lambda_i}{\partial \psi} \dot{\psi} \\ &= \bar{\omega}_i \end{aligned} \tag{49.17}$$

Comparing (49.17) with (49.12) we see that

$$\frac{\partial \Lambda_1}{\partial \phi} = \sin \psi \sin \theta, \qquad \frac{\partial \Lambda_1}{\partial \theta} = \cos \psi \tag{49.18}$$

But then we would have

$$\frac{\partial^2 \Lambda_1}{\partial \theta\, \partial \phi} = \sin \psi \cos \theta, \qquad \frac{\partial^2 \Lambda_1}{\partial \phi\, \partial \theta} = 0 \tag{49.19}$$

and so no such functions Λ_i exist. This result is a general one and in no way depends on the particular parameters used to describe rotations. The angular velocity components are nonintegrable combinations of the time derivatives of the parameters of the rotation group. They are frequently called the derivatives of *quasi coordinates*. (See Sec. 77.) This result is intimately connected with the fact that the rotations describing the orientation of the body do not commute, so that not only the magnitudes of the rotations, but also their order, must be given for the orientation to be specified.

The above analysis may also be used to relate the velocity and acceleration of a moving point with respect to the body axes with its velocity and acceleration relative to the space axes. Let us consider a particle which is not necessarily fixed with respect to either set of axes and denote its position vector relative to body and space axes by

$\bar{\mathbf{r}}$, $\mathbf{r}$ respectively. Thus [cf. (48.1)]

$$\bar{\mathbf{r}} = S\mathbf{r}$$

so that

$$\dot{\bar{\mathbf{r}}} = S\dot{\mathbf{r}} + \dot{S}\mathbf{r}$$

$$\ddot{\bar{\mathbf{r}}} = S\ddot{\mathbf{r}} + 2\dot{S}\dot{\mathbf{r}} + \ddot{S}\mathbf{r}$$

or

$$S^{\mathsf{T}}\ddot{\bar{\mathbf{r}}} = \ddot{\mathbf{r}} + 2S^{\mathsf{T}}\dot{S}\mathbf{r} + S^{\mathsf{T}}\ddot{S}\mathbf{r} \tag{49.20}$$

Using the results

$$\Omega = S^{\mathsf{T}}\dot{S}, \qquad \dot{\Omega} = \dot{S}^{\mathsf{T}}\dot{S} + S^{\mathsf{T}}\ddot{S}$$

$$\Omega^2 = -(\dot{S}^{\mathsf{T}}S)(S^{\mathsf{T}}\dot{S}) = -\dot{S}^{\mathsf{T}}\dot{S}$$

we have then

$$\begin{aligned} S^{\mathsf{T}}\dot{\bar{\mathbf{r}}} &= \dot{\mathbf{r}} + \Omega\mathbf{r} \\ S^{\mathsf{T}}\ddot{\bar{\mathbf{r}}} &= \ddot{\mathbf{r}} + 2\Omega\dot{\mathbf{r}} + (\dot{\Omega} + \Omega^2)\mathbf{r} \end{aligned} \tag{49.21}$$

If we choose the axes of the two coordinate systems to coincide instantaneously, $S^{\mathsf{T}} = I$ and the components of the acceleration of the particle relative to the body axes are given as the sum of four terms

$$\begin{aligned} \dot{\bar{\mathbf{r}}} &= \dot{\mathbf{r}} - \boldsymbol{\omega} \times \mathbf{r} \\ \ddot{\bar{\mathbf{r}}} &= \ddot{\mathbf{r}} - 2\boldsymbol{\omega} \times \dot{\mathbf{r}} - \dot{\boldsymbol{\omega}} \times \mathbf{r} + \boldsymbol{\omega} \times (\boldsymbol{\omega} \times \mathbf{r}) \end{aligned} \tag{49.22}$$

where $\boldsymbol{\omega}$ is the angular velocity of the body system relative to the space system and $\bar{\mathbf{r}}$, $\mathbf{r}$ denote distances measured with respect to body and space systems respectively.

The first of equations (49.22) states that the velocity of the particle relative to the rotating body axes is equal to its velocity relative to the space axes minus the velocity relative to the space axes of the point in the body axes instantaneously occupied by the particle. It is merely a statement of the law of composition of relative velocities. It may alternatively be written

$$\dot{\mathbf{r}} = \dot{\bar{\mathbf{r}}} + \boldsymbol{\omega} \times \bar{\mathbf{r}} \tag{49.23}$$

since the vector to the particle is the same relative to the two sets of axes and instantaneously possesses the same components.

The second of equations (49.22) may, using the first, be written

$$\ddot{\mathbf{r}} = \ddot{\bar{\mathbf{r}}} + 2\boldsymbol{\omega} \times \dot{\bar{\mathbf{r}}} + \dot{\boldsymbol{\omega}} \times \bar{\mathbf{r}} + \boldsymbol{\omega} \times (\boldsymbol{\omega} \times \bar{\mathbf{r}}) \tag{49.24}$$

giving the acceleration in the space system in terms of quantities measured in the body system. When we multiply throughout by the mass of the particle, the left-hand side becomes the force if the space axes form an inertial system, for which the coordinates may be taken as cartesian. In nonrelativistic mechanics, with which we are here concerned, an inertial system is one which is at rest with respect to, or moving with constant velocity relative to, the average position of the distant stars.*

Thus, for example, if $\boldsymbol{\omega}$ denotes the angular velocity of the earth relative to this system, (49.24) gives the mass times acceleration of the particle relative to the earth as the sum of the four terms

$m\ddot{\mathbf{r}}$ = the mass times the acceleration relative to the inertial system, i.e., the external force

$-2m\boldsymbol{\omega} \times \mathring{\mathbf{r}}$ = the Coriolis force in a direction at right angles to $\boldsymbol{\omega}$ and to the velocity $\mathring{\mathbf{r}}$ of the particle relative to the earth

$-m\dot{\boldsymbol{\omega}} \times \bar{\mathbf{r}}$ = in this case a practically negligible term, since $\boldsymbol{\omega}$ is approximately constant

$-m\boldsymbol{\omega} \times (\boldsymbol{\omega} \times \bar{\mathbf{r}})$ = the centrifugal force

For a particle at the equator moving in the plane of the equator, for example, the ratio of the magnitudes of the Coriolis and centripetal forces is approximately $4.4 \times 10^{-5}v$, where v is its velocity in centimeters per second.

50. THE MOMENTAL ELLIPSOID

In order to discuss the dynamics of a rigid body, as opposed to the kinematic description of its motion, it is necessary to return to (14.8), (14.9), and (14.10), which relate the spin angular momentum $\boldsymbol{\sigma}$, the angular velocity $\boldsymbol{\omega}$, the kinetic energy T, and the inertial tensor J of a rigid body rotating about a fixed point. In terms of the matrix notation described in Appendix II, these equations may be written:

$$\boldsymbol{\sigma} = J\boldsymbol{\omega}, \qquad T = \tfrac{1}{2}(\boldsymbol{\omega}, J\boldsymbol{\omega}) \tag{50.1}$$

* A much more satisfactory definition on an inertial system is provided by the general theory of relativity, in which it is a system in the neighborhood of which the gravitational field vanishes, whether this field is produced by matter or by acceleration or rotation of the axes relative to the coordinate system mentioned above.

with

$$J = \begin{pmatrix} A & -H & -G \\ -H & B & -F \\ -G & -F & C \end{pmatrix} \tag{50.2}$$

The transformation law (II.36) for the tensor J then becomes

$$\bar{J} = SJS^{\mathsf{T}} \tag{50.3}$$

The components of J in the space coordinate system are not constant, because in this system the values of the x's change with time. For this reason it is convenient to work in the body coordinate system when discussing the motion of rigid rotators.

The inertia tensor $\bar{J}$ is symmetric and therefore its matrix has three real eigenvalues, and the corresponding eigenvectors are mutually orthogonal. If these three vectors are introduced as the coordinate axes of the body coordinate system, then $\bar{J}$ is diagonal when referred to these axes, i.e., the products of inertia vanish in this coordinate system. From the definition (14.9) of J_{ij}, it follows that the diagonal elements of J are positive for any coordinate system, so that the eigenvalues A, B, C of J are positive.

The directions of the eigenvectors of J are called the *principal axes* of the rotator relative to the fixed point. In this system, the kinetic energy can be written:

$$T = \tfrac{1}{2}(A\bar{\omega}_1{}^2 + B\bar{\omega}_2{}^2 + C\bar{\omega}_3{}^2) \tag{50.4}$$

and the angular momentum has components $(A\bar{\omega}_1, B\bar{\omega}_2, C\bar{\omega}_3)$. If the angular velocity happens to be along one of the principal axes, the angular velocity and angular momentum are therefore parallel to each other.

Associated with a symmetric matrix $\bar{J}$ there is a quadratic form, and a quadric surface defined by

$$(\mathbf{x}, J\mathbf{x}) = \bar{J}_{ij}\bar{x}_i\bar{x}_j = 1 \tag{50.5}$$

which, if J is the inertia matrix, is necessarily an ellipsoid. This surface is called the *momental ellipsoid* because the moment of inertia of the rotator about an axis through the fixed point is easily expressed in terms of this ellipsoid. The moment of inertia I about an axis is defined by the equation

$$T = \tfrac{1}{2}I\omega^2 \qquad (\boldsymbol{\omega} \text{ along the axis}) \tag{50.6}$$

If s_ρ denotes the distance of the ρth particle of the rotator from the axis, then

$$I = \sum_{\rho} m_\rho s_\rho^{\ 2} \tag{50.7}$$

Now let $\mathbf{x}$ be parallel to $\boldsymbol{\omega}$. Then, from (50.5) and (50.1),

$$\boldsymbol{\omega} = (2T)^{1/2}\mathbf{x} \tag{50.8}$$

Insertion of this in (50.6) then yields

$$T = \tfrac{1}{2}I(2T\mathbf{x}^2), \qquad I = \frac{1}{\mathbf{x}^2} \tag{50.9}$$

Thus the moment of inertia about an axis is the square of the reciprocal of the distance from the origin to the momental ellipsoid in the direction of the axis.

51. THE FREE ROTATOR

The motion of a rigid rotator with no forces acting can be described geometrically by use of the momental ellipsoid just discussed. In such a motion the angular momentum and the kinetic energy must both remain constant. These conditions suffice to determine the motion. From (50.1) we have

$$T = \tfrac{1}{2}(\boldsymbol{\omega}, J\boldsymbol{\omega}), \qquad \boldsymbol{\sigma} = J\omega \tag{51.1}$$

The momental ellipsoid is defined by (50.5) as

$$(\mathbf{x}, J\mathbf{x}) = 1 \tag{51.2}$$

Let us find the equation of the plane tangent to the momental ellipsoid at the point where the instantaneous axis cuts the ellipsoid. Here $\mathbf{x} = \boldsymbol{\omega}/(2T)^{1/2}$ and the equation of the tangent plane is

$$(\mathbf{r}, J\mathbf{x}) = [\mathbf{r}, J\boldsymbol{\omega}/(2T)^{1/2}] = 1 \tag{51.3}$$

where $\mathbf{r}$ is the radius vector of a point in the tangent plane. Equation (51.3) can be written

$$(\mathbf{r}, \boldsymbol{\sigma}) = (2T)^{1/2} \tag{51.4}$$

which shows that the equation of the tangent plane depends only on the kinetic energy and angular momentum, and this plane is therefore fixed in space. Since the point I of tangency between the ellipsoid and the plane is on the instantaneous axis, this point of the ellipsoid has zero velocity and the momental ellipsoid rolls without sliding on the tangent plane. This result, first pointed out by Poinsot, is illustrated in Fig. 51-1.

The motion of the rotator can now be visualized. The center of the ellipsoid is fixed and the ellipsoid rolls on a fixed plane. There are three parts to the general motion: (1) rotation about the instantaneous axis, (2) precession of the instantaneous axis about the direction of the

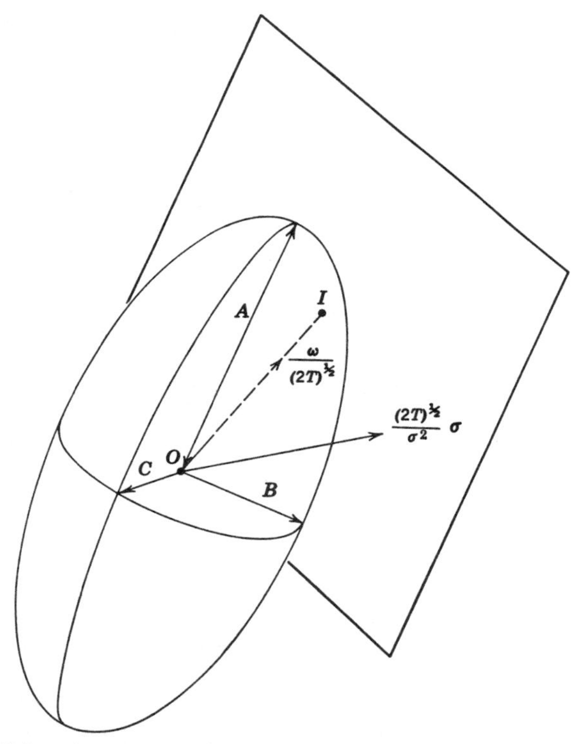

Fig. 51-1. Poinsot's representation of the motion of a free rotator. The fixed tangent plane is at right angles to the angular momentum and the momental ellipsoid is in contact with it instantaneously at point I.

total angular momentum S, (3) nutation, or change in angle between the instantaneous axis and the angular momentum. The frequency of nutation is twice the frequency of rotation.

52. EULER'S EQUATIONS OF MOTION

The rate of change of angular momentum of a rigid body about a point is the moment $\mathbf{M}$ of the applied forces about that point. Thus for a rigid rotator we may write

$$\dot{\boldsymbol{\sigma}} = \mathbf{M} \tag{52.1}$$

As always, $\dot{\boldsymbol{\sigma}}$ must be measured in an inertial coordinate system, which here means the space system.

In the space system, however, the inertia tensor itself is not constant, and so it is advantageous to use the body coordinate system representation of J. The simplest body coordinate system is that which coincides with the space system instantaneously, so that $S = 1$ although $\dot{S} \neq 0$. We have, since $\bar{J}$ is constant,

$$\dot{\bar{J}} = \dot{S}JS^{\mathsf{T}} + S\dot{J}S^{\mathsf{T}} + SJ\dot{S}^{\mathsf{T}} = 0$$

or

$$\begin{aligned} \dot{J} &= -S^{\mathsf{T}}\dot{S}J - J\dot{S}^{\mathsf{T}}S \\ &= J\Omega - \Omega J \end{aligned} \tag{52.2}$$

by (49.3). Thus

$$\begin{aligned} \dot{\boldsymbol{\sigma}} &= J\dot{\boldsymbol{\omega}} + (J\Omega - \Omega J)\boldsymbol{\omega} \\ &= J\dot{\boldsymbol{\omega}} - \Omega\boldsymbol{\sigma} \\ &= J\dot{\boldsymbol{\omega}} + \boldsymbol{\omega} \times \boldsymbol{\sigma} \end{aligned} \tag{52.3}$$

by use of (49.5).

Combining (52.1) and (52.3), we obtain

$$\mathbf{M} = J\dot{\boldsymbol{\omega}} + \boldsymbol{\omega} \times \boldsymbol{\sigma} \tag{52.4}$$

The only vector which appears differentiated in (52.4) is $\boldsymbol{\omega}$, and $\dot{\boldsymbol{\omega}} = \dot{\bar{\boldsymbol{\omega}}} + \Omega\boldsymbol{\omega} = \dot{\bar{\boldsymbol{\omega}}}$ since $S = 1$ instantaneously and $\Omega\boldsymbol{\omega}$ vanishes. Thus (52.4) may be written in the body coordinate system. For simplicity we choose the principal axis system and obtain, since J is diagonal,

$$\begin{aligned} \bar{M}_1 &= A\dot{\bar{\omega}}_1 + (C - B)\bar{\omega}_2\bar{\omega}_3 \\ \bar{M}_2 &= B\dot{\bar{\omega}}_2 + (A - C)\bar{\omega}_3\bar{\omega}_1 \\ \bar{M}_3 &= C\dot{\bar{\omega}}_3 + (B - A)\bar{\omega}_1\bar{\omega}_2 \end{aligned} \tag{52.5}$$

These are the celebrated *Euler equations of motion.*

The Euler equations can be integrated in the case where all the force moments are zero, yielding an analytical description of the motion represented by Poinsot's geometrical construction. They also give immediate information about the steady motion of a rotator, when all the $\dot{\omega}$'s are zero. Let us consider a symmetric rotator, i.e., one with two of its principal moments of inertia equal, and inquire what torques are needed to obtain a steady motion. Suppose that $A = B$. Then $M_3 = 0$ and there can be no torque about the unique axis of the rotator. Since $A = B$, the inertia matrix is degenerate and any axis through the fixed point perpendicular to the unique axis is also a principal axis. We choose the 2 axis of the space system in such a direction that $\omega_2 = 0$. Then M_1, the torque about the 1 axis, must also vanish. The only permissible torque is that about the 2 axis, and it must have the magnitude given by the second of equations (52.5):

$$M_2 = (A - C)\omega_3\omega_1 \tag{52.6}$$

This is a possible situation for a heavy top which is symmetric and which has its center of mass on the unique axis. This situation is the opposite of that of the free rotator. There the angular velocity vector precessed about the constant angular momentum vector. Here a moment of force is applied which makes the angular momentum vector precess about the constant angular velocity vector. If all three principal moments of inertia are equal, the angular velocity and angular momentum vectors are necessarily parallel, and, if the former is to be constant, the latter must also be constant and there can be no applied force moments.

The equations of motion for a rigid rotator can be written in Lagrangian form, the Euler angles being used as coordinates. This proves to be practical only if the rotator has two equal principal moments of inertia. If this is so, the kinetic energy can be written as

$$T = \tfrac{1}{2}[A(\dot{\theta}^2 + \sin^2\theta\dot{\phi}^2) + C(\dot{\psi} + \cos\theta\dot{\phi})^2] \tag{52.7}$$

where the expressions (49.12) have been inserted in (50.4).

Many applications of this method are discussed in detail in the references listed at the end of this chapter, and one application—to the motion of a spinning particle in an electromagnetic field—is developed in Chapter 16.

GENERAL REFERENCES

F. Klein and A. Sommerfeld, *Uber die Theorie des Kreisels.* Teubner, Leipzig (1910–23).

H. Lamb, *Higher Mechanics.* Cambridge University Press (1929).

J. L. Synge and B. A. Griffith, *Principles of Mechanics.* McGraw-Hill, New York (1942).

EXERCISES

1. Derive (49.11) from (49.3), (48.2).

2. If successive rotations S, T are equivalent to a rotation $U = TS$, and if $\Omega_S = S^{\mathsf{T}}\dot{S}$, $\Omega_T = T^{\mathsf{T}}\dot{T}$, $\Omega_U = U^{\mathsf{T}}\dot{U}$, and $\bar{\Omega}_S = S\Omega_S S^{\mathsf{T}}$, show that

$$\Omega_U = \Omega_S + S^{\mathsf{T}}\Omega_T S$$

and

$$\dot{\Omega}_U = \dot{\Omega}_S + S^{\mathsf{T}}\dot{\Omega}_T S + S^{\mathsf{T}}(\Omega_T\bar{\Omega}_S - \bar{\Omega}_S\Omega_T)S$$

Hence show that, if the elements of Ω_S, Ω_T are constant, the elements of Ω_U are constant in general only if Ω_T commutes with $\bar{\Omega}_S$. Interpret this physically.

3. A particle is projected downward from a point h above the equator (a) radially in the space coordinate system, (b) radially in the earth coordinate system. Assuming that the particle moves with constant speed v, find the angles of incidence on the surface of the earth in the two cases and sketch both paths in both coordinate systems. Show that this cannot account for the east-west effect of cosmic radiation.

4. A uniform solid circular cylinder of radius a rests on a horizontal plane, and an identical cylinder rests on it touching it along the highest generator. If no slipping occurs, show that as long as the cylinders remain in contact

$$\dot{\theta}^2 = \frac{12g(1 - \cos\theta)}{a(17 + 4\cos\theta - 4\cos^2\theta)}$$

where θ is the angle which the plane containing the axes makes with the vertical. Show also that the path of a point on the axis of the upper cylinder may be written

$$x = \tfrac{1}{3}a(\theta + 4\sin\theta)$$

$$y = 2a(1 - \cos\theta)$$

relative to horizontal and vertical axes through the initial position of the point.

5. Two smooth fixed rods Ox, Oy in a vertical plane slope downward from O at angles 45° to the vertical. Equal uniform rods AC, BC are smoothly hinged at C, and the other ends A, B slide by means of rings on the fixed rods. Obtain the equations of motion and find the times of small oscillations about equilibrium.

6. A uniform right circular cone of semivertical angle α rolls without sliding on a plane inclined at an angle β with the horizontal, being released from rest with the line of contact horizontal. Prove that the cone will remain in contact with the plane provided that

$$9\tan\beta < \cot\alpha + 4\tan\alpha$$

7. A uniform, thin, hollow circular cylinder of mass M and radius b moves on a smooth horizontal plane. A second uniform, thin, hollow cylinder of mass m and radius $a(<b)$ rolls, without slipping, on the inner surface of the

first cylinder. Show that

$$\dot{\theta}^2 = \frac{g}{b-a}\frac{2(M+m)}{2M+m\sin^2\theta}(\cos\theta - \cos\theta_0)$$

where θ is the angle between the vertical plane and the plane containing the axes of the cylinders, and the motion starts from rest with $\theta = \theta_0$.

8. Three similar uniform straight rods AB, BC, CD, each of length $2l$, mass M, smoothly hinged at B and C, lie on a smooth horizontal table. A light inextensible taut string attached to the mid-point of BC passes perpendicularly over a smooth edge of the table and carries a particle, also of mass M, at the end of the portion which hangs vertically. Initially all three rods are in a straight line parallel to the edge of the table. Apply Lagrange's equations to show that the time required for the free ends A and D to meet is

$$\left(\frac{l}{3g}\right)^{1/2}\int_0^{2\pi/3}\left(\frac{5+3\sin^2\theta}{\sin\theta}\right)^{1/2}d\theta$$

9. A body is rotating about its center of mass under no forces, and A, B, C $(A > B > C)$ are the principal moments of inertia at the center of mass. If initially ω_3 is positive and ω_1 negative and if $\sigma^2 = 2BT$ (σ = angular momentum, T = kinetic energy), show that

$$\omega_2 = \frac{\sigma}{B}\tanh\tau$$

$$\omega_1 = -\frac{\sigma}{B}\left[\frac{B(B-C)}{A(A-C)}\right]^{1/2}\operatorname{sech}\tau$$

$$\omega_3 = \frac{\sigma}{B}\left[\frac{B(A-B)}{C(A-C)}\right]^{1/2}\operatorname{sech}\tau$$

where

$$\tau = \frac{\sigma}{B}\left[\frac{(A-B)(B-C)}{AC}\right]^{1/2}(t-t_0)$$

What happens when t increases indefinitely?

10. Investigate the motion of a symmetrical top ($A = B$) in a gravitational field, one point on the axis of the top being held fixed. Show that the total energy E and the angular momenta p_ϕ, p_ψ about the vertical axis and about the symmetry axis of the top are constants of the motion, and without solving the equations of motion determine the general nature of the motion from the condition that $\dot{\theta}^2 \geq 0$.

10 HAMILTONIAN THEORY

53. HAMILTON'S EQUATIONS

We turn now to an investigation of the beautiful superstructure on Lagrangian theory built chiefly by Sir W. R. Hamilton (1805–1865) though not without important contributions from Poisson (1781–1840), Jacobi (1804–1851), and, later, Lie (1842–1899) and many others. This development yields a deeper insight into the equations of motion of a mechanical system which can be described by a Lagrangian, making it possible to write these equations in a very elegant form (53.6) or in terms of a very simple principle (57.4). Furthermore, contact transformations (Sec. 58), the partial differential equation of Hamilton and Jacobi (61.7), and perturbation theory (Sec. 74) which emerge from this analysis are sometimes of practical use in obtaining approximations to the solutions of complicated dynamic problems. Perhaps most important of all, however, is the fact that the generalization of Hamilton's ideas to the theories of relativity, quantum mechanics, and statistical mechanics is fundamental to these branches of theoretical physics.

As usual, we shall restrict ourselves to the consideration of a system which is described by a Lagrangian L which is a function of the generalized coordinates q_j, the generalized velocities $\dot{q}_j$, and the time t:

$$L = L(q_j, \dot{q}_j, t) \qquad (j = 1 \cdots f) \tag{53.1}$$

For a single particle referred to rectangular cartesian coordinates x_i and acted upon by a force derivable from a potential function $V(x_i)$ we have already obtained

$$L = T - V = \tfrac{1}{2}m\dot{x}_i\dot{x}_i - V(x_j)$$

so that

$$\frac{\partial L}{\partial \dot{x}_j} = m\dot{x}_j = p_j$$

the component of momentum in the j direction. We therefore adopt the definition (34.3) for the more general problem.

$$p_j = \frac{\partial L}{\partial \dot{q}_j} \tag{53.2}$$

is the *generalized momentum* corresponding to, or conjugate to, the generalized coordinate q_j. It is not necessary, of course, that the momenta p_j should have the dimensions MLT^{-1} any more than it was necessary that the coordinates q_j should all be lengths. However, the product of any momentum and its conjugate position coordinate must be of dimensions ML^2T^{-1}, which are the dimensions of angular momentum or of the dynamic variable "action" defined later (57.10).

With the definition (53.2), Lagrange's equations (32.2) may be written in the form

$$\dot{p}_j = \frac{\partial L}{\partial q_j} \tag{53.3}$$

We suppose that it is possible to solve equations (53.2) for the $\dot{q}_i$ explicitly as functions of the q_j, p_j, t, and thus to use the variables q_j, p_j, t rather than q_j, $\dot{q}_j$, t to specify the system. Hence if δ denotes the increment in any function of the variables q_j, p_j, t or q_j, $\dot{q}_j$, t due to infinitesimal changes in these variables, we have

$$\begin{aligned}\delta L &= \frac{\partial L}{\partial q_j}\,\delta q_j + \frac{\partial L}{\partial \dot{q}_j}\,\delta \dot{q}_j + \frac{\partial L}{\partial t}\,\delta t \\ &= \dot{p}_j\,\delta q_j + p_j\,\delta \dot{q}_j + \frac{\partial L}{\partial t}\,\delta t \qquad \text{using (53.3)} \\ &= \delta(p_j\dot{q}_j) + \dot{p}_j\,\delta q_j - \dot{q}_j\,\delta p_j + \frac{\partial L}{\partial t}\,\delta t\end{aligned}$$

or

$$\delta(p_j\dot{q}_j - L) = -\dot{p}_j\,\delta q_j + \dot{q}_j\,\delta p_j - \frac{\partial L}{\partial t}\,\delta t \tag{53.4}$$

We have already seen in (33.1) the importance of the quantity

$$H = p_j\dot{q}_j - L \tag{53.5}$$

which in a system described by a Lagrangian not involving the time explicitly is the Jacobian integral. For a natural system (33.6), this integral is the total energy of the system. Even in the more general case in which L involves t and the system is not necessarily natural, the combination (53.5) plays an important role. Since variations in the velocities $\dot{q}$ do not appear on the right-hand side of (53.4), H may be written as a function of the q's and the p's and the time. To obtain this function $H(q, p, t)$, the $\dot{q}$'s are eliminated between (53.2) and (53.5). This function $H(q, p, t)$ is called the *Hamiltonian* of the system. It is to be stressed that the Hamiltonian is always to be written thus, so that, for example $\partial H/\partial q_i$ means that the other q's, all the p's, and t are to be held fixed in the differentiation; the $\dot{q}$'s are not involved in the process explicitly. This is in contrast to the situation in Lagrange's equations where the p's are not explicitly involved, but the Lagrangian is always a function of the q's, the $\dot{q}$'s, and t.

Then we have

$$\delta H = \frac{\partial H}{\partial q_j}\,\delta q_j + \frac{\partial H}{\partial p_j}\,\delta p_j + \frac{\partial H}{\partial t}\,\delta t$$

so that, since the variations δq_j, δp_j, δt are mutually independent, it follows by comparison with (53.4) that

$$\dot{q}_j = \frac{\partial H}{\partial p_j} \tag{53.6}$$

$$\dot{p}_j = -\frac{\partial H}{\partial q_j}$$

and

$$\frac{\partial H}{\partial t} = -\frac{\partial L}{\partial t} \tag{53.7}$$

Equations (53.6) are called *Hamilton's canonical equations*, and the variables q_j, p_j are said to be *canonically conjugate*. Hamilton's equations constitute a set of $2f$ differential equations of the first order to give the $2f$ variables q_j, p_j as functions of the time t and $2f$ arbitrary constants. Whereas the f Lagrange's equations (32.2) (of the second order) describe the motion of the system by the motion of a representative point in f-dimensional configuration space, the $2f$ canonical equations describe the motion in terms of that of a representative point in a $2f$-dimensional space specified by the f variables q_j together with the f variables p_j. Such a space is called the *phase space* of the system. Boltzmann and Gibbs, in particular, have applied this concept extensively to the kinetic theory of gases and to statistical mechanics. On the other hand, the variables p_j describe the state of motion of the system, but do not specify its configuration. The state of motion may thus be represented by a point in an f-dimensional space p_j, which is called the *momentum space* of the system. Equations analogous to those of Lagrange, describing the motion of the system in terms of that of a representative point in momentum space, are derived in Sec. 60.

The complete solution of the f Lagrange equations of motion requires $2f$ constants for its specification. These constants may be taken to be the values of the coordinates and the velocities at some time t_0. Giving the velocities is equivalent to giving the values of the coordinates at time $t_0 + \delta t$ in addition to those at t_0. We may say, then, that the complete specification of a solution of Lagrange's equations requires the values of the coordinates at two times. (These times need not be infinitesimally separated.) With Hamilton's equations the situation is quite different. We may consider the coordinates and momenta as being the independent elements describing the system. The velocities are then derived from Hamilton's equations. A state of motion is described completely by giving the values of the f coordinates and the f momenta at one time t_0.

There are other ways of obtaining first-order equations of motion from the second-order Lagrange equations. One way is to introduce the $\dot{q}$'s as new variables in the Lagrange equations. If $\dot{q}_i = u_i$, then $\ddot{q}_i = \dot{u}_i$ and we have $2f$ first-order equations. The form of the equations is, in general, not such as to lead to interesting results in a simple way.

Through each point in configuration space there is an f-fold infinity of possible paths of a given system moving under given forces. Through each point in momentum space there is, similarly, an f-fold infinity of possible paths. On the other hand, through each point in phase space there is, for a given problem, only one path, which we call the *trajectory* of the system. The $2f$ canonical equations of the first order require, for their complete solution, the specification of $2f$ arbitrary constants. Once a point in phase space is prescribed, $2f$ such constants are given, and there is no more freedom of choice. We may say that the aggregate of values of q_j, p_j at a certain instant specifies the state of the system at that instant. When the state is given at any time, it may then be determined for all times, both later and earlier, for which the Hamiltonian is known.

According to the Heisenberg uncertainty principle, it is fundamentally impossible to give a coordinate and its conjugate momentum precise values at a common time, since measurement of the one interferes with measurement of the other. In quantum mechanics it is therefore necessary to adopt a different definition of the state of a system, although one which reduces to the above definition for situations in which the effects of the uncertainty principle are negligible.

The rate of increase of H along the trajectory of the system is

$$\dot{H} = \frac{\partial H}{\partial q_j}\dot{q}_j + \frac{\partial H}{\partial p_j}\dot{p}_j + \frac{\partial H}{\partial t}$$

or

$$\dot{H} = \frac{\partial H}{\partial t} \qquad \text{by (53.6)} \tag{53.8}$$

$$= -\frac{\partial L}{\partial t} \qquad \text{by (53.7)}$$

If L does not contain t explicitly, neither does H, and then, as already indicated, H is a constant of the motion

$$H(q_j, p_j) = E$$

If L is expressible as in Sec. 33 as the sum of terms L_2, L_1, L_0, respectively quadratic and linear in $\dot{q}_j$ and independent of $\dot{q}_j$, then, from (53.5),

$$E = 2L_2 + L_1 - L$$

$$= L_2 - L_0$$

The part of L linear in $\dot{q}_j$ does not contribute to the *value* of H but of course does modify the definition of p_j, hence the *form* of H as a function of q_j, p_j, t.

Writing $L_2 = T, L_0 = -V$, where T, V are the kinetic and potential energies of the system, we have $H = T + V$. Usually the terms quadratic in q_j do not involve t explicitly, so that under these conditions (53.8) becomes

$$\frac{d}{dt}(T + V) = \frac{\partial H}{\partial t} = \frac{\partial V}{\partial t} \tag{53.9}$$

which is the relation derived earlier (11.12).

We note that, from (53.3) and (53.6),

$$\dot{p}_j = \frac{\partial L}{\partial q_j}(q, \dot{q}, t) = -\frac{\partial H}{\partial q_j}(q, p, t) \tag{53.10}$$

Thus, if a particular coordinate does not appear explicitly in L, it does not appear explicitly in H. Such coordinates are called *cyclic* or *ignorable* (Sec. 34).

It might at first be imagined that, since the canonical equations are of the first order, they represent great mathematical simplicity when compared with the second-order equations of Lagrange. In a particular problem, however, the kinetic and potential energies are given, or can be found, so that there is usually little difficulty in writing down the Lagrangian for the problem. Once this is done, in order to set up the Hamiltonian it is necessary to (i) obtain p_j from (53.2); (ii) solve the equations derived under (i) to derive the $\dot{q}_j$; (iii) substitute these $\dot{q}_j$ in both terms of (53.5) to obtain H. When this has been carried out, the first set of equations (53.6) reproduces the equations derived under (i) and the second set of equations (53.6) yields the equations of motion. To solve these we may eliminate p_j between both equations (53.6), and the resulting equations are then nothing but Lagrange's equations. The fact that Hamilton's equations are of the first order does not make them easier to integrate as they stand. The advantage lies in their simple form and in the properties of the variables q_j, p_j in terms of which they describe the behavior of the system. In particular, it is sometimes possible to choose some of these so that their values do not change during the motion of the system.

54. HAMILTON'S EQUATIONS IN VARIOUS COORDINATE SYSTEMS

To illustrate the form of Hamilton's equations, we consider a system consisting of a particle moving under the influence of a potential which is independent of the velocity, so that the momentum p_i conjugate to the coordinate q_i is $\partial T/\partial \dot{q}_i$.

(a) Cylindrical Coordinates. See Sec. 5.

$$x_1 = \rho \cos \phi$$

$$x_2 = \rho \sin \phi$$

$$x_3 = z$$

In these coordinates the kinetic energy of a particle is

$$T = \tfrac{1}{2}m(\dot{\rho}^2 + \rho^2\dot{\phi}^2 + \dot{z}^2)$$

so that

$$(54.1) \qquad p_\rho = m\dot{\rho}, \qquad p_\phi = m\rho^2\dot{\phi}, \qquad p_z = m\dot{z}$$

p_ρ is a linear momentum. It is the component of linear momentum along a line through the origin and lying in the 1-2 plane. p_z is also a linear momentum, the 3 component of the linear momentum of the particle. p_ϕ is an angular momentum, the 3 component of the angular momentum of the particle. We have

$$H = (2m)^{-1}(p_\rho^{\,2} + \rho^{-2}p_\phi^{\,2} + p_z^{\,2}) + V$$

Hamilton's equations are

$$\begin{aligned}
\dot{p}_\rho &= -\frac{\partial H}{\partial \rho} = (m\rho^3)^{-1}p_\phi^{\,2} - \frac{\partial V}{\partial \rho}, & \dot{\rho} &= m^{-1}p_\rho \\
\dot{p}_\phi &= -\frac{\partial H}{\partial \phi} = -\frac{\partial V}{\partial \phi}, & \dot{\phi} &= (m\rho^2)^{-1}p_\phi \\
\dot{p}_z &= -\frac{\partial H}{\partial z} = -\frac{\partial V}{\partial z}, & \dot{z} &= m^{-1}p_z
\end{aligned} \tag{54.2}$$

p_ϕ is a constant if V is independent of ϕ, and now, if p_ϕ is zero, p_ρ is a constant if V is also independent of ρ. Potential functions which are independent of ϕ are said to possess cylindrical symmetry. These coordinates are particularly applicable to problems which possess such symmetry.

(b) Spherical Coordinates. See (5.5).

$$\begin{aligned}
x_1 &= r \sin\theta \cos\phi \\
x_2 &= r \sin\theta \sin\phi \\
x_3 &= r \cos\theta
\end{aligned}$$

The kinetic energy of the particle is

$$T = \frac{m}{2}(\dot{r}^2 + r^2\dot{\theta}^2 + r^2 \sin^2\theta\dot{\phi}^2)$$

so that

$$\begin{aligned}
p_r &= \frac{\partial T}{\partial \dot{r}} = m\dot{r} \\
p_\theta &= \frac{\partial T}{\partial \dot{\theta}} = mr^2\dot{\theta} \\
p_\phi &= \frac{\partial T}{\partial \dot{\phi}} = mr^2 \sin^2\theta\dot{\phi}
\end{aligned} \tag{54.3}$$

p_r is the linear momentum along the radial direction. p_θ is the angular momentum about an axis through the origin and perpendicular to the plane containing the particle and the x_3 axis. Since here the value of the Hamiltonian is the energy $E = T + V$, to obtain H we must express T in terms of coordinates and momenta using (53.2). Thus

$$H = \frac{1}{2m}\left(p_r^2 + \frac{p_\theta^2}{r^2} + \frac{p_\phi^2}{r^2 \sin^2 \theta}\right) + V(r, \theta, \phi) \tag{54.4}$$

Hamilton's equations are now

$$\dot{r} = \frac{\partial H}{\partial p_r} = \frac{p_r}{m}$$

$$\dot{\theta} = \frac{\partial H}{\partial p_\theta} = \frac{p_\theta}{mr^2}$$

$$\dot{\phi} = \frac{\partial H}{\partial p_\phi} = \frac{p_\phi}{mr^2 \sin^2 \theta}$$

which are just the solutions of (54.3) for the velocities, and

$$\begin{aligned} \dot{p}_r &= -\frac{\partial H}{\partial r} = \frac{1}{mr^3}\left(p_\theta^2 + \frac{p_\phi^2}{\sin^2 \theta}\right) - \frac{\partial V}{\partial r} \\ \dot{p}_\theta &= -\frac{\partial H}{\partial \theta} = \frac{1}{mr^2}\left(\frac{p_\phi^2 \cos \theta}{\sin^3 \theta}\right) - \frac{\partial V}{\partial \theta} \\ \dot{p}_\phi &= -\frac{\partial H}{\partial \phi} = -\frac{\partial V}{\partial \phi} \end{aligned} \tag{54.5}$$

If V is independent of ϕ, p_ϕ is a constant. If V is also independent of θ, and if p_ϕ vanishes, p_θ is a constant (cf. Sec. 36).

(c) Rotating Coordinates. If a set of rectangular axes $\bar{x}$, $\bar{y}$, $\bar{z}$ is rotating with angular velocity ω about the z axis of a set of rectangular coordinates x, y, z, so that the origins and the z, $\bar{z}$ axes coincide,

$$x = \bar{x} \cos \omega t - \bar{y} \sin \omega t$$

$$y = \bar{x} \sin \omega t + \bar{y} \cos \omega t$$

$$z = \bar{z}$$

The kinetic energy of a particle relative to x, y, z is

$$\begin{aligned} T &= \tfrac{1}{2}m(\dot{x}^2 + \dot{y}^2 + \dot{z}^2) \\ &= \tfrac{1}{2}m(\dot{\bar{x}}^2 + \dot{\bar{y}}^2 + \dot{\bar{z}}^2) + m\omega(\bar{x}\dot{\bar{y}} - \dot{\bar{x}}\bar{y}) + \tfrac{1}{2}m\omega^2(\bar{x}^2 + \bar{y}^2) \end{aligned}$$

Hence, in the presence of forces independent of the velocities,

$$\bar{p}_x = \frac{\partial L}{\partial \dot{\bar{x}}} = \frac{\partial T}{\partial \dot{\bar{x}}} = m(\dot{\bar{x}} - \omega \bar{y})$$

$$\bar{p}_y = \frac{\partial L}{\partial \dot{\bar{y}}} = \frac{\partial T}{\partial \dot{\bar{y}}} = m(\dot{\bar{y}} + \omega \bar{x}) \tag{54.6}$$

$$\bar{p}_z = m\dot{\bar{z}}$$

In terms of these momenta, the Hamiltonian is simply

$$H = \frac{1}{2m}(\bar{p}_x{}^2 + \bar{p}_y{}^2 + \bar{p}_z{}^2) + \omega(\bar{p}_x\bar{y} - \bar{p}_y\bar{x}) + V$$

and the equations of motion are

$$\dot{\bar{p}}_x = \omega\bar{p}_y - \frac{\partial V}{\partial \bar{x}}, \text{ etc.} \qquad \text{and} \qquad \dot{\bar{x}} = m^{-1}\bar{p}_x + \omega\bar{y}, \text{ etc.}$$

If V is a constant, the equations of motion for the coordinates are

$$\ddot{\bar{x}} - 2\omega\dot{\bar{y}} - \omega^2\bar{x} = 0$$

$$\ddot{\bar{y}} + 2\omega\dot{\bar{x}} - \omega^2\bar{y} = 0$$

giving the expressions for the Coriolis and centripetal accelerations derived in Sec. 49.

55. CHARGED PARTICLE IN AN ELECTROMAGNETIC FIELD

We saw in Sec. 35 that the motion of a charged particle in an electromagnetic field is described by the Lagrangian

$$L = \frac{1}{2}mv^2 + \frac{e}{c}\mathbf{v}\cdot\mathbf{A} - e\phi \tag{55.1}$$

If cartesian coordinates are used to describe the vectors $\mathbf{v}$ and $\mathbf{A}$, the momenta can be written as

$$p_i = \frac{\partial L}{\partial v_i} = mv_i + \frac{e}{c}A_i \tag{55.2}$$

The velocities are thus easily expressed in terms of coordinates and momenta

$$\mathbf{v} = \frac{1}{m}\left(\mathbf{p} - \frac{e}{c}\mathbf{A}\right) \tag{55.3}$$

which yields the Hamiltonian

$$H(\mathbf{r}, \mathbf{p}, t) = \frac{1}{2m}\left[\mathbf{p} - \frac{e}{c}\mathbf{A}(\mathbf{r}, t)\right]^2 + e\phi(\mathbf{r}, t) \tag{55.4}$$

The canonical equations are then (55.3) together with

$$\dot{\mathbf{p}} = -\nabla H = -e\,\nabla\phi + \frac{e}{c}\nabla(\mathbf{v}\cdot\mathbf{A}) \tag{55.5}$$

Equation (55.5) is identical with (35.6) and thus leads to the Lorentz force equation (35.3).

If ϕ and $\mathbf{A}$ do not involve t explicitly the Jacobian energy integral exists. We see that (55.4) may then be written in the form

$$E = \tfrac{1}{2}mv^2 + e\phi \tag{55.6}$$

so that E is easily interpreted as the sum of the kinetic and electrostatic potential energies of the particle. The effect of the vector potential on the equations of motion appears in the definition of the momentum $\mathbf{p}$, which is not, in general, a vector in the direction of the velocity. It is not surprising that the vector potential $\mathbf{A}$ does not appear in the Jacobian integral (55.6), for the magnetic field, derived from $\mathbf{A}$, does not contribute to the energy of the particle.

The above formulation of the problem is not by any means the only one, although for most purposes it is the most convenient. In Sec. 58, for instance, we give a formulation in which the momenta p_j turn out to be $mv_j - (e/c)x_i\,\partial_j A_i$. The fact that this differs from (55.2) is of no physical significance, any more than in the case of rotating coordinates (54.6) where the momenta $\bar{p}_j$ were not just $m\dot{x}_j$. The momenta are defined as $\partial L/\partial\dot{q}_j$ and therefore depend on the choice of L. As we shall see in Sec. 58, the Lagrangian describing a given dynamic system is not unique, hence the momenta p_j are not unique either.

If we write the kinetic energy as

$$T = \tfrac{1}{2}mv^2 = E - e\phi$$

from (55.5), i.e., as the difference between the total energy and the electrostatic potential energy, we can write similarly

$$P_j = m\dot{x}_j = p_j - \frac{e}{c}A_j$$

for the "kinetic momentum" as the difference between the total momentum and what we might call the electromagnetic momentum. The kinetic momentum is, of course, related to the kinetic energy in the usual way

$$T = \frac{1}{2m} \sum_j P_j{}^2 = \frac{P^2}{2m}$$

but it is not canonically conjugate to the position coordinate; i.e.,

$$\dot{P}_j \neq -\frac{\partial H}{\partial x_j}, \qquad \dot{x}_j \neq \frac{\partial H}{\partial P_j}$$

As in the case of the equivalent Lagrangian treatment (Sec. 35), the Hamiltonian (55.4) contains terms which refer either to the particle or to the coupling between the particle and the electromagnetic field. There are no terms which refer to the field alone. Such terms could be incorporated into (55.4) to yield not only the equation of motion (55.5) but also Maxwell's equations. We do not attempt to treat such terms here; cf., however, Chapter 15.

56. THE VIRIAL THEOREM

We define the quantity $\lambda = p_j q_j$, which for any system for which the coordinates and momenta stay finite must also remain bounded. From Hamilton's equations we have

$$\dot{\lambda} = p_j \frac{\partial H}{\partial p_j} - q_j \frac{\partial H}{\partial q_j}$$

so that

$$\frac{\lambda(t) - \lambda(0)}{T} = \frac{1}{T} \int_0^T \dot{\lambda}\, dt = \frac{1}{T} \int_0^T dt \left(p_j \frac{\partial H}{\partial p_j} - q_j \frac{\partial H}{\partial q_j} \right) \tag{56.1}$$

For a periodic system the left-hand side vanishes if T is chosen to equal the period. For a nonperiodic system for which λ is bounded, the left-hand side tends to zero as $T \to \infty$. In either case, we have

$$\left\langle p_j \frac{\partial H}{\partial p_j} \right\rangle = \left\langle q_j \frac{\partial H}{\partial q_j} \right\rangle \tag{56.2}$$

where the brackets denote the time average values of the included quantities, taken from a certain instant until (if ever) the representative point in phase space returns to its initial position.

The quantity

$$\mathcal{V} = -q_j \frac{\partial H}{\partial q_j} \tag{56.3}$$

is a generalization of the *virial function* introduced into the kinetic theory of gases by Clausius. Thus

$$\langle \mathcal{V} \rangle = -\left\langle p_j \frac{\partial H}{\partial p_j} \right\rangle \tag{56.4}$$

For a system of N particles, the right-hand side of (56.4) becomes

$$\begin{aligned} -\left\langle \sum_{\rho=1}^{N} p_\rho \cdot \dot{x}_\rho \right\rangle &= -2\left\langle \sum_{\rho=1}^{N} T_\rho \right\rangle \\ &= -2N\bar{T} \end{aligned} \tag{56.5}$$

where the bar denotes the process of averaging not only over the time but also over the N particles of the system. On the other hand, for such a system (56.3) becomes

$$\langle \mathcal{V} \rangle = \langle \Sigma \dot{p}_\rho \cdot x_\rho \rangle = I + \oint P\mathbf{r} \cdot \mathbf{n}\, dS$$

where I arises from the contribution to $\dot{p}_\rho$ due to the interparticle forces, and the last term, integrated over the surface of the containing vessel, represents the contribution from the pressure P. Since $\mathbf{n}$ is the unit normal to the surface, directed inward, it follows that

$$\langle \mathcal{V} \rangle = I - VP \operatorname{div} \mathbf{r} = I - 3VP \tag{56.6}$$

where V is the total volume. Comparison of (56.5) and (56.6) yields

$$\begin{aligned} PV &= \frac{2}{3} N\bar{T} + \frac{I}{3} \\ &= Nk\theta + \frac{I}{3} \end{aligned}$$

where θ is the absolute temperature ($\bar{T} = \frac{3}{2}k\theta$).

The departure of the equation of state from that of the perfect gas law is therefore seen to be associated with the term I which comes from the interparticle forces.

For the special case of a single particle moving in a potential $\phi(r)$, (56.4) becomes

$$\langle \mathcal{V} \rangle = -\langle \mathbf{p} \cdot \dot{\mathbf{r}} \rangle = \langle \dot{\mathbf{p}} \cdot \mathbf{r} \rangle$$

or

$$2\langle T \rangle = \langle \mathbf{r} \cdot \nabla\phi \rangle$$

For $\phi = Kr^{-n}$, this becomes

$$\langle T \rangle = -\tfrac{1}{2}n\langle \phi \rangle \tag{56.7}$$

so that for a planet moving around the sun, a satellite moving around the earth, or an electron in a classical orbit in a hydrogen atom ($n = 1$) the average value of the kinetic energy is $-\frac{1}{2}$ of the average value of the potential energy. In particular, for a circular orbit, the time averaging process is no longer necessary and the total energy $E = T + \phi$ is minus the kinetic energy. In general, from (56.7), we have

$$E = \langle T \rangle + \langle \phi \rangle = \frac{n-2}{n} \langle T \rangle$$

so that for $n > 2$ the total energy is positive.

57. VARIATIONAL PRINCIPLES

The equations of motion of a system in Lagrangian or Hamiltonian form are differential equations. As such, they describe the temporal development of the system from instant to instant. In this section we show that the motion of the system can be described as that motion which makes a certain integral have a stationary value.

It was noted in Sec. 53 that the path of the system in configuration space is completely specified by giving the values of the coordinates at two times. We now take these two times to be t_1 and t_2, with a finite interval between them. Thus if we are given the values $q_i(t_1)$ and $q_i(t_2)$ ($i = 1, 2, \cdots, f$), the Lagrange equations of motion determine the functions $q_i(t)$ for all t. [It can happen that no path of the system going through the point $q_i(t_1)$ in configuration space goes through the point $q_i(t_2)$. For example, this is the situation for a harmonic oscillation if $t_2 - t_1$ is an integral number of periods and if $q(t_2) \neq q(t_1)$. We exclude such cases.]

Suppose that we are given a set of functions $q_i(t)$ which take on the prescribed values at $t = t_1$ and at $t = t_2$. This means that we are given a curve A in configuration space connecting the initial and final configurations of the system. With this curve we may associate the quantity

$$\Phi = \int_{t_1}^{t_2} L(q, \dot{q}, t)\, dt \tag{57.1}$$

where L is the Lagrangian of the system, as the integrand is a known function of t. Φ is a *functional* of the q_i, being determined by the functional dependence of the q_i on t, and may be written as $\Phi[q_i \cdots q_f]$. Now consider a curve B defined by the set of functions $q_i(t) + \delta q_i(t)$. At each time t this second curve differs from the first by $\delta q_i(t)$ which we take to be a small quantity (cf. Fig. 57-1). With curve B we associate

$$\Phi_B = \int_{t_1+\Delta t_1}^{t_2+\Delta t_2} L(q + \delta q, \dot{q} + \delta \dot{q}, t)\, dt \tag{57.2}$$

where the times at which the two configurations are specified are also varied. Then the difference $\Delta\Phi = \Phi_B - \Phi$ can be evaluated by

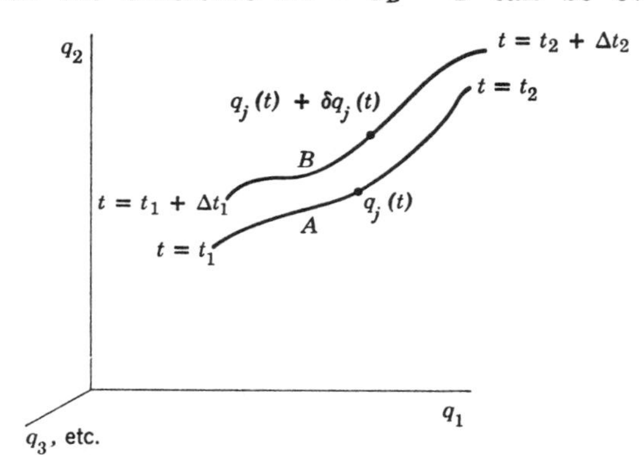

Fig. 57-1. Path A in configuration space and a neighboring curve B.

expanding the integrand of Φ_B in a Taylor series, keeping only linear terms in the δq_i. We obtain

$$\Delta\Phi = \int_{t_1}^{t_2} \left\{ \frac{\partial L}{\partial q_i}\, \delta q_i + \frac{\partial L}{\partial \dot{q}_i}\, \delta \dot{q}_i \right\} dt + [L\, \Delta t]_{t_1}^{t_2}$$

The second term in the integral may be integrated by parts since $\delta \dot{q}_i = d\, \delta q_i/dt$. This yields

$$\Delta\Phi = \int_{t_1}^{t_2} \left\{ \frac{\partial L}{\partial q_i} - \frac{d}{dt}\frac{\partial L}{\partial \dot{q}_i} \right\} \delta q_i\, dt + \left[\frac{\partial L}{\partial \dot{q}_i}\, \delta q_i + L\, \Delta t \right]_{t_1}^{t_2} \tag{57.3}$$

The variation in Φ receives contributions from the entire curve through the integral, and from the variation of the curve at the end times and from variations of the end times themselves. If we do not vary the end times ($\Delta t_1 = \Delta t_2 = 0$) or the curve at the end times [$\delta q_i(t_1) = \delta q_i(t_2) = 0$], then

$$\Delta\Phi = \int_{t_1}^{t_2} \left\{ \frac{\partial L}{\partial q_i} - \frac{d}{dt}\frac{\partial L}{\partial \dot{q}_i} \right\} \delta q_i\, dt. \tag{57.4}$$

If now $\Delta\Phi = 0$ for arbitrary $\delta q_i(t)$, then the quantity in braces vanishes, so that Lagrange's equations are satisfied. Conversely, if Lagrange's equations are satisfied $\Delta\Phi$ vanishes for arbitrary $\delta q_i(t)$. Thus we may say that *the actual path in configuration space between two configurations* $q_i(t_1)$ *and* $q_i(t_2)$ *at times* t_1 *and* t_2 *respectively is that which makes the time integral of the Lagrangian stationary with respect to variations of the path which vanish at the end points.* This statement is known as *Hamilton's principle.*

We return now to the general expression (57.3). The integral vanishes for the actual path and only the end point variations remain. It is convenient to introduce the total variation Δq_i at the end points:

$$\Delta q_i = \delta q_i + \dot{q}_i \, \Delta t \tag{57.5}$$

Here δq_i and $\dot{q}_i$ are evaluated at t_1 or at t_2. Then (57.3) becomes

$$\Delta\Phi = \left[\left(-\frac{\partial L}{\partial \dot{q}_i}\dot{q}_i + L\right)\Delta t + \frac{\partial L}{\partial \dot{q}_i}\Delta q_i\right]_{t_1}^{t_2} \tag{57.6}$$

or

$$\Delta\Phi = [-H\,\Delta t + p_i\,\Delta q_i]_{t_1}^{t_2} \tag{57.7}$$

Equation (57.7) contains Hamilton's principle, but it is useful also because it gives an explicit expression for $\Delta\Phi$ even when Δq, Δt do not vanish. We shall use this expression extensively in the following discussion of the principle of stationary action.

If the path between times t_1 and t_2 satisfies Lagrange's equations, the functional Φ, obtained by integrating the Lagrangian along this path, becomes just a function of the end points:

$$\Phi = \phi[q_j(t_1), t_1, q_j(t_2), t_2]$$

This function ϕ is called *Hamilton's principal function* for the path. It is the value of the functional Φ obtained by integrating along the curve that makes Φ an extreme, i.e., along the actual path of the system.

For particular cases, Hamilton's principle remains valid when a wider class of curves is considered. Thus, if a particular coordinate q_i is cyclic, the corresponding momentum p_i is a constant of the motion, and the contribution to expression (57.7) from variations in this coordinate is

$$[p_i\,\Delta q_i]_{t_1}^{t_2} = p_i[\Delta q_i]_{t_1}^{t_2} \qquad \text{(not summed)} \tag{57.8}$$

This term vanishes even for curves which commence from the configuration for which the cyclic coordinate has been changed by Δq_i, provided that the curves end on the configuration in which this same coordinate has been changed by the same amount, i.e., for all curves for which the total change in q_i is the same. Thus, if there exist a number of cyclic coordinates, it is not necessary to specify that the initial and final values of these coordinates should be the same for each curve considered in order that Hamilton's principle should apply; it is sufficient to state that the total change in each coordinate is the same for each curve.

Similarly, if H does not involve t explicitly we have $H = E$, and (57.7) becomes, for vanishing Δq_i,

$$\Delta\Phi = -E[\Delta t]_{t_1}^{t_2} \tag{57.9}$$

If the total time required to get from one configuration to the other is specified, the right-hand side of this expression vanishes. This may be stated: *Given an initial and final configuration of a system described by a Hamiltonian which is* independent of the time, *and given the* time *required for the system to change from one configuration to the other, the path of the system in configuration space is that along which* Φ *is stationary.*

If H does not depend on t, i.e., for conservative systems, it is convenient to introduce another integral which is characteristic of the path of the dynamic system. We define Σ by

$$\Sigma = \int_{t_1}^{t_2} \frac{\partial L}{\partial \dot{q}_j} \dot{q}_j \, dt = \int_{t_1}^{t_2} p_j \dot{q}_j \, dt \tag{57.10}$$

Like Φ, Σ is a functional of the q_j, and in evaluating the integral the values of q_j, $\dot{q}_j$ as functions of t must be known; i.e., the integral depends on the equation to the curve, expressed parametrically in terms of t. Equation (57.10) may be written

$$\Sigma = \int_{t_1}^{t_2} (H + L) \, dt = \int_{t_1}^{t_2} H \, dt + \Phi$$

If we restrict the variation in Σ to curves along each of which H is constant, it follows that

$$\begin{aligned} \Delta\Sigma &= \Delta\Phi + \Delta[Ht]_{t_1}^{t_2} \\ &= [t \, \Delta H + p_j \, \Delta q_j]_{t_1}^{t_2} \end{aligned} \tag{57.11}$$

by (57.7). If now we consider only those curves for which the initial and final configurations are the same ($\Delta q_j = 0$) and only those curves which are characterized by the same value of the energy

$E = H(\Delta H = 0)$, it follows that

$$\Delta \Sigma = \Delta \int p_j \dot{q}_j \, dt = 0 \tag{57.12}$$

This result is known as the *principle of stationary action.*

If, as in (32.1), $L = T - V$, where T is quadratic in the velocities and V is independent of the velocities, then

$$\Sigma = 2 \int_{t_1}^{t_2} T \, dt \tag{57.13}$$

where T is the kinetic energy of the system. The principle of stationary action may then be stated: *Given the initial and final configuration of a system which is described by a Hamiltonian which is* independent of the time, *and given the* energy *of the system, the path of the system in configuration space is that along which the integral of the kinetic energy is stationary, when compared with that along neighboring curves which satisfy the given conditions.*

In the application of Hamilton's principle to conservative systems, the class of curves considered is such that the time to move along each is the same, although the energies associated with them can differ. In the principle of stationary action, which is applicable only to conservative systems, the times required to move along the curves considered may be different, but the energy, defined by $E = p_j \dot{q}_j - L$, is the same for each curve. As given above, this principle merely states that Σ is stationary for the actual motion of the system. For conditions under which the principle may be refined to a principle of least action—a statement that Σ is a minimum for the actual motion—the reader is referred elsewhere.* Sometimes, however, the statement in italics following (57.13) is referred to loosely as a least action principle.

If the path of the system between times t_1 and t_2 is given, the functional Σ, like Φ, becomes an ordinary function of the end points and times. As indicated above, the introduction of Σ is useful only for systems for which H, hence L, does not contain the time explicitly. For such systems the momenta p_j, derivable from L, may be expressed as functions of the coordinates and their time derivatives, without involving t explicitly, and the integral

$$\Sigma = \int p_j \, dq_j$$

becomes a function depending only on the *coordinates* of the end points,

$$\Sigma = S[q_j(t_1), q_j(t_2)]$$

* E. T. Whittaker, *Analytical Dynamics*, Dover, New York (1944), p. 250.

when the integration in Σ is taken along the actual path of the system. The function S is called the *action* for the path. Thus, for $H = E = p_j\dot{q}_j - L = \text{const}$,

$$S = \int_{t_1}^{t_2} (E + L)\,dt = \phi + E(t_2 - t_1)$$

or

$$\phi[q_j(t_1), t_1, q_j(t_2), t_2] = -E(t_2 - t_1) + S[q_j(t_1), q_j(t_2)]$$

For given initial conditions, we have then, writing $t_2 = t$,

$$\phi(q_j, t) = -Et + S(q_j) + \text{const} \tag{57.14}$$

When the system is conservative, Hamilton's principal function may therefore be separated into a function of the coordinates and a linear function of the time.

The configuration space of the system is described by generalized coordinates q_i. It is sometimes convenient to consider the arc length along a curve in configuration space. For a system consisting of more than one particle this arc length has no immediate interpretation in physical space. A natural definition is given by the kinetic energy. We define the element of arc ds by the equation

$$T = \frac{1}{2}\left(\frac{ds}{dt}\right)^2 \tag{57.15}$$

as though our system consisted of a single particle of unit mass. Then*

$$ds = (2T)^{1/2}\,dt \tag{57.16}$$

and

$$S = \int (2T)^{1/2}\,ds$$

If the system is conservative, Hamilton's principle may be written

$$\Delta S = \Delta \int [2(E - V)]^{1/2}\,ds = 0 \tag{57.17}$$

As stated here, Hamilton's principle and the principle of stationary action have the advantage of being independent of the coordinate system used to describe the system, but have the disadvantage of referring to the configuration at two distinct times rather than to the state of the system at a single time. It is possible to avoid this by asserting that the motion of a dynamic system is describable by the development of a contact transformation in time, i.e., by taking the subject of the next section as fundamental and deriving from it the form of the

* See (10.8) and Appendix I.

equations of motion. This order of procedure is advantageous in a quantum theory but in classical particle mechanics Hamilton's principle seems preferable.

58. CONTACT TRANSFORMATIONS

According to Hamilton's principle, the path of a system in configuration space between two fixed points and two fixed times is that which makes the time integral of the Lagrangian stationary:

$$\Delta \int_{t_1}^{t_2} L(q, \dot{q}, t)\, dt = 0 \tag{58.1}$$

The content of (58.1) is not changed if the coordinates q are replaced by other coordinates $\bar{q}_i = \bar{q}_i(q, t)$. This leads to the known invariance of Lagrange's equations under coordinate transformation. There is a much larger class of transformations which also do not affect the content of (58.1) but which change the *value* as well as the *form* of the Lagrangian. These more general transformations are called *contact transformations.*

Let $L(q, \dot{q}, t)$ and $\bar{L}(\bar{q}, \dot{\bar{q}}, t)$ be two Lagrangians involving the same number f of degrees of freedom. If the q and $\bar{q}$ are related so that a path in q space giving a stationary value to the time integral of L corresponds to a path in $\bar{q}$ space giving the time integral of $\bar{L}$ a stationary value, then these two Lagrangians can be regarded as giving two different descriptions of the same system. This correspondence will hold provided that

$$\bar{L}(\bar{q}, \dot{\bar{q}}, t) = L(q, \dot{q}, t) - \frac{d}{dt}\phi(q, \bar{q}, t) \tag{58.2}$$

because under the conditions of Hamilton's principle the total time derivative cannot contribute to the variation of the time integral.

In (58.2) $\bar{L}$ depends only on barred coordinates and L only on unbarred coordinates, in addition to the time. Differentiating (58.2) yields

$$\begin{aligned}\frac{\partial \bar{L}}{\partial \bar{q}_i}\,\delta\bar{q}_i + \bar{p}_i\,\delta\dot{\bar{q}}_i + \frac{\partial \bar{L}}{\partial t}\,\delta t = {} & \frac{\partial L}{\partial q_i}\,\delta q_i + p_i\,\delta\dot{q}_i + \frac{\partial L}{\partial t}\,\delta t \\ & - \frac{d}{dt}\left(\frac{\partial \phi}{\partial q_i}\,\delta q_i + \frac{\partial \phi}{\partial \bar{q}_i}\,\delta\bar{q}_i + \frac{\partial \phi}{\partial t}\,\delta t\right)\end{aligned} \tag{58.3}$$

where the momenta have their usual definitions. Equating coefficients of $\delta\dot{\bar{q}}_i$ and $\delta\dot{q}_i$ shows that

$$p_i = \frac{\partial\phi}{\partial q_i}, \qquad \bar{p}_i = -\frac{\partial\phi}{\partial\bar{q}_i} \tag{58.4}$$

These equations provide the necessary and sufficient conditions that L and $\bar{L}$ describe the same system. The reader may verify that equating coefficients of δq_i, $\delta\bar{q}_i$, and δt leads to nothing new.

If H and $\bar{H}$ are constructed from L and $\bar{L}$ according to the usual prescription (53.5), we see that

$$\begin{aligned} \bar{H}(\bar{q}, \bar{p}, t) &= \bar{p}_i\dot{\bar{q}}_i - \bar{L} \\ &= p_i\dot{q}_i - \phi + \frac{\partial\phi}{\partial t} - L + \phi \\ &= H(q, p, t) + \frac{\partial\phi(q, \bar{q}, t)}{\partial t} \end{aligned} \tag{58.5}$$

The value of the barred Hamiltonian differs from the unbarred only if ϕ depends explicitly on the time.

Equations (58.4) and (58.5) are conveniently combined into

$$p_i\,dq_i - \bar{p}_i\,d\bar{q}_i - (H - \bar{H})\,dt = d\phi \tag{58.6}$$

It can happen that $\phi(q, \bar{q}, t)$ does not depend explicitly on all of its arguments so that certain differentials are missing from the right side of (58.6). At first this seems to lead to the impossible conclusion that certain momenta vanish. However, we should interpret this as implying the existence of relations between the barred and unbarred coordinates independent of any momenta. If there are $n \leq f$ such independent relations

$$g_l(q_1 \cdots q_f, \bar{q}_1 \cdots \bar{q}_f) = 0 \qquad (l = 1, 2, \cdots, n) \tag{58.7}$$

then the equations between the differentials dq and $d\bar{q}$

$$\frac{\partial g_l}{\partial q_i}\,dq_i + \frac{\partial g_l}{\partial\bar{q}_i}\,d\bar{q}_i = 0$$

can be solved for n of the differentials, leaving $2f - n$ independent differentials on the left side of (58.6). In this way the number of independent differentials on the left can be reduced to the number which appear on the right or to f, whichever is larger. In either case, then, the equating of coefficients can be carried out. We give an example.

Example 1. Take $f = 2$ and take $\phi = \frac{1}{2}q_1^2$. Then

$$d\phi = q_1\,dq_1$$

and three differentials are missing on the right side of (58.6). Assume therefore that $n = 2$ and write

$$\bar{q}_1 = \bar{q}_1(q_1, q_2), \qquad \bar{q}_2 = \bar{q}_2(q_1, q_2)$$

so that (58.6) becomes

$$p_1\,dq_1 + p_2\,dq_2 - \bar{p}_1\left(\frac{\partial \bar{q}_1}{\partial q_1}\,dq_1 + \frac{\partial \bar{q}_1}{\partial q_2}\,dq_2\right) - \bar{p}_2\left(\frac{\partial \bar{q}_2}{\partial q_1}\,dq_1 + \frac{\partial \bar{q}_2}{\partial q_2}\,dq_2\right) = q_1\,dq_1$$

Equating coefficients of differentials yields

$$p_1 - \bar{p}_1\frac{\partial \bar{q}_1}{\partial q_1} - \bar{p}_2\frac{\partial \bar{q}_2}{\partial q_1} = q_1$$

$$p_2 - \bar{p}_1\frac{\partial \bar{q}_1}{\partial q_2} - \bar{p}_2\frac{\partial \bar{q}_2}{\partial q_2} = 0$$

which can be solved for $\bar{p}_1$ and $\bar{p}_2$ in terms of q_1, q_2, p_1, p_2. The solution is especially simple if $\bar{q}_1 = q_1$, $\bar{q}_2 = q_2$. Then

$$\bar{p}_1 = p_1 - q_1, \qquad \bar{p}_2 = p_2$$

The function ϕ does not determine the functions g_l, but only their number, and does not therefore provide a convenient description of a point transformation. This can be done easily, as is shown in the next section. We see, however, that all transformations derived in this way, i.e., all contact transformations, leave both the Lagrangian and Hamiltonian forms of the equations of motion unchanged, as both follow from Hamilton's principle which is unchanged. ϕ is called a *generator* of the transformation.

We give two more examples as illustrations of contact transformations.

Example 2. Let the Lagrangian be that of a charged particle in an electromagnetic field (55.1)

$$L = \frac{1}{2}m\mathbf{v}^2 + \frac{e}{c}\mathbf{v}\cdot\mathbf{A} - e\phi_M$$

where ϕ_M is the scalar potential of the field. Taking

$$\phi = \frac{e}{c}\mathbf{r}\cdot\mathbf{A} \tag{58.8}$$

so that

$$\bar{L} = \frac{1}{2} m\mathbf{v}^2 - \frac{e}{c}\mathbf{r} \cdot \dot{\mathbf{A}} - e\phi_M$$

we see that only the three unbarred coordinates are involved and we may choose $f = 3$ relations (58.7) which we take to be $\bar{\mathbf{r}} = \mathbf{r}$. Then we obtain

$$p_i = m\dot{x}_i - \frac{e}{c} x_j \frac{\partial A_j}{\partial x_i}$$

The barred Hamiltonian is

$$\bar{H} = \frac{1}{2m} \sum_{i=1}^{3} \left(\bar{p}_i + \frac{e}{c} x_j \frac{\partial A_j}{\partial x_i}\right)^2 + e\phi_M + \frac{e}{c}\mathbf{r} \cdot \frac{\partial \mathbf{A}}{\partial t} \tag{58.9}$$

which differs from H in value by the last term, which is $\partial\phi/\partial t$. Lagrange's equations are unchanged as no coordinate change has been made.

Example 3. The harmonic oscillator is described by the Hamiltonian

$$H = \tfrac{1}{2}(p^2 + \omega^2 q^2) \tag{58.10}$$

We make a contact transformation generated by

$$\phi = \tfrac{1}{2}\omega q^2 \cot 2\pi\bar{q} \tag{58.11}$$

which does depend explicitly on all the barred and unbarred coordinates and which therefore specifies the transformation uniquely. Then, from (58.4),

$$p = \frac{\partial\phi}{\partial q} = \omega q \cot 2\pi\bar{q}$$

$$\bar{p} = -\frac{\partial\phi}{\partial\bar{q}} = \pi\omega q^2 \operatorname{cosec}^2 2\pi\bar{q}$$

Thus

$$q = \left(\frac{\bar{p}}{\pi\omega}\right)^{1/2} \sin 2\pi\bar{q}$$

$$p = \left(\frac{\bar{p}\omega}{\pi}\right)^{1/2} \cos 2\pi\bar{q}$$

$$H = \nu\bar{p}$$

where $\nu = \omega/2\pi$ is the frequency of the oscillation. The transformation therefore throws the Hamiltonian into an exceptionally simple form, when expressed in terms of $\bar{q}$, $\bar{p}$, actually making $\bar{q}$ a cyclic variable and the Hamiltonian a constant times $\bar{p}$. The equations of motion are then easily integrated.

$$\bar{p} = \frac{E}{\nu}, \qquad \bar{q} = \nu(t - t_0)$$

The new position coordinate $\bar{q}$, conjugate to $\bar{p}$, is ν times the time. Thus the expressions for q, p become

$$q = a \sin \omega(t - t_0)$$

$$p = a\omega \cos \omega(t - t_0)$$

where $a = (2E)^{1/2}/\omega$. In Sec. 62 we shall show how the generating function (58.11) is obtained.

Contact transformations possess the properties which characterize a group.

(i) The result of two successive contact transformations is a contact transformation. Let

$$p_i\, dq_i - \bar{p}_i\, d\bar{q}_i + (\bar{H} - H)\, dt = d\phi_1(q, \bar{q}, t)$$

and

$$\bar{p}_i\, d\bar{q}_i - \bar{\bar{p}}_i\, d\bar{\bar{q}}_i + (\bar{\bar{H}} - \bar{H})\, dt = d\phi_2(\bar{q}, \bar{\bar{q}}, t)$$

Adding, we obtain

$$p_i\, dq_i - \bar{\bar{p}}_i\, d\bar{\bar{q}}_i + (\bar{\bar{H}} - H)\, dt = d\phi(q_i, \bar{\bar{q}}_i, t)$$

Since, on adding, the terms in $d\bar{q}_i$ cancel, the right-hand side does not depend on the $\bar{q}_i$. The generator of the resultant transformation is the sum of the generators of the separate transformations expressed as a function of the initial and final coordinates.

(ii) The associative law holds. The proof is left to the reader.

(iii) The identity transformation is a contact transformation and its generator is zero.

(iv) There exists an inverse to every contact transformation. Exchanging the roles of the barred and unbarred variables merely changes the sign of the generator.

Any differentiable function $\phi(q_i, q_j, t)$ may be regarded as the generator of a contact transformation. Our result may then be stated thus: *The form of the canonical equations is invariant with respect to the group of contact transformations.* For this reason contact transformations are sometimes referred to as *canonical* transformations. Just as Hamilton's principle was seen to lay no stress on the particular coordinate system adopted, so here Hamilton's equations are seen to lay no stress on the particular coordinates and their conjugate momenta employed. We are attempting to formulate the laws of motion of dynamic systems relative to some observer, who is characterized by means of a coordinate system. It is satisfactory to discover that the form of the equations which describe this motion does not single out one particular observer but is the same over a wide set (actually a nonenumerable infinity) of such observers, and over a wide choice of coordinates and momenta for each observer.

59. ALTERNATIVE FORMS OF CONTACT TRANSFORMATIONS

The defining equation (58.2) for a contact transformation is not always the most convenient to use. It becomes awkward when some of the new coordinates are functions of the old coordinates but not of the old momenta. In this case the $d\bar{q}_i$ are not independent of the dq_i and (58.6) becomes difficult to interpret. This difficulty can be eliminated by using one of three alternative forms of generator for contact transformations.

Let

$$\phi(q, \bar{q}, t) = \psi + p_j q_j \tag{59.1}$$

Differentiating (59.1), we obtain with the use of (58.6)

$$p_i\, dq_i - \bar{p}_i\, d\bar{q}_i + (\bar{H} - H)\, dt = d\psi + p_i\, dq_i + q_i\, dp_i \tag{59.2}$$

which shows that ψ is a function of $\bar{q}_j$, p_j, t. Rewriting this, we get

$$-q_j\, dp_j - \bar{p}_j\, d\bar{q}_j + (\bar{H} - H)\, dt = d\psi(p, \bar{q}, t) \tag{59.3}$$

$\psi(p, \bar{q}, t)$ thus generates a contact transformation according to

$$q_i = -\frac{\partial \psi(p, \bar{q}, t)}{\partial p_i}, \qquad \bar{p}_i = -\frac{\partial \psi(p, \bar{q}, t)}{\partial \bar{q}_i} \tag{59.4}$$

$$\bar{H}(\bar{q}, \bar{p}, t) = H(q, p, t) + \frac{\partial \psi(p, \bar{q}, t)}{\partial t} \tag{59.5}$$

Similarly, we may write

$$\phi(q, \bar{q}, t) = \phi'(p, \bar{p}, t) + p_j q_j - \bar{p}_j \bar{q}_j \tag{59.6}$$

where ϕ' generates a contact transformation according to

$$q_i = -\frac{\partial \phi'(p, \bar{p}, t)}{\partial p_i}, \qquad \bar{q}_i = \frac{\partial \phi'(p, \bar{p}, t)}{\partial \bar{p}_i} \tag{59.7}$$

$$\bar{H}(\bar{q}, \bar{p}, t) = H(q, p, t) + \frac{\partial \phi'(p, \bar{p}, t)}{\partial t} \tag{59.8}$$

and lastly

$$\phi(q, \bar{q}, t) = \psi'(q, \bar{p}, t) - \bar{p}_j\bar{q}_j \tag{59.9}$$

where ψ' generates a contact transformation according to

$$p_i = \frac{\partial \psi'(q, \bar{p}, t)}{\partial q_i}, \qquad \bar{q}_i = \frac{\partial \psi'(q, \bar{p}, t)}{\partial \bar{p}_i} \tag{59.10}$$

$$H(\bar{q}, \bar{p}, t) = H(q, p, t) + \frac{\partial \psi'(q, \bar{p}, t)}{\partial t} \tag{59.11}$$

The last form is particularly convenient for writing a coordinate transformation as a contact transformation. For let

$$\psi'(q, \bar{p}, t) = f_i(q)\bar{p}_i \tag{59.12}$$

Then according to (59.10),

$$p_i = \frac{\partial f_j(q)}{\partial q_i}\bar{p}_j, \qquad \bar{q}_i = f_i(q) \tag{59.13}$$

The new coordinates are functions of the old coordinates only. The new momenta are linear combinations of the old momenta. The identity transformation is obviously generated by

$$\psi'(q, \bar{p}, t) = q_i\bar{p}_i$$

which gives

$$\phi(q, \bar{q}, t) = (q_i - \bar{q}_i)\bar{p}_i$$
$$= 0$$

since $q_i = \bar{q}_i$. However, $\phi(q, \bar{q}, t) = 0$ does not lead only to the identity transformation. The ψ' of (59.12) also yields $\phi = 0$. $\phi = 0$ (or $=$ const) merely states that the dq's and the $d\bar{q}$'s are not independent, hence functional relationships between the q's and $\bar{q}$'s must exist; indeed, enough must exist to define a coordinate transformation.

As an example of the coordinate transformation (59.12) we may take the familiar case of spherical polar coordinates. Let

$$\bar{p}_j = p_x, p_y, p_z \qquad (j = 1, 2, 3)$$
$$\bar{q}_j = x, y, z$$
$$x = r \sin \theta \cos \phi$$
$$y = r \sin \theta \sin \phi$$
$$z = r \cos \theta$$

Then

$$\psi' = p_x r \sin\theta \cos\phi + p_y r \sin\theta \sin\phi + p_z r \cos\theta$$

and

$$p_1 = p_r = \bar{p}_k \frac{\partial \bar{q}_k}{\partial r} = p_x \sin\theta \cos\phi + p_y \sin\theta \sin\phi + p_z \cos\theta$$

$$p_2 = p_\theta = \bar{p}_k \frac{\partial \bar{q}_k}{\partial \theta} = p_x r \cos\theta \cos\phi + p_y r \cos\theta \sin\phi - p_z r \sin\theta$$

$$p_3 = p_\phi = \bar{p}_k \frac{\partial \bar{q}_k}{\partial \phi} = -p_x r \sin\theta \sin\phi + p_y r \sin\theta \cos\phi$$

Thus

$$p_x^2 + p_y^2 + p_z^2 = p^2 = p_r^2 + r^{-2} p_\theta^2 + r^{-2} \operatorname{cosec}^2 \theta p_\phi^2$$

Cf. (54.4).

In the special case in which the functions f_j are linear inhomogeneous functions of the q_k,

$$\bar{q}_j = f_j = S_{jk} q_k + b_j \tag{59.14}$$

where the S_{jk} and b_j are constants, the transformation is a *linear coordinate transformation*. For the momenta this yields

$$p_j = \bar{p}_k S_{kj} \tag{59.15}$$

The momenta then also transform linearly. The generator of this transformation is

$$\psi' = \bar{p}_j S_{jk} q_k + b_j \bar{p}_j \tag{59.16}$$

If $b_j = 0$, the q_j, p_j are transformed linearly and homogeneously into the $\bar{q}_j$, $\bar{p}_j$.

60. ALTERNATIVE FORMS OF THE EQUATIONS OF MOTION

Starting from a Lagrangian $L(q, \dot{q}, t)$, we may write down Lagrange's equations of motion in configuration space and Hamilton's equations in phase space. There is a contact transformation which interchanges the roles of coordinates and momenta, namely that generated by

$$\phi(q, \bar{q}) = q_i \bar{q}_i \tag{60.1}$$

from which

$$p_i = \bar{q}_i, \qquad \bar{p}_i = -q_i$$

One would expect, therefore, that it is possible to write equations of Lagrange's form in momentum space.

To accomplish this we proceed as in the previous section, where we went from a generator $\phi(q, \bar{q}, t)$ to a generator $\phi'(p, \bar{p}, t)$ by (59.6). We write

$$L(q, \dot{q}, t) = K(p, \dot{p}, t) + p_i\dot{q}_i + q_i\dot{p}_i \tag{60.2}$$

The arguments of K are necessarily those written, as may be seen by differentiation and use of the Lagrange equations in the q's. Inserting this form of L in Hamilton's principle yields

$$\frac{\partial K(p, \dot{p}, t)}{\partial p_i} - \frac{d}{dt}\frac{\partial K(p, \dot{p}, t)}{\partial \dot{p}_i} = 0 \tag{60.3}$$

The equations of motion in this form are of little practical use in classical problems because the Lagrangian $L(q, \dot{q}, t)$ is generally a simpler function of $\dot{q}$ than of q. Their quantum analog is useful.

Finally there exists a curious form of the equations of motion somewhat like Hamilton's equations. Write

$$L(q, \dot{q}, t) = q_i\dot{p}_i - F(\dot{q}, \dot{p}, t) \tag{60.4}$$

where the arguments of F are necessarily those given. Hamilton's principle then leads to

$$q_i = \frac{\partial F(\dot{q}, \dot{p}, t)}{\partial \dot{p}_i}, \qquad p_i = -\frac{\partial F(\dot{q}, \dot{p}, t)}{\partial \dot{q}_i} \tag{60.5}$$

The nature of the independent variables, velocities and forces, makes these equations appear of little practical value. We note the identity

$$L(q, \dot{q}, t) + H(q, p, t) + K(p, \dot{p}, t) + F(\dot{q}, \dot{p}, t) = 0$$

GENERAL REFERENCES

W. Yourgrau and S. Mandelstam, *Variational Principles in Dynamics and Quantum Theory.* Pitman, London (1955).

P. M. Morse and H. Feshbach, *Methods of Theoretical Physics,* Vol. I. McGraw-Hill, New York (1953).

C. Lanczos, *The Variational Principles of Mechanics.* University of Toronto (1949).

H. Margenau and G. M. Murphy, *The Mathematics of Physics and Chemistry.* Van Nostrand, New York (1943).

EXERCISES

1. A particle is free to slide inside a smooth straight tube which is held at an angle α to the horizontal and constrained to move parallel to itself with an acceleration α in the direction of the projection of the tube onto a horizontal plane. Write down and solve Hamilton's equations for the motion of the particle, and show that they are equivalent to the solution of the problem by elementary methods.

2. Examine the motion of a particle in a potential field

$$V = -\frac{1}{2mr^2} p_\phi{}^2 \operatorname{cosec}^2 \theta$$

where p_ϕ is the angular momentum of the particle around the z axis. Interpret physically the conclusion that $p_\theta = \text{const}$ (equation 54.5).

3. Similarly, examine the motion of a particle in a potential field

$$V = -\frac{p_\phi{}^2}{2m\rho^2}$$

(equation 54.2).

4. Show by comparison of (54.6) and (55.2) the result of Sec. 40 that a small constant magnetic field B may be eliminated from the problem of the motion of a charge e by viewing the particle from a coordinate system which is rotating with the Larmor frequency $\omega = eB/2mc$. What is the physical significance of the terms neglected by the provision that the field be small?

5. Show that if the variables q_r are cyclic ($r = 1 \cdots l < f$) and if

$$R = L - \sum_{r=1}^{l} \beta_r \dot{q}_r \qquad \left(\beta_r = \frac{\partial L}{\partial \dot{q}_r}\right)$$

is expressed thus

$$R = R(q_s, \dot{q}_s, \beta_r, t) \qquad (s = l + 1 \cdots f, r = 1 \cdots l)$$

then

$$\frac{d}{dt}\frac{\partial R}{\partial \dot{q}_s} = \frac{\partial R}{\partial q_s}$$

thereby reducing the problem to one of $f - l$ degrees of freedom. R is known as *Routh's function*.

6. A particle is projected vertically upward in an elevator which is moving upward with acceleration a. Treating the gravitational field of the earth as a known function of height, write down the Hamiltonian for the motion of the particle (*a*) relative to the earth, (*b*) relative to the elevator.

7. The minimum potential energy of a uniform flexible cable of fixed length held between two fixed supports in a uniform gravitational field is achieved in a configuration which satisfies a variational principle. Set up the integral to be minimized with the constraint on the length included by means of a Lagrange multiplier. Find and solve the equation for the curve. Use the horizontal coordinate x as independent variable.

8. A system described by a Lagrangian $L(q_i, \dot{q}_i, t)$ $(i = 1, 2, \cdots, f)$ is subjected to the constraint $g(q_1, \cdots, q_f) = 0$, $\partial g/\partial q_f \neq 0$. Show that the equations obtained from (57.4) (with $\Delta\Phi = 0$) by eliminating δq_f are not identical with those obtained from the Lagrangian $L(q_i, q_f(q_i), \dot{q}_i, \dot{q}_f(q_i), t)$ $(i = 1, 2, \cdots, f - 1)$. Which are the equations of motion and why?

9. A particle moves in one dimension under a potential $V = Kx^n$. Using $\dot{p} = -(\partial V/\partial x)$, eliminate x to find the F function (60.4) and thus show that equations (60.5) reduce to the usual equations of motion.

10. Construct the function (60.4) for the simple harmonic oscillator, and write the corresponding equations of motion.

11. A harmonic oscillator is described by the Lagrangian $L = \frac{1}{2}m(\dot{x}^2 - \omega^2 x^2)$. Construct the momentum space Lagrangian $K(p, \dot{p})$ and write out the equations of motion.

12. A particle moves vertically in a uniform gravitational field g, the Lagrangian being $L = \frac{1}{2}\dot{z}^2 - gz$. Construct the momentum space Lagrangian $K(p, \dot{p})$. (*Hint:* Add a total time derivative such as $\frac{1}{2}d/dt(\lambda z^2) = \lambda z\dot{z}$ to the Lagrangian.)

11 THE HAMILTON-JACOBI METHOD

61. THE HAMILTON-JACOBI EQUATION

A contact transformation derived from a generating function which depends explicitly on the time not only results in a transformation in the phase space of a system and the functional form of the Hamiltonian, but also changes the value of the Hamiltonian. Thus on contact transformation from the variables q, p to $\bar{q}$, $\bar{p}$, we have according to (58.5)

$$\bar{H}(\bar{q}, \bar{p}, t) = H(q, p, t) + \frac{\partial \phi(q, \bar{q}, t)}{\partial t} \tag{61.1}$$

The variables appearing on the right-hand side of this equation must all be transformed into the barred variables according to (58.4), namely

$$\bar{p}_i = -\frac{\partial \phi(q, \bar{q}, t)}{\partial \bar{q}_i}, \qquad p_i = \frac{\partial \phi(q, \bar{q}, t)}{\partial q_i} \tag{61.2}$$

The canonical equations of motion are now

$$\dot{\bar{q}}_i = \frac{\partial \bar{H}(\bar{q}, \bar{p}, t)}{\partial \bar{p}_i}, \qquad \dot{\bar{p}}_i = -\frac{\partial \bar{H}(\bar{q}, \bar{p}, t)}{\partial \bar{q}_i} \tag{61.3}$$

We now consider one way in which to make the canonical equations (61.3) easy to integrate—to choose ϕ so that the *new Hamiltonian $\bar{H}$ is independent of the $\bar{q}_j$, $\bar{p}_j$, and t.* These new coordinates and momenta are then constants of the motion which we write thus:

$$\bar{q}_j = \alpha_j, \qquad \bar{p}_j = -\beta_j \tag{61.4}$$

The α_j, β_j are $2f$ constants, the specification of which is sufficient to describe the $2f$ initial conditions of the motion.

Since $\bar{H}$ is independent of $\bar{q}_j$, $\bar{p}_j$, and t, it must be a constant, and for definiteness we take this constant as zero.* From (61.2) we write

$$\phi = \phi(q_j, \alpha_j, t)$$

with

$$p_j = \frac{\partial \phi}{\partial q_j} \tag{61.5}$$

and

$$\beta_j = \frac{\partial \phi}{\partial \alpha_j} \tag{61.6}$$

The condition (61.1) now becomes

$$\frac{\partial \phi}{\partial t} + H\left(q_j, \frac{\partial \phi}{\partial q_j}, t\right) = 0 \tag{61.7}$$

where in the Hamiltonian the momenta have been expressed in terms of the derivatives of ϕ by means of (61.5).

Equation (61.7) is known as the *Hamilton-Jacobi partial differential equation.* It is an equation of the first order, but, in all cases of interest, of the second degree, in the $f + 1$ independent variables q, t. A complete integral of this equation requires for its specification $f + 1$ constants, but since only derivatives of ϕ appear in (61.7) it follows that one of these constants is additive. The other f constants may be taken as the α_j. Thus, if we can find a complete integral of (61.7) involving f arbitrary constants, none of which is an additive constant, the equations of motion of the dynamic system are simply (61.6); i.e., the partial derivative of ϕ with respect to any one of these constants is itself a constant.

Since $\bar{H} = 0$ and the $\bar{q}_j$ are constants of the motion, it follows that the new Lagrangian

$$\bar{L} = \bar{p}_j\dot{\bar{q}}_j - \bar{H}$$

is also zero. Hence, from (58.2),

$$\frac{d\phi}{dt} = L \qquad \text{or} \qquad \phi = \int L \, dt$$

Thus ϕ is Hamilton's principal function for the motion. It is the value of the function Φ of (57.1) obtained by integrating along the

* If the generating function ϕ_1 yields $\bar{H} = a = \text{const}$, the function $\phi_1 + at$ makes the new Hamiltonian zero.

actual path of the system. ϕ may be regarded as a function of the end coordinates and times of the motion, and the change in it due to infinitesimal changes in these is given by (57.7).

$$d\phi = [-H\,dt + p_j\,dq_j]_{t_1}^{t}$$

Writing H, p_j, q_j for the values of these quantities at time t, and $\bar{H}$, $\bar{p}_j$, $\bar{q}_j$ for their values at $t = t_1$, and restricting the variation in t at the end points so that $[dt]_{t_2} = [dt]_{t_1} = dt$, we have

$$d\phi = p_j\,dq_j - \bar{p}_j\,d\bar{q}_j - (H - \bar{H})\,dt$$

which reproduces (58.6). *The principal function between time t_1 and t is therefore the generating function of the contact transformation from the initial values of the coordinates and momenta to the values of the coordinates and momenta at time t.* The state of the system at any time is thus obtained from the initial state by the contact transformation generated by the principal function up to that time.

In the special case in which the Hamiltonian is independent of the time, we saw in Sec. 57 that it was convenient to introduce the action S, which is related to the principal function ϕ for a path starting on an arbitrary but definite configuration and ending on the configuration at time t. By (57.14)

$$\phi = -Et + S \tag{61.8}$$

apart from an arbitrary additive constant. We should expect that substitution of this expression into the Hamilton-Jacobi equation should lead to a simplification of that equation for a system described by a Hamiltonian that does not involve the time explicitly, and that the constant E in (61.8) should then represent the constant total energy for a natural system (33.6).

$$H\left(q_j, \frac{\partial S}{\partial q_j}\right) = E \tag{61.9}$$

where

$$S = S(q_j, \alpha_j), \qquad p_j = \frac{\partial S}{\partial q_j} \qquad \text{[by (61.5)]} \tag{61.10}$$

and the time variable has been separated out of the $H - J$ equation. The function S is then the generator of a time-independent contact transformation, which, in the many cases in which H does not involve t, is sufficient for the purpose. Since ϕ is a function of the q_j, t, and f arbitrary constants α_j, we see from (61.10) that the extra constant E introduced into (61.8) cannot be independent of the α_j, for otherwise

ϕ would contain $f + 1$ arbitrary constants. There is therefore some relation

$$E = E(\alpha_j) \tag{61.11}$$

the explicit form of which depends on the manner in which the f constants α_j have been chosen.

A symmetrical way of choosing the relation (61.11) is

$$E = \sum_{j=1}^{f} \alpha_j$$

so that equations (61.6) become

$$\frac{\partial S}{\partial \alpha_j} = \beta_j + t$$

These f equations describe the path of the system parametrically as functions of the time. Sometimes it is more convenient to derive the equation of the path itself, without reference to the particular time at which any given point on the path is occupied. To do this we choose as the form of (61.11)

$$E = \alpha_f$$

Then equations (61.6) become

$$\begin{aligned} \frac{\partial S}{\partial \alpha_j} &= \beta_j \qquad [j = 1 \cdots (f-1)] \\ \frac{\partial S}{\partial E} &= \beta + t \qquad (\beta_f = \beta) \end{aligned} \tag{61.12}$$

The first $f - 1$ equations then describe the path, and the last equation tells what point on the path is occupied at any given time. We note that it is not necessary to solve for ϕ or S explicitly in any given problem, since only derivatives of these functions appear in the equations of motion, (61.6) or (61.12). This fact usually simplifies the analysis considerably.

It is easily verified without the use of transformation theory that the equations of motion, (61.5) and (61.6), and the $H - J$ equation (61.7) are consistent with the canonical equations. Differentiating (61.6) with respect to the time,

$$0 = \frac{\partial^2 \phi}{\partial t\, \partial \alpha_j} + \frac{\partial^2 \phi}{\partial \alpha_j\, \partial q_k}\, \dot{q}_k$$

and, differentiating the $H - J$ equation with respect to α_j,

$$\frac{\partial^2 \phi}{\partial t\, \partial \alpha_j} + \frac{\partial H}{\partial \left(\dfrac{\partial \phi}{\partial q_k}\right)} \frac{\partial^2 \phi}{\partial \alpha_j\, \partial q_k} = 0$$

Since $\partial \phi / \partial q_k = p_k$, these equations are consistent with the canonical equation $\dot{q}_k = \partial H / \partial p_k$. We leave to the reader to verify the consistency of the equations with $\dot{p}_k = -\partial H / \partial q_k$ by differentiating (61.5) with respect to the time and (61.7) with respect to q_j.

Reference to Sec. 59 shows that it is possible to formulate similar Hamilton-Jacobi theories in terms of ψ, ψ', ϕ'. Since ψ and ϕ' are expressed in terms of the p_j, the differential equations for these generators to make the new Hamiltonian vanish are

$$\frac{\partial \psi}{\partial t} + H\left(-\frac{\partial \psi}{\partial p_j}, p_j, t\right) = 0$$

or

$$\frac{\partial \phi'}{\partial t} + H\left(-\frac{\partial \phi'}{\partial p_j}, p_j, t\right) = 0$$

Since in general the Hamiltonian involves q_j in a much more complicated manner than it does p_j, these equations are of little interest. The function $\psi'(q_j, \bar{p}_j, t)$, however, will generate a contact transformation so that $\bar{H} = 0$ if

$$\frac{\partial \psi'}{\partial t} + H\left(q_j, \frac{\partial \psi'}{\partial q_j}, t\right) = 0$$

where, in H, p_j is replaced by $\partial \psi' / \partial q_j$ according to (59.10). Then, as before, $\bar{q}_j$ and $\bar{p}_j$ are constants, since $\bar{H} = 0$, and the rest of the above theory follows. It is obvious that this formulation merely repeats what has been given above, since, from (59.9),

$$\psi' = \phi + \bar{p}_j \bar{q}_j$$

i.e., the functions ψ' and ϕ merely differ by an additive constant.

If H does not involve t, we may write

$$\psi' = -Et + S'(q_j, \bar{p}_j) \tag{61.13}$$

as in (61.8), with

$$S' = S + \bar{p}_j \bar{q}_j$$

Indeed, this result is valid even if the $\bar{p}_j$ and $\bar{q}_j$ are not constants. Then

$$H\left(q_j, \frac{\partial S'}{\partial q_j}\right) = E \tag{61.14}$$

Thus starting with the function ψ', by means of a contact transformation we may pass to a system for which the Hamiltonian is zero and the new coordinates and momenta are constants of the motion, with ψ' determined by (61.13) and (61.14).

If the time does not appear in H, there is another simple transformation which is of great interest, and which also leads to the Hamilton-Jacobi equation (61.14). This is a time-independent transformation generated by $S'(q_j, \bar{p}_j)$ alone, such that as before the $\bar{p}_j$ are constant, but now the new Hamiltonian is not zero but equal to its original value: $\bar{H}(\bar{p}) = H(q, p)$. As a consequence the $\bar{q}_j$ are not constants. This transformation, generated by the time-independent part of ψ' in (61.13), leads naturally to *action and angle variables* discussed in the next two sections.

62. ACTION AND ANGLE VARIABLES—PERIODIC SYSTEMS

To illustrate the ideas which follow, let us consider the special case of a system with one degree of freedom described by a Hamiltonian $H(q, p)$ which does not contain the time explicitly. We may apply the contact transformation generated by $S'(q, \bar{p})$ so that, from (59.10),

$$\begin{aligned} p &= \frac{\partial S'}{\partial q}, \qquad \bar{q} = \frac{\partial S'}{\partial \bar{p}}, \qquad \bar{H} = H \\ \dot{\bar{p}} &= -\frac{\partial \bar{H}}{\partial \bar{q}}, \qquad \dot{\bar{q}} = \frac{\partial \bar{H}}{\partial \bar{p}} \end{aligned} \tag{62.1}$$

Let us choose S' so that $\bar{p}$ is a constant of the motion, $\bar{p} = J$ say. Then, when we write $\bar{q} = w$, it follows that w is a cyclic coordinate and

$$\bar{H} = \bar{H}(J) = H(q, p) = E \tag{62.2}$$

with

$$w = \frac{\partial S'}{\partial J}, \qquad \dot{w} = \frac{d\bar{H}}{dJ} = \nu \tag{62.3}$$

Since $\bar{H}$ and J are both constants, ν is also. Thus the cyclic coordinate w conjugate to the constant of the motion J increases linearly with the time:

$$w = \nu t + \delta \tag{62.4}$$

If, for instance, J has the dimensions of action (energy $\times$ time), it follows that w is dimensionless, ν is a frequency, and w increases by unity in time ν^{-1}.

The constant new coordinates $\bar{q}_j$ of the last section were given by

$$\bar{q}_j = \frac{\partial \psi'}{\partial \bar{p}_j}$$

or, for the case considered here, with one degree of freedom

$$\bar{q} = \frac{\partial \psi'}{\partial \bar{p}}$$

The new coordinate $\bar{q}$, which in the present section we have called w, is given by

$$w = \frac{\partial S'}{\partial \bar{p}}$$

From (61.13) it follows that

$$w = t\frac{\partial E}{\partial \bar{p}} + \bar{q}$$

or

$$w = \frac{d\bar{H}}{dJ}t + \delta$$

by (61.4) which reproduces (62.4). The arbitrary constant δ is nothing but the constant of integration α of the last section. The time-dependent generator ψ' leads to a time-independent new coordinate $\bar{q}$. The time-independent part of ψ' generates a transformation which leads to a new coordinate which increases linearly with time.

Let us suppose now, as would be the case in many systems, that q and p are periodic functions of the time, of period τ, and let us evaluate the action over one period. This is a constant of the motion, and we shall choose J equal to it

$$J = \oint p\, dq$$

the integration proceeding over the values of q from time t to time $t + \tau$. Of course, in order to evaluate this integral, it is necessary to know the value of p as a function of q by solving the equation $H = E$.

During one period of the motion, w increases by an amount

$$\begin{aligned}\Delta w &= \oint dw = \oint \frac{\partial}{\partial q}\left(\frac{\partial S'}{\partial J}\right) dq \\ &= \frac{\partial}{\partial J}\oint \frac{\partial S'}{\partial q}\, dq \\ &= \frac{\partial J}{\partial J} = 1\end{aligned}$$

The change in w during a complete period is unity, and, as it has been shown that w increases by unity in time ν^{-1}, it follows that $\nu = \tau^{-1}$ is the frequency of the periodic motion. For such a problem J is called the *action variable* and w the corresponding *angle variable.*

The transformation has been generated by $S'(q, \bar{p}) = S'(q, J)$, where S' corresponds to the time-independent part of ψ' in (61.13). The generating function could be equally well expressed as $S(q, \bar{q}) = S(q, w)$ with

$$S' = S + \bar{p}\bar{q} = S + Jw \tag{62.5}$$

from (59.9). Then, instead of (62.1), we have

$$p = \frac{\partial S}{\partial q}, \qquad J = -\frac{\partial S}{\partial w}$$

Since, by definition, J is the increase in S' during one period, it follows from (62.5), by taking the increase of both sides over a period, that S is a periodic function of w with period 1.

Thus, for example, for the simple oscillator discussed in Sec. 58,

$$\begin{aligned}J &= \oint (2E - \omega^2 q^2)^{\frac{1}{2}}\, dq \\ &= \pi\omega a^2 = \frac{2\pi E}{\omega}\end{aligned} \tag{62.6}$$

where $a = (2E)^{\frac{1}{2}}/\omega$. Thus

$$E = \frac{\omega J}{2\pi} = \nu J = \bar{H} \tag{62.7}$$

The generator of the transformation to angle variables is

$$S' = \int (2E - \omega^2 q^2)^{\frac{1}{2}}\, dq$$

or

$$S'(q, J) = \int \left(\frac{\omega J}{\pi} - \omega^2 q^2\right)^{\frac{1}{2}} dq$$

The angle variable w is given by

$$w = \frac{\partial S'}{\partial J} = \frac{\omega}{2\pi} \int \left(\frac{\omega J}{\pi} - \omega^2 q^2 \right)^{-1/2} dq$$

$$= \frac{1}{2\pi} \sin^{-1} \left(\frac{\omega \pi}{J} \right)^{1/2} q$$

so that $dw = (\omega/2\pi p)\, dq$, a result that follows directly from (62.4), provided that ν is the frequency. The momentum is

$$p = \frac{\partial S'}{\partial q} = (2E)^{1/2} \cos 2\pi w$$

In terms of the constant a, which is the amplitude of the motion,

$$J = \pi \omega a^2$$

$$w = \frac{1}{2\pi} \sin^{-1} \frac{q}{a}$$

or

$$q = a \sin 2\pi w$$

Thus J is simply ω times the area enclosed in the corresponding circular motion, and w is $(2\pi)^{-1}$ times the phase angle. w increases by unity for each complete oscillation of the system.

We note that these results are identical with those derived at the end of Sec. 58. There the generating function is expressed in terms of q, w, and it corresponds to S above. If we express S' in terms of q, w, we have

$$S' = \tfrac{1}{2}\omega q^2 \cot 2\pi w + \pi \omega q^2 w \operatorname{cosec}^2 2\pi w$$

i.e.,

$$S' = \tfrac{1}{2}\omega q^2 \cot 2\pi \bar{q} + Jw$$

Thus, from (62.5),

$$S = \tfrac{1}{2}\omega q^2 \cot 2\pi \bar{q}$$

This is the mysterious generating function (58.11) which was introduced earlier without any apparent reason. It is, of course, periodic in $\bar{q}$, with period unity. More simply, since $E = \frac{1}{2}\omega^2 q^2 \operatorname{cosec}^2 2\pi w$, the function S is obtained from $S = \int (2E - \omega^2 q^2)^{1/2}\, dq$, with E expressed in terms of q and w, and w kept constant during the integration.

It should be re-emphasized that the introduction of action and angle variables does not assist us in the solution of a problem as simple as this, for before J could be found it was already necessary to know p

as a function of q (62.6). Furthermore, since the Hamiltonian is quadratic in p, the expression for p in terms of q is double-valued, corresponding, of course, to the two possible directions of the velocity when the oscillator is at a particular point. In evaluating (62.6), the positive sign of the square root was taken as q increased, the negative as q decreased. Indeed, the graph of p against q is the path of the system in phase space, and as the system oscillates the representative point in phase space traverses an ellipse. The integral J is clearly the area of this ellipse.

More generally, if the motion of a periodic system of one degree of freedom is represented by a closed curve in phase space, the action variable corresponding to this motion is the area enclosed by the curve. Such a motion is called a *libration* or oscillation. Thus, for the anharmonic oscillator (see Chapter 3),

$$p = m\dot{x} = \pm\{2m[E - V(x)]\}^{1/2}$$

For this to be represented by a closed loop in phase space, there must exist two values x', x'' of x between which $E > V$ and at which $E = V$. Furthermore, dp/dx must also be infinite at $x = x'$, $x = x''$. The last condition is satisfied provided that x', x'' are single roots of the equation

$$V(x) = E$$

The motion is then a libration, and, since $p\,dq = p\,dx = 2T\,dt$, the expression $p\,dq$ is positive throughout the motion. The loop in phase space is then traversed in the clockwise sense. Other possible types of motion of the oscillator have been discussed in Chapter 3. One other possibility is of special interest here. It may happen that the potential function $V(x)$ is itself a periodic function of x. If, then, $E > V$ for all values of x, the system assumes another form of periodicity. Without loss of generality, we may choose the variable $q = q(x)$ so that the period of V is 2π; i.e.,

$$V(q) = V(q + 2\pi)$$

Then clearly

$$p(q) = p(q + 2\pi)$$

so that the momentum is periodic in q. The path in phase space is then no longer a loop but a periodic curve, but nevertheless for this case, too, we may define an action variable. The integration in (62.6) is then over values of q corresponding to the period of p, i.e., from q_0 to $q_0 + 2\pi$. Such a motion is called a *rotation*. As an example, we note that for the plane pendulum discussed in Sec. 18 the equation for p_θ as a function of θ may be written

$$p_\theta^2 = 2ml^2[E - mgl(1 - \cos\theta)]$$

It is therefore periodic in θ, and, if $E > 2mgl$, the motion is a rotation. The action associated with the motion is then

$$(2m)^{1/2} l \int_0^{2\pi} [E - mgl(1 - \cos\theta)]^{1/2} \, d\theta$$

and the corresponding angle variable is $w = \theta/2\pi$, which clearly increases by unity during each period.

Action and angle variables are of particular interest in quantum theory. According to the original treatment of the latter, the values which J may assume for a periodic system are not arbitrary but are limited to be integral multiples of Planck's constant h; i.e.,

$$J = nh$$

where n is an integer. Thus, for example, the energy of a harmonic oscillator is not arbitrary but is limited by (62.7) to the values $E = nh\nu$. For a detailed discussion of this, and indeed for an exhaustive treatment of action and angle variables in classical and quantum theory, we refer the reader to a treatise by Professor Born.†

It may be noted that the state of a system with one degree of freedom may be specified by the complex number

$$z = 2^{-1/2}(q + iap) \tag{62.8}$$

where a is a real constant. To every point in phase space there corresponds a point in the z plane, and vice versa. The Hamiltonian $H(q, p)$ may be expressed in terms of z and its complex conjugate z^* by means of the relations

$$\begin{aligned} q &= 2^{-1/2}(z + z^*) \\ iap &= 2^{-1/2}(z - z^*) \end{aligned} \tag{62.9}$$

The canonical equations may be written

$$\begin{aligned} \frac{\partial H'}{\partial z^*} &= i\dot{z} \\ \frac{\partial H'}{\partial z} &= -i\dot{z}^* \end{aligned} \tag{62.10}$$

† M. Born, *The Mechanics of the Atom*, Bell, London (1927).

where $aH(q, p) = H'(z, z^*)$ and in the differentiation the variables z, z^* are treated as **independent**. The second canonical equation is then just the complex conjugate of the first.

For a motion of libration, the action variable

$$J = \oint p\, dq$$

evaluated along a closed curve in phase space taken in the clockwise sense may be written as a contour integral

$$J = \frac{i}{2a} \oint (z - z^*)\, d(z + z^*)$$

$$= \frac{i}{2a} \oint (z\, dz^* - z^*\, dz)$$

evaluated in the usual counterclockwise sense. Thus

$$J = a^{-1}\, \mathrm{Im} \oint z^*\, dz \tag{62.11}$$

where Im denotes "imaginary part of." The contour of integration in the z plane corresponds, of course, to the closed curve in the $p - q$ plane through the relations (62.9).

If E is the energy of the particle,

$$H'(z, z^*) = aE$$

so that z^* may be expressed as a function of aE and z:

$$z^* = z^*(aE, z) \tag{62.12}$$

In general, z^* does not appear as a single-valued function of z, so that the integrand in (62.11) possesses one or more branch points (cf. Exercise 3, Chapter 3). In the example given below, however, z^* may be made a single-valued function by appropriate choice of the parameter a.

We shall consider the case of the simple oscillator described by

$$H = \tfrac{1}{2}p^2 + \tfrac{1}{2}\omega^2 q^2 \tag{62.13}$$

Then, using (62.9),

$$H' = a[\tfrac{1}{4}(\omega^2 - a^{-2})(z^2 + z^{*2}) + \tfrac{1}{2}(\omega^2 + a^{-2})zz^*]$$

If we choose $a = \omega^{-1}$, this becomes

$$H' = \omega z z^* \tag{62.14}$$

The Hamiltonian equations of motion are then

$$\dot{z} = -i\omega z$$

and its conjugate complex, or

$$z = Ae^{-i\omega t}$$

where $|A| = E^{1/2}/\omega$ from (62.13). The representative point in the complex plane thus moves in a circle of radius $E^{1/2}\omega^{-1}$ with angular velocity ω. The action variable for this motion is, from (62.11),

$$J = \frac{E}{\omega}\,\mathrm{Im}\oint\frac{dz}{z} = \frac{2\pi E}{\omega}$$

in agreement with (62.6).

If we have a periodic system that is slightly disturbed by some external force, we may under certain conditions still introduce action and angle variables and use them to estimate the rate of change of the amplitude of the resulting almost periodic motion. We shall write the Hamiltonian of the system as $H = H[q, p, a(t)]$, where the effect of the external force is exhibited by the explicit dependence of H on t through $a(t)$, and the fact that this effect is small implies that $\dot{a}$ is small. At any given instant throughout the motion we shall suppose that, if the external influence were suddenly removed ($\dot{a} = 0$), the system would become truly periodic, with action and angle variables J and w introduced as above. One would think that the action variable J derived in this way would depend on the instant at which the external force were supposed removed, but it may be shown that,* provided $a(t)$ is unrelated to the period of the system,

$$\frac{\Delta J}{\Delta t} \sim \dot{a}^2$$

where ΔJ is the change in J in the time interval Δt. In the limit of infinitesimal $\dot{a}$, then, the change in J goes to zero even over an extremely long interval of time for which $\Delta a = \dot{a}\,\Delta t$ is finite. The type of external influence considered above, which is sufficiently slow so that the change in energy of the system per period is small compared with the energy itself, and which does not produce resonance, is known as an *adiabatic change*, and the resulting invariance of J is known as *adiabatic invariance*.

For a simple harmonic oscillator, we have, from equations (62.7)

* Born, *loc. cit.*, p. 56.

et seq.,

$$q = \left(\frac{J}{\pi\omega}\right)^{1/2} \sin (2\pi w)$$

so that, since J is an invariant for adiabatic changes, the amplitude of the oscillation is proportional to $\omega^{-1/2}$ for such changes. If the frequency of oscillation is slowly increased by an external influence, the amplitude then slowly decreases. This result may be derived more directly thus: If

$$H = \tfrac{1}{2}(p^2 + \omega^2 q^2)$$

where ω is a given slowly varying function of time, and if we look for a solution of the equations of motion of the form

$$q = a \sin \int \omega \, dt$$

then, neglecting terms involving second time derivatives of a,

$$\ddot{q} = -a\omega^2 \sin \int \omega \, dt + (a\dot{\omega} + 2\dot{a}\omega) \cos \int \omega \, dt$$

so that

$$a\dot{\omega} + 2\dot{a}\omega = 0$$

or

$$a \sim \omega^{-1/2}$$

Action and angle variables are therefore useful for finding the rate of change of the amplitude of the motion under the influence of an adiabatic perturbation.

For a simple pendulum the length of which is being increased adiabatically, or for any other oscillator the natural frequency of which is being slowly increased, the total energy of the oscillator is proportional to $a^2\omega^2$ and thus to ω. According to quantum mechanics the energy is $(n + \frac{1}{2})\hbar\omega$, where n is an integer, and therefore also is proportional to ω for an adiabatic change which keeps n constant.

63. SEPARABLE MULTIPLY-PERIODIC SYSTEMS

It is relatively easy to generalize the ideas of the previous section to systems with more than one degree of freedom. Introducing a time-independent generating function $S'(q_j, \bar{p}_j)$, we suppose that it has been chosen so that the $\bar{p}_j$ are constants:

$$\bar{p}_j = J_j$$

Then, if H is independent of the time, we have, writing $\bar{q}_j = w_j$,

$$p_j = \frac{\partial S'}{\partial q_j}, \qquad w_j = \frac{\partial S'}{\partial J_j} \tag{63.1}$$
$$H(q_j, p_j) = \bar{H}(w_j, J_j) = E$$

As before,

$$\dot{w}_j = \frac{\partial \bar{H}}{\partial J_j}, \qquad \dot{J}_j = 0 = -\frac{\partial \bar{H}}{\partial w_j} \tag{63.2}$$

Thus the w_j are cyclic, $\bar{H} = \bar{H}(J_j)$, and, as $\bar{H}$, J_j are constants of the motion,

$$w_j = \nu_j t + \delta_j \tag{63.3}$$

where ν_j, δ_j are constants.

All the above relations hold independently of any periodicity properties of the dynamic system in question, but they are of little interest unless it is possible to choose the constants J_j so introduced as equal to the action variables of the system. We shall therefore suppose that for each degree of freedom there exists an interval $t_0 < t \leq t_0 + T_j$ over which the integral

$$J_j = \oint p_j \, dq_j \qquad \text{(not summed)} \tag{63.4}$$

may be determined, the value for each j being independent of t_0. This condition is satisfied if, for instance, q_j and p_j are periodic functions of the time, with period T_j:

$$\begin{aligned} q_j(t) &= q_j(t + nT_j) \\ p_j(t) &= p_j(t + nT_j) \end{aligned} \qquad (n \text{ integer})$$

The motion is then one of libration. If, on the other hand, the motion is one of rotation, p_j is a periodic function of q_j and the integral in (63.4) is to be taken over this period. In order to evaluate each term in (63.4) each p must be known in terms of the corresponding q, a condition that by (63.1) is satisfied if S' is separable:

$$S' = \Sigma S_k'(q_k)$$

During the motion of the system, the J_j remain constant, so that changes in the angle variables w_j arise only from changes in the coordinates q_k, i.e., from (63.1)

$$dw_j = \sum_k \frac{\partial}{\partial q_k}\frac{\partial S'}{\partial J_j}\,dq_k \tag{63.5}$$

$$= \sum_k \frac{\partial}{\partial J_j}\,p_k\,dq_k$$

Thus, from (63.3),

$$\nu_j = \sum_k \frac{\partial p_k}{\partial J_j}\,\dot{q}_k \tag{63.6}$$

giving an expression for the frequencies of the motion that may be computed if S' is known as a function of the q's and J's.

If the frequencies are commensurable, the system will return to its initial state after some finite time T where

$$T = n_j T_j \qquad \text{(not summed)}$$

and the n_j are integers. If we integrate (63.5) over this complete period of the system, we obtain

$$\int_{t=0}^{T} dw_j = \frac{\partial}{\partial J_j}\sum_k \int_{t=0}^{T} p_k\,dq_k \tag{63.7}$$

$$= \frac{\partial}{\partial J_j}\sum_k n_k J_k$$

$$= n_j$$

From (63.3), the left-hand side is equal to

$$\nu_j T = n_j \nu_j T_j \qquad \text{(not summed)}$$

Thus, for each j, $\nu_j T_j = 1$, i.e., ν_j is in fact the frequency with which the motion of the jth degree of freedom repeats itself.

If the frequencies are not commensurable, strictly speaking T is infinite, i.e., the system will never return exactly to its initial state. However, for large T the system comes back to a state which is arbitrarily close to that from which it departed, so that in this case (63.7) contains an error that may be made as small as desired.

The function S' is not completely defined as a function of the q_k, for each one of the terms in the sum

$$S' = \sum_k p_k \dot{q}_k\,dt \tag{63.8}$$

is defined only when the time over which the integration is to extend is specified. Over a complete period T of the system, the generating function S' increases by an amount $\sum_k n_k J_k$, even although after that time all the q's and p's have returned to their initial values. During this time, however, the alternative generating function

$$S = S' - \Sigma J_k w_k \tag{63.9}$$

increases by an amount

$$\Delta S = \Delta S' - \Sigma J_k \, \Delta w_k$$
$$= 0 \qquad \text{by (63.7)}$$

Thus S is periodic with the period T.

The ambiguity in S' as a function of q_k also reveals the fact that S is *multiply-periodic*. If the kth term in

$$S' = \sum_k \int p_k \, dq_k$$

is taken over m_k extra periods of the motion of q_k, and at the same time w_k in (63.9) is replaced by $w_k + m_k$, the value of the function S is unaltered, i.e.,

$$S(q_k, w_k) = S(q_k, w_k + m_k) \tag{63.10}$$

for all integers m_k. Since this is valid for any value of k, $S(q_k, w_k)$ is in this sense multiply-periodic in w_k with fundamental period unity.

It may happen that S' is not separable when expressed in terms of the variables q_j but may become separable when expressed in terms of some other variables: $\bar{q}_j = \bar{q}_j(q_k)$. For this reason the choice of a coordinate system is of great importance in dealing with such problems. The coordinate transformation above corresponds to the momentum transformation (59.10) with $\psi' = f_i \bar{p}_i + g(q)$

$$p_j = \bar{p}_k \frac{\partial \bar{q}_k}{\partial q_j} + \frac{\partial g}{\partial q_j}$$

so that

$$\oint p_j \, dq_j = \oint \bar{p}_k \frac{\partial \bar{q}_k}{\partial q_j} \, dq_j$$

since the second term on the right-hand side gives no contribution. Hence

$$\oint p_j \, dq_j = \oint \bar{p}_j \, d\bar{q}_j \tag{63.11}$$

These integrals may be evaluated only if the p_j are known as functions of the q_j, and the $\bar{p}_j$ as functions of the $\bar{q}_j$. Thus, if the $H - J$ equation is separable and if f integrals of it are known, the sum of the action variables is invariant under a coordinate transformation which leaves the equation separable.

If, independently of any initial conditions, the fundamental periods T_j are mutually incommensurable, it may be shown that all transformations which leave the $H - J$ equation separable are of the form

$$\bar{q}_j = \bar{q}_j(q_j) \qquad \text{(each } j\text{)}$$

i.e., each new coordinate is a function of a particular old coordinate. The generator of such transformations is

$$\psi' = \sum_j \bar{p}_j f_j(q_j) + g$$

It then follows that (63.11) is valid *for each* j, and the phase integrals J_j are invariant with respect to all transformations which leave the $H - J$ equation separable.

64. APPLICATIONS

In the following applications, equation numbers on the right-hand side of the page refer to the general equations of the text of which the equality is a particular case.

(a) Central Motion. A particle moves in the field of a fixed center of force, the potential being $V(r)$ (cf. Sec. 36). Using spherical coordinates,

$$L = \tfrac{1}{2}\mu(\dot{r}^2 + r^2\dot{\theta}^2 + r^2 \sin^2 \theta\dot{\phi}^2) - V \tag{32.1}$$

$$p_r = \mu\dot{r}, \qquad p_\theta = \mu r^2\dot{\theta}, \qquad p_\phi = \mu r^2 \sin^2 \theta\dot{\phi} \tag{54.3}$$

$$H = \frac{1}{2\mu}(p_r{}^2 + r^{-2}p_\theta{}^2 + r^{-2} \operatorname{cosec}^2 \theta p_\phi{}^2) + V$$

The Hamiltonian does not involve t, so we have

$$\frac{1}{2\mu}\left[\left(\frac{\partial S}{\partial r}\right)^2 + r^{-2}\left(\frac{\partial S}{\partial \theta}\right)^2 + r^{-2} \operatorname{cosec}^2 \theta \left(\frac{\partial S}{\partial \phi}\right)^2\right] + V = E \tag{61.9}$$

with

$$S = S(r, \theta, \phi, \alpha_j) \qquad (j = 1, 2, 3)$$

The $H - J$ equation is separable, for if we write

$$S = S_r(r) + S_\theta(\theta) + S_\phi(\phi)$$

we have

$$\left(\frac{dS_r}{dr}\right)^2 + r^{-2}\left(\frac{dS_\theta}{d\theta}\right)^2 + r^{-2}\operatorname{cosec}^2\theta\left(\frac{dS_\phi}{d\phi}\right)^2 = 2\mu(E - V)$$

or

$$r^2 \sin^2\theta\left[2\mu(E - V) - \left(\frac{dS_r}{dr}\right)^2 - r^{-2}\left(\frac{dS_\theta}{d\theta}\right)^2\right] = \left(\frac{dS_\phi}{d\phi}\right)^2$$

The right-hand side of this equation is a function of ϕ alone, the left-hand side a function of r and θ. But r, θ, ϕ are independent variables. Each side of this equation must, therefore, be a constant, and we write

$$\frac{dS_\phi}{d\phi} = p_\phi = m \tag{61.10}$$

The constant m here introduced is the same as that specified earlier (36.3), and it may be regarded here as one of the constants α_j. The $H - J$ equation may now be written

$$\left(\frac{dS_\theta}{d\theta}\right)^2 + m^2\operatorname{cosec}^2\theta = 2r^2\mu(E - V) - r^2\left(\frac{dS_r}{dr}\right)^2$$

and, as before, each side must be equal to a constant l^2. Thus we may write

$$p_\theta = \frac{dS_\theta}{d\theta} = (l^2 - m^2\operatorname{cosec}^2\theta)^{1/2} \tag{61.10}$$

[cf. (36.4)], hence

$$p_r = \frac{dS_r}{dr}[2\mu(E - V) - l^2 r^{-2}]^{1/2} \tag{61.10}$$

The three momenta are then given explicitly as functions of r, θ, ϕ and three constants m, l, E, and then S is given by

$$S = \int(p_r\,dr + p_\theta\,d\theta + p_\phi\,d\phi)$$

The equations of motion then follow in any one of the following three ways:

1. Express the momenta above in terms of the velocities by (53.2). This gives three differential equations which are, of course, Lagrange's equations and are not in an integrated form.

2. Use the relations

$$\beta + t = \frac{\partial S}{\partial E} = \frac{\partial S_r}{\partial E} = \mu \int [2\mu(E - V) - l^2 r^{-2}]^{-1/2}\, dr$$

$$\begin{aligned} \beta_1 &= \frac{\partial S}{\partial m} = \frac{\partial S_\phi}{\partial m} + \frac{\partial S_\theta}{\partial m} \\ &= \phi - m \int \operatorname{cosec}^2 \theta (l^2 - m^2 \operatorname{cosec}^2 \theta)^{-1/2}\, d\theta \qquad (61.12) \\ \beta_2 &= \frac{\partial S}{\partial l} = \frac{\partial S_\theta}{\partial l} + \frac{\partial S_r}{\partial l} \\ &= l \int (l^2 - m^2 \operatorname{cosec}^2 \theta)^{-1/2}\, d\theta - l \int r^{-2} [2\mu(E - V) - l^2 r^{-2}]^{-1/2}\, dr \end{aligned}$$

The evaluation of these integrals is facilitated by choosing the axes so that $m = 0$ [cf. (36.5)]. The problem is then one of two degrees of freedom, with $\phi = \beta_1 = \text{const.}$ Thus

$$\beta + t = \left(\frac{\mu}{2}\right)^{1/2} I_1, \qquad \theta - \beta_2 = \frac{l}{(2\mu)^{1/2}} I_2$$

where

$$I_1 = \int (E - V^*)^{-1/2}\, dr, \qquad I_2 = \int r^{-2} (E - V^*)^{-1/2}\, dr$$

and where

$$V^* = V + \frac{1}{2} \frac{l^2}{\mu r^2}$$

is the potential function (36.13) for the one-dimensional problem equivalent to the radial motion.

3. For general orientation of axes ($m \neq 0$) the problem is one of three degrees of freedom, and there exist three first integrals of the equations of motion, expressible in terms of the coordinates and momenta thus:

$$\begin{aligned} m &= p_\phi \\ l &= (m^2 \operatorname{cosec}^2 \theta + p_\theta^2)^{1/2} \\ E &= \frac{1}{2\mu} (p_r^2 + l^2 r^{-2}) + V = \frac{p_r^2}{2\mu} + V^* \end{aligned}$$

These three constants—the angular momentum about the polar axis, the total angular momentum, and the energy—correspond here to the three constants α_j in the complete integral of the $H - J$ equation. As written above, each constant expresses a relation between one coordinate alone and the corresponding momentum. The problem is thus broken into three equivalent one-dimensional problems for the ϕ, θ, r motions respectively, corresponding to the fact that the variables in the $H - J$ equation are separable. Attention has already been drawn to this in Sec. 36 for the case of the radial motion. In the case of the θ motion, the equation

$$p_\theta^2 + m^2 \operatorname{cosec}^2 \theta = l^2$$

may be regarded as giving a Hamiltonian H_θ such that

$$\dot{p}_\theta = -\frac{\partial H_\theta}{\partial \theta}, \qquad \dot{\theta} = \frac{\partial H_\theta}{\partial p_\theta}$$

If $H_\theta = V^* - V = l^2/2\mu r^2$, constant as far as variations of θ are concerned, it is easily verified that these canonical equations are satisfied. The potential energy of the centrifugal force when expressed as a function of θ and p_θ may therefore be regarded as the Hamiltonian which describes the θ motion.

If the system is periodic, the action variables may be evaluated as follows:

$$J_\phi = \oint p_\phi \, d\phi = m \oint d\phi$$

and, as the integration is over a complete range of ϕ from 0 to 2π,

$$J_\phi = 2\pi m$$

The condition that p_θ be real is that

$$|\sin \theta| \geq \frac{m}{l}$$

where, since the angular momentum about a particular axis cannot exceed the total angular momentum,

$$m \leq l$$

Thus

$$J_\theta = \oint (l^2 - m^2 \operatorname{cosec}^2 \theta)^{1/2} \, d\theta \tag{64.1}$$

the positive sign of the square root being taken as θ increases from $\theta_{\min} = \sin^{-1}(m/l)$ to $\theta_{\max} = \pi - \theta_{\min}$, the negative sign as θ decreases. Hence

$$\begin{aligned} J_\theta &= 2 \int_{\theta_{\min}}^{\theta_{\max}} (l^2 - m^2 \operatorname{cosec}^2 \theta)^{1/2} \, d\theta \\ &= 4l \int_{\theta_{\min}}^{\pi/2} \left(1 - \frac{m^2}{l^2} \operatorname{cosec}^2 \theta\right)^{1/2} d\theta \end{aligned}$$

Writing $\sin^2 \theta = u$, we have

$$J_\theta = 2 \int_{m^2/l^2}^{1} \left(\frac{l^2 u - m^2}{1 - u}\right)^{1/2} \frac{du}{u}$$

Replacement of u by v^{-1} in the integral changes the sign and interchanges l and m, so that J is a function of $(l - m)$. Furthermore, for $m = 0$,

$$(J_\theta)_{m=0} = 2l \int_0^1 \frac{du}{[u(1 - u)]^{1/2}}$$

$$= 2\pi l$$

Thus

$$J_\theta = 2\pi(l - m)$$

Finally

$$J_r = \oint [2\mu(E - V) - l^2 r^{-2}]^{1/2}\, dr \tag{64.2}$$

if the potential V and the constants E and $l^2\mu^{-1}$ are such that the radial motion is periodic.

In the special case of a potential $V = \kappa/r$, the condition that the integrand should vanish is

$$2\mu E r = \mu\kappa \pm (\mu^2\kappa^2 + 2\mu E l^2)^{1/2}$$

These two roots are both positive only if both κ and E are negative, i.e., for an attractive potential with the particle having insufficient energy to escape to infinity. The radial motion is then periodic, and

$$J_r = 2 \int_{r_{\min}}^{r_{\max}} [2\mu(E - \kappa r^{-1}) - l^2 r^{-2}]^{1/2}\, dr$$

with

$$\left.\begin{matrix} r_{\max} \\ r_{\min} \end{matrix}\right\} = \frac{\kappa}{2E}\left[1 \pm \left(1 + \frac{2El^2}{\mu\kappa^2}\right)^{1/2}\right]$$

The value of this integral is* $J_r = -2\pi l + (2\pi\mu\kappa/(-2\mu E)^{1/2})$

*See, for example, Born, *loc. cit.*, p. 308. The analysis requires that the term $-2\pi l$ should be negative, and here we can define l as positive because the sense of rotation is irrelevant. To allow for both signs of l, the term $-2\pi l$ must be replaced by $-2\pi|l| = -|J_\theta + J_\phi|$. This distinction is essential if the sense of the rotation is significant, as for the Foucault pendulum, for example. See Alexander Khein and D. F. Nelson, *Am. J. Phys.*, 61, 170, 175 (1993).

so that

$$E = -\frac{2\pi^2\mu\kappa^2}{(J_r + J_\theta + J_\phi)^2} \tag{64.3}$$

giving the energy in terms of the action variables, hence the Hamiltonian in terms of these variables. The frequencies of the r, θ, ϕ motion are then given by (63.2)

$$\nu_r = \frac{\partial \bar{H}}{\partial J_r}, \qquad \nu_\theta = \frac{\partial \bar{H}}{\partial J_\theta}, \qquad \nu_\phi = \frac{\partial \bar{H}}{\partial J_\phi}$$

but since the energy depends only on the sum $J_r + J_\theta + J_\phi$, these frequencies are all equal to

$$\left(\frac{2}{\mu}\right)^{1/2} \frac{(-E)^{3/2}}{\pi|\kappa|} \tag{64.4}$$

The orbit is closed in the special case considered here, and the frequencies of the r, θ, and ϕ motion are identical.

Systems possessing two or more commensurable frequencies are called *degenerate*. The degeneracy $\nu_\theta = \nu_\phi$ arises as a general consequence of the central nature of the force; that ν_r also assumes the same value is a consequence of the special form $V = \kappa/r$ of the central force.

Thus, for an electron of charge $-e$ moving in the field of a nucleus of charge Ze, we have $\kappa = -Ze^2$ and (64.3) becomes

$$E = -\frac{2\pi^2\mu Z^2 e^4}{J^2}$$

where

$$J = J_r + J_\theta + J_\phi$$

If the generating function S is expressed in the form of the function $S'(q_j, J_j)$ of Sec. 63, we have

$$S'(r, \theta, \phi, J_r, J_\theta, J_\phi) = \int (p_r\, dr + p_\theta\, d\theta + p_\phi\, d\phi)$$

where

$$p_r = \mu\dot{r} = \left[-2\mu\kappa\left(\frac{2\pi^2\mu\kappa}{(J_r + J_\theta + J_\phi)^2} + \frac{1}{r}\right) - \frac{(J_\theta + J_\phi)^2}{4\pi^2 r^2}\right]^{1/2}$$

$$p_\theta = \mu r^2\dot{\theta} = \frac{1}{2\pi}[(J_\theta + J_\phi)^2 - J_\phi^2 \operatorname{cosec}^2 \theta]^{1/2}$$

$$p_\phi = \mu r^2 \sin^2 \theta\dot{\phi} = \frac{J_\phi}{2\pi}$$

Thus, for example, (63.6) gives, for ν_ϕ,

$$\nu_\phi = \dot{r}\frac{\partial p_r}{\partial J_\phi} + \dot{\theta}\frac{\partial p_\theta}{\partial J_\phi} + \dot{\phi}\frac{\partial p_\phi}{\partial J_\phi}$$

$$= \frac{4\pi^2\mu\kappa^2}{(J_r + J_\theta + J_\phi)^2}$$

which reduces to the result (64.4).

It was postulated by Bohr and Sommerfeld that the action variables are not arbitrary but may assume only values of the form

$$J_r = n_r h, \qquad J_\theta = n_\theta h, \qquad J_\phi = n_\phi h$$

where n_r, n_θ, n_ϕ are integers and h is Planck's constant. This additional assumption imposed on classical mechanics was successful in describing the observed energy levels of an electron in an atom, at least approximately. Here, for instance, one obtains for the energy levels of a hydrogen-like atom

$$E = -\frac{2\pi^2\mu Z^2 e^4}{n^2 h^2}$$

where

$$n = n_r + n_\theta + n_\phi$$

Coupled with the condition $E = h\nu$ for the frequency of light emitted when an electron makes a transition from one energy level to another, this leads to

$$\nu = \frac{2\pi^2\mu Z^2 e^4}{h^3}\left(\frac{1}{n_1{}^2} - \frac{1}{n_2{}^2}\right)$$

for the frequency spectrum, where n_1 and n_2 are integers. For $n_1 = n \gg 1$, $n_2 = n + 1$, the frequency associated with the change of n by an integer is

$$\nu \approx \frac{2\pi^2\mu Z^2 e^4}{h^3}\frac{2}{n^3}$$

$$= \left(\frac{2}{\mu}\right)^{1/2}\frac{(-E)^{3/2}}{\pi Z e^2}$$

which is identical with (64.4). Thus, in the limit of large quantum numbers n, the frequency of the light emitted when n changes by unity is the same as the frequency of the electron in its orbit. This result is a particular case of the correspondence principle of Bohr.

(b) The Problem of Two Centers of Gravitation. We turn now to the problem of the motion of a particle under the influence of forces

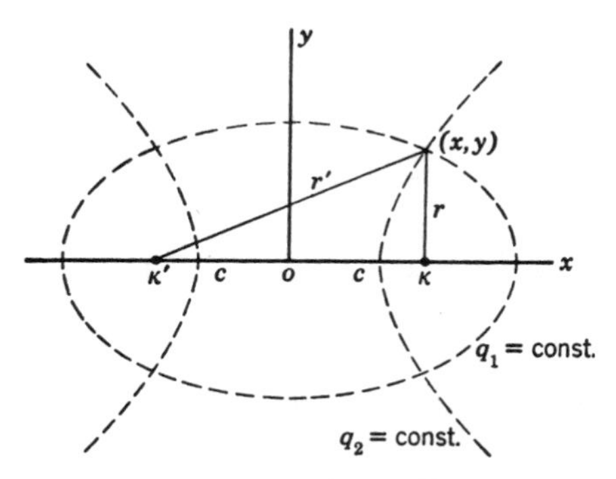

Fig. 64-1. Elliptic coordinates.

toward two fixed centers, the forces being inversely proportional to the square of the distance of the particle from these centers; i.e.,

$$V = \kappa[(x - c)^2 + y^2]^{-1/2} + \kappa'[(x + c)^2 + y^2]^{-1/2}$$

where $2c$ is the distance between the centers. We consider only the two-dimensional problem, the initial conditions of the motion being such that the particle remains in a plane. We introduce elliptic coordinates q_1, q_2 (see Fig. 64-1), defined by

$$x = c \cosh q_1 \cos q_2$$

$$y = c \sinh q_1 \sin q_2$$

The equations $q_1 = \text{const}$ and $q_2 = \text{const}$ then represent respectively ellipses and hyperbolas whose focuses are at the centers of force. Then

$$V = \frac{\kappa}{c} (\cosh q_1 - \cos q_2)^{-1} + \frac{\kappa'}{c} (\cosh q_1 + \cos q_2)^{-1}$$

and the kinetic energy is

$$\begin{aligned} T &= \tfrac{1}{2}(\dot{x}^2 + \dot{y}^2) \\ &= \frac{c^2}{2} (\cosh^2 q_1 - \cos^2 q_2)(\dot{q}_1{}^2 + \dot{q}_2{}^2) \end{aligned}$$

Hence

$$\begin{aligned} p_1 &= \frac{\partial L}{\partial \dot{q}_1} = c^2(\cosh^2 q_1 - \cos^2 q_2)\dot{q}_1 \\ p_2 &= \frac{\partial L}{\partial \dot{q}_2} = c^2(\cosh^2 q_1 - \cos^2 q_2)\dot{q}_2 \end{aligned} \tag{53.2}$$

and

$$\begin{aligned}H &= T + V \\ &= \frac{1}{2c^2}(\cosh^2 q_1 - \cos^2 q_2)^{-1}(p_1{}^2 + p_2{}^2) \\ &\qquad + \frac{\kappa}{c}(\cosh q_1 - \cos q_2)^{-1} + \frac{\kappa'}{c}(\cosh q_1 + \cos q_2)^{-1} \\ &= E\end{aligned}$$

The Hamilton-Jacobi equation is, then,

$$\left(\frac{\partial S}{\partial q_1}\right)^2 + \left(\frac{\partial S}{\partial q_2}\right)^2 = 2c^2E(\cosh^2 q_1 - \cos^2 q_2) - 2\kappa c(\cosh q_1 + \cos q_2) - 2\kappa' c(\cosh q_1 - \cos q_2) \tag{61.9}$$

or

$$\begin{aligned}\left(\frac{\partial S}{\partial q_1}\right)^2 - 2c^2E\cosh^2 q_1 + 2(\kappa + \kappa')c\cosh q_1 &= \alpha_1 \\ &= -\left(\frac{\partial S}{\partial q_2}\right)^2 - 2c^2E\cos^2 q_2 - 2(\kappa - \kappa')c\cos q_2\end{aligned}$$

The equation is therefore separable, and we may write

$$S = S_1(q_1) + S_2(q_2)$$

where

$$\begin{aligned}S_1 &= \int [2c^2E\cosh^2 q_1 - 2c(\kappa + \kappa')\cosh q_1 + \alpha_1]^{1/2}\,dq_1 \\ S_2 &= \int [-2c^2E\cos^2 q_2 - 2c(\kappa - \kappa')\cos q_2 - \alpha_1]^{1/2}\,dq_2\end{aligned}$$

The equations of motion are

$$\begin{aligned}t + \beta &= \frac{\partial S}{\partial E} \\ \beta_1 &= \frac{\partial S}{\partial \alpha_1}\end{aligned} \tag{61.12}$$

As usual, the first of these equations gives the time at which a point on the orbit is occupied; the second gives the equation to the orbit.

$$\begin{aligned}2\beta_1 = \int &[2c^2E\cosh^2 q_1 - 2c(\kappa + \kappa')\cosh q_1 + \alpha_1]^{-1/2}\,dq_1 \\ &- \int [-2c^2E\cos^2 q_2 - 2c(\kappa - \kappa')\cos q_2 - \alpha_1]^{-1/2}\,dq_2\end{aligned}$$

These integrals are of the elliptic type.

The analysis of this problem by the Hamilton-Jacobi method is considerably simpler than the direct integration of Lagrange's equation of motion. This results essentially from the fact that the $H - J$ equation, when expressed in terms of the elliptic coordinates, is separable.

Two special cases are of interest.

Case 1. $\kappa' \to -\kappa$, $c \to 0$, $2c\kappa \to \mu$ (finite). The problem is then that of *the motion of a particle in the field of a dipole* of strength μ and axis along the x axis. We write $q_2 = \theta$ and introduce

$$r = c \cosh q_1$$

Then as $c \to 0$, $\cosh q_1 \to \infty$, so that also

$$r \to c \sinh q_1$$

Then $x = r \cos \theta$, $y = r \sin \theta$, and the elliptic coordinates reduce to plane polar coordinates. The equation to the orbit becomes

$$2\beta_1 = \int (2Er^2 + \alpha_1)^{-1/2} \frac{dr}{r} - \int (-2\mu \cos \theta - \alpha_1)^{-1/2} \, d\theta$$

and the time at which a point on it is occupied is given by

$$t + \beta = \int (2Er^2 + \alpha_1)^{-1/2} r \, dr$$

In this limit, the momenta become

$$p_1 = r\dot{r}, \qquad p_2 = r^2\dot{\theta}$$

(note that p_1 is conjugate to $q_1 = \ln r + \text{const}$, not to r).

These results may, of course, be obtained independently by starting from first principles:

$$V = \frac{\mu}{r^2} \cos \theta$$

so that

$$H = \frac{1}{2r^2} ({p_1}^2 + {p_2}^2) + \frac{\mu}{r^2} \cos \theta = E$$

Thus

$$\left(\frac{dS_1}{dq_1}\right)^2 + \left(\frac{dS_2}{dq_2}\right)^2 = 2Er^2 - 2\mu \cos \theta$$

or

$$S_1 = \int (2Er^2 + \alpha_1)^{1/2} \frac{dr}{r}$$

$$S_2 = \int (-2\mu \cos \theta - \alpha_1)^{1/2} \, d\theta$$

as before.

Case 2. The special case $c \to \infty$, $\kappa'/4c^2 \to F$ (finite). In this limit the motion is that of *a particle in a Coulomb field upon which is superimposed a uniform parallel field.* The solution should then describe approximately the motion of a charged electron around a nucleus under the influence of a uniform external electrostatic field, as in the *Stark effect.*

We shall first move the origin to the point occupied by the center κ and, for convenience in generalizing this case to three dimensions, replace the variable x by z; i.e.,

$$y = c \sinh q_1 \sin q_2$$

$$z = c(-1 + \cosh q_1 \cos q_2)$$

Let us now take the limit $c \to \infty$, replacing q_1, q_2 by $q_1 c^{-1/2}$, $q_2 c^{-1/2}$ respectively. Then

$$y = q_1 q_2$$

$$z = \tfrac{1}{2}(q_1^2 - q_2^2)$$

the parabolic coordinates introduced earlier (Exercise 6, Chapter 1). If we wish to consider this motion in three dimensions, these coordinates are easily generalized to

$$x = q_1 q_2 \cos q_3$$

$$y = q_1 q_2 \sin q_3$$

$$z = \tfrac{1}{2}(q_1^2 - q_2^2)$$

Thus

$$2z = q_1^2 - \frac{x^2 + y^2}{q_1^2} = -q_2^2 + \frac{x^2 + y^2}{q_2^2}$$

so that the surfaces $q_1 = $ const are paraboloids of revolution about the z axis which intersect a plane containing the z axis in parabolas with focus at the origin. This origin is now the center of the force of potential κr^{-1}, and the center κ' has been moved to a distance large compared with the separation of the particle from the origin and its strength correspondingly increased so that it produces a uniform field F in the z direction, of potential $-Fz$. For definiteness we shall take F as positive.

The kinetic energy is

$$T = \tfrac{1}{2}[(q_1^2 + q_2^2)(\dot{q}_1^2 + \dot{q}_2^2) + q_1^2 q_2^2 \dot{q}_3^2]$$

so that

$$p_1 = (q_1^2 + q_2^2)\dot{q}_1$$
$$p_2 = (q_1^2 + q_2^2)\dot{q}_2$$
$$p_3 = q_1^2 q_2^2 \dot{q}_3$$

Hence

$$H = [2(q_1^2 + q_2^2)]^{-1}\left[p_1^2 + p_2^2 + \left(\frac{1}{q_1^2} + \frac{1}{q_2^2}\right)p_3^2 + 4\kappa - F(q_1^4 - q_2^4)\right] = E$$

and the $H - J$ equation is separable:

$$\left(\frac{dS_1}{dq_1}\right)^2 + \left(\frac{dS_2}{dq_2}\right)^2 + \left(\frac{1}{q_1^2} + \frac{1}{q_2^2}\right)\left(\frac{dS_3}{dq_3}\right)^2 + 4\kappa - F(q_1^4 - q_2^4) = 2E(q_1^2 + q_2^2)$$

Hence

$$\frac{dS_3}{dq_3} = \alpha_3$$

$$\left(\frac{dS_1}{dq_1}\right)^2 + \alpha_3^2 q_1^{-2} - 2Eq_1^2 - Fq_1^4 = -\left(\frac{dS_2}{dq_2}\right)^2 - \alpha_3^2 q_2^{-2} + 2Eq_2^2 - Fq_2^4 - 4\kappa = \alpha_1$$

Thus

$$S_1 = \int [f_1(q_1)]^{1/2}\, dq_1$$
$$S_2 = \int [f_2(q_2)]^{1/2}\, dq_2$$

with

$$\alpha_3 = \frac{J_3}{2\pi}$$

$$f_1(q) = \alpha_1 + 2Eq^2 + Fq^4 - \frac{J_3^2}{4\pi^2 q^2}$$

$$f_2(q) = -\alpha_1 - 4\kappa + 2Eq^2 - Fq^4 - \frac{J_3^2}{4\pi^2 q^2}$$

We shall be interested in the case in which the particle is bound to the center of force ($E < 0$), so that for it to move into the region $z > 0$ the central force must be attractive ($\kappa < 0$). Without performing the above integrations we may learn something about the nature of the motion from the conditions $f_1(q_1) \geq 0$, $f_2(q_2) \geq 0$. Writing $q^2 = s$,

$J_3{}^2/4\pi^2 = \alpha$, we have

$$\frac{df_1}{ds} = 2E + 2Fs + \frac{\alpha}{s^2}$$

$$\frac{d^2f_1}{ds^2} = 2F - \frac{2\alpha}{s^3}$$

so that $d^2f_1/ds^2 = 0$ for $s = s_0 = (\alpha/F)^{1/3}$. This is a minimum of df_1/ds, at which point $df_1/ds = 2E + 3\alpha^{1/3}F^{2/3}$. Hence if

$$E < -\frac{3}{2}\alpha^{1/3}F^{2/3} = -\frac{3}{2}\left(\frac{J_3F}{2\pi}\right)^{2/3}$$

f_1 has two turning points, and initial conditions may be chosen (corresponding to choice of α_1) so that the graph of f_1 against s is as indicated in Fig. 64-2. The particle, once confined between the limits $q_{1\min} \leq q_1 \leq q_{1\max}$, will remain within these limits. We leave to the reader to verify that the function f_2 always possesses one and only one maximum, so that, as its value at this maximum must be positive, q_2 also is confined between limits $q_{2\min} \leq q_2 \leq q_{2\max}$. In this motion, then, since $J_3/2\pi$, the angular momentum about the axis of F, is different from zero, the particle executes a rotation about this axis. Its path is thus confined to the interior of a ring symmetrical about the line through the force center parallel to the direction of F, the cross section of the ring being bounded by parabolas as indicated in Fig. 64-3. Note that the nonvanishing of J_3 implies that $q_{1\min}$ and $q_{2\max}$ are both greater than zero.

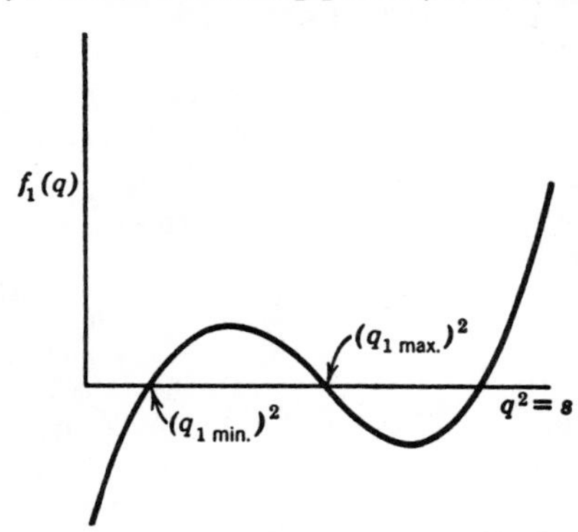

Fig. 64-2. The range of values available to variable q_1 when $J_3 \neq 0$.

For $J_3 = 0$, the motion takes place in a plane parallel to the field F. It then follows that for f_2 ever to be positive, $-\alpha - 4\kappa > 0$, and then $0 \leq q_2 \leq q_{2\max}$, where

$$(q_{2\max})^2 = F^{-1}\{E + [E^2 - (\alpha_1 + 4\kappa)F]^{1/2}\}$$

The graph of f_1 against $q_1{}^2$ has a minimum at $q_1{}^2 = -E/F$, and at that point $f_1 = \alpha_1 - E^2/F$. For the particle to be bound near the

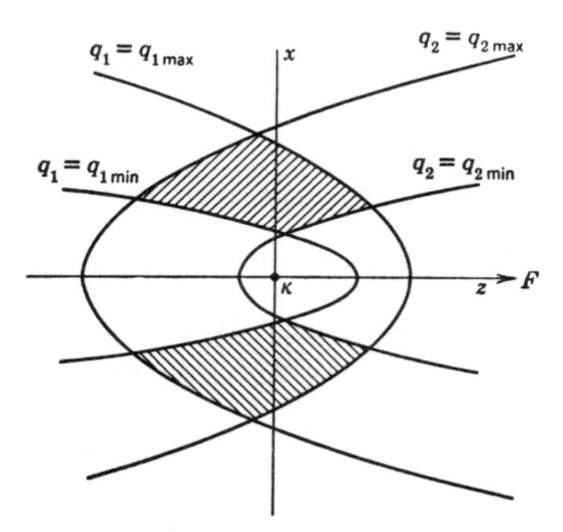

Fig. 64-3. Cross section of region in which particle may move ($J_3 \neq 0$).

origin it is necessary that $\alpha_1 > 0$, and, if the parallel field is not too large ($F < E^2/\alpha_1$), then q_1 is also bound between limits $0 \leq q_1 \leq q_{1\max}$. The motion is then confined to the region bounded between the parabolas $q_1 = q_{1\max}$ and $q_2 = q_{2\max}$.

GENERAL REFERENCES

M. Born, *The Mechanics of the Atom.* Bell, London (1927).
R. P. Feynman, *Rev. Mod. Phys.*, 20, 367 (1948).
E. T. Whittaker, *Analytical Dynamics.* Dover, New York (1944).

EXERCISES

1. A homogeneous bar is free to slide on a smooth vertical plane which is constrained to rotàte with constant angular velocity ω about a vertical axis fixed in the plane. Describe the motion (*a*) by using Lagrange's equations, (*b*) by using the method of Hamilton and Jacobi.

2. Write down the Hamilton-Jacobi equation for the motion of a charged particle in a uniform magnetic field.

3. Show that the $H - J$ equation for a particle moving in a lamellar field of force may be written

$$i\hbar \frac{\partial \psi}{\partial t} = -\frac{\hbar^2}{2m} \nabla^2 \psi + V\psi + \underline{\frac{i\hbar}{2m} \psi \nabla^2 \phi}$$

where

$$\psi = e^{i\phi/\hbar}$$

and $\hbar$ is a constant of dimensions of action. If V is independent of t, this becomes

$$\nabla^2 u + \frac{2m}{\hbar^2}(E - V)u = \underline{\frac{i}{\hbar} u \nabla^2 S}$$

with

$$\psi = e^{-iEt/\hbar}u(\mathbf{r})$$

If $\hbar = h/2\pi$, these equations, with the underlined nonlinear terms absent, become formally Schrödinger's fundamental equations of wave mechanics.

4. Evaluate the integrals (64.1) and (64.2) by contour integration.

5. Develop the problem of two centers of gravitation by using Lagrange's equations.

6. Write down the action variables J_1, J_2, J_3 for the motion of a particle of mass m in a central field upon which is superimposed a uniform parallel field. If the central field is a Coulomb attraction ($\kappa = -Ze^2$) and the uniform field is an electric field $E(F = -eE)$, show that the energy of the particle is, to terms of first order in E,

$$W = \frac{-2\pi^2 mZ^2e^4}{(J_1 + J_2 + J_3)^2} - \frac{3E}{8\pi^2 mZe}(J_1 + J_2 + J_3)(J_2 - J_1)$$

(Born, *loc. cit.*, p. 212.)

7. Show directly that, if during the small oscillations of a simple pendulum the thread is slowly drawn up through the point of support, the action variable associated with the motion is invariant. Use this result to estimate the rate of change of the amplitude of the oscillations.

8. Show that for a particle of mass m moving in a circle with speed $v(\ll c)$ under an attractive force to a fixed center such that the potential energy at a radius r is $g(e^{-\kappa r}/r)$ (g, κ constants), the Bohr theory gives, for a fixed n, two states or none according as κ is less than or greater than

$$\kappa_0 = \frac{gm}{\hbar^2 n^2}(2 + \sqrt{5}) \exp\left[-\tfrac{1}{2}(1 + \sqrt{5})\right]$$

Show further that if $\kappa > (2mg)/(en^2\hbar^2)$, both states have positive energy, but that if $\kappa < (2mg)/(en^2\hbar^2)$ one state is bound and the other has positive energy. Discuss the case $\kappa \to 0$ and derive the condition on the parameters such that nonrelativistic theory is applicable.

9. A neutral particle of mass m moves in a circle under its gravitational attraction to a fixed center of mass M. Find the Bohr radii of the orbits and assuming ordinary densities determine the order of magnitude of the largest central mass M such that the inner orbit of a circulating neutron would lie outside it. Estimate the temperature above which such a system would be unstable.

(*Ans.* $\sim$100 grams, 10^{-13} °K!)

12 INFINITESIMAL CONTACT TRANSFORMATIONS

65. TRANSFORMATION THEORY OF CLASSICAL DYNAMICS

We have seen that the form of the equations of motion of a system is unchanged if the set of variables q, p describing the state of the system is replaced by another set of variables $\bar{q}$, $\bar{p}$ connected with the original set by contact transformation. The functional form of the Hamiltonian in general changes when this replacement is made, and the value of the Hamiltonian changes if the contact transformation depends upon the time.

Any function $f(q, p, t)$ has a value determined by the state of the system at a given time. Before we can call this a dynamic variable, however, we must prescribe how this function is to transform under contact transformation. The most convenient prescription is that the *value* of f must not change. Thus under the contact transformation $\bar{q}, \bar{p} \rightarrow q, p$ we say that if f is a dynamical variable

$$\begin{aligned} f(q, p, t) &= f(q(\bar{q}, \bar{p}, t), p(\bar{q}, \bar{p}, t), t) \\ &= \bar{f}(\bar{q}, \bar{p}, t) \end{aligned} \tag{65.1}$$

According to this prescription the Hamiltonian is not a dynamical variable when time-dependent transformations are considered. This is reasonable because we have seen that the Hamiltonian of any system can be made to vanish. The kinetic energy, however, is a dynamical variable.

If a system is described in a fixed coordinate system, the useful dynamical variables (kinetic energy, momentum and angular momentum, etc.) do not depend *explicitly* on the time. When this same system is described in a moving coordinate system these quantities may become explicitly time-dependent, hence we cannot omit the explicit time dependence of dynamical variables.

It is difficult to discuss the effects of finite contact transformations on dynamical variables, just as it is difficult to discuss the change in value of a function when the argument is changed by a finite amount. We therefore consider *infinitesimal contact transformations* (ICT's) in which the new variables q, p differ from the old variables $\bar{q}$, $\bar{p}$ by small quantities whose products are negligible in a limiting process. Thus we write

$$q_i = \bar{q}_i + \delta q_i, \qquad p_i = \bar{p}_i + \delta p_i \tag{65.2}$$

Inserting this into the defining equation for a contact transformation (58.6) yields

$$\begin{aligned} \delta p_i\, dq_i - \delta q_i\, dp_i + (\bar{H}(\bar{q}, \bar{p}, t) - H(q, p, t))\, dt \\ = d(\phi(q, \bar{q}, t) - p_i\, \delta q_i) \end{aligned} \tag{65.3}$$

Because only the differentials of the q_i, the p_i, and t appear on the left side of (65.3), the right side is necessarily a function of these variables, which we denote by $-\epsilon X(q, p, t)$. Here ϵ is a small parameter and the minus sign is introduced for future convenience. Thus

$$\delta q_i = \frac{\partial X(q, p, t)}{\partial p_i}, \qquad \delta p_i = -\epsilon \frac{\partial X(q, p, t)}{\partial q_i} \tag{65.4}$$

$$H(q, p, t) = \bar{H}(\bar{q}, \bar{p}, t) + \epsilon \frac{\partial X(q, p, t)}{\partial t} \tag{65.5}$$

$X(q, p, t)$ is called the *generator* of the ICT. It is a function of the coordinates and momenta of the system and of the time, hence of the state of the system at the time t. We may take $X(q, p, t)$ to be a dynamical variable. If q, p and $\bar{q}$, $\bar{p}$ are related by a finite contact transformation, we may say that the ICT's generated by $X(q, p, t)$ and by $\bar{X}(\bar{q}, \bar{p}, t)$ respectively are the same ICT described in two different coordinate systems. It will be shown immediately below that the z component of the angular momentum generates a rotation of coordinates about the z axis. In any other coordinate system this same component of angular momentum will generate a rotation about this same axis although the description of the rotation will be different. We examine some specific examples of ICT's.

Example 1. Let X be the Hamiltonian H of the system, and let us write $\epsilon = \delta t$. Then

$$\frac{\delta q_j}{\delta t} = \frac{\partial H}{\partial p_j}$$

$$\frac{\delta p_j}{\delta t} = -\frac{\partial H}{\partial q_j}$$

$$H = \bar{H} + \delta t \frac{\partial H}{\partial t} = \bar{H} + \delta H$$

Hence, from Hamilton's equations

$$\frac{\delta q_j}{\delta t} = \dot{q}_j, \qquad \frac{\delta p_j}{\delta t} = \dot{p}_j, \qquad \frac{\delta H}{\delta t} = \frac{\partial H}{\partial t} = \dot{H} \tag{65.6}$$

This therefore represents a transformation from the variables q_j, p_j at time t to the variables $q_j + \delta q_j$, $p_j + \delta p_j$ at time $t + \delta t$, and the motion of the system during this interval of time may be regarded as an ICT generated by the Hamiltonian. The motion of the system may then be regarded as a succession of ICT's. If H is independent of the time, the generator of the successive transformations remains constant; if H depends on the time, the new generator at time $t + \delta t$ is just $\bar{H} + \delta H$, i.e., the new Hamiltonian at that later time, since δH is defined as $(\partial H/\partial t)\,\delta t$.

Example 2. We may consider a set of particles described by rectangular cartesian coordinates x_ρ, y_ρ, z_ρ, and let

$$X = -\sum_\rho p_{x,\rho} = -p_x$$

Then,

$$\delta x_\rho = \epsilon \frac{\partial X}{\partial p_{x,\rho}} = -\epsilon$$

$$\delta y_\rho = 0, \qquad \delta z_\rho = 0, \qquad \delta \mathbf{p}_\rho = 0$$

or

$$\begin{aligned} x &= \bar{x} - \epsilon \\ y &= \bar{y}, \qquad z = \bar{z}, \qquad \mathbf{p} = \bar{\mathbf{p}} \end{aligned} \tag{65.7}$$

for each particle. This transformation, generated by the x component of the momentum, therefore corresponds to a translation of the coordinate system parallel to itself by a distance ϵ in the positive x direction.

As is to be expected, the momentum components are unchanged by this transformation.

Example 3. Let X be the negative of the orbital angular momentum of the system of particles about the z axis:

$$X = -l_z = -\sum_\rho (x_\rho p_{y,\rho} - y_\rho p_{x,\rho})$$

Then

$$\delta x_\rho = \epsilon \frac{\partial X}{\partial p_{x,\rho}} = \epsilon y_\rho, \qquad \delta p_{x,\rho} = -\epsilon \frac{\partial X}{\partial x_\rho} = \epsilon p_{y,\rho}$$

$$\delta y_\rho = \epsilon \frac{\partial X}{\partial p_{y,\rho}} = -\epsilon x_\rho, \qquad \delta p_{y,\rho} = -\epsilon \frac{\partial X}{\partial y_\rho} = -\epsilon p_{x,\rho}$$

$$\delta z_\rho = \epsilon \frac{\partial X}{\partial p_{z,\rho}} = 0, \qquad \delta p_{z,\rho} = -\epsilon \frac{\partial X}{\partial z_\rho} = 0$$

Thus, for each ρ, i.e., for the coordinates and momenta of each particle,

$$\begin{aligned} x &= \bar{x} + \epsilon y, & p_x &= \bar{p}_x + \epsilon p_y \\ y &= \bar{y} - \epsilon x, & p_y &= \bar{p}_y - \epsilon p_x \\ z &= \bar{z}, & p_z &= \bar{p}_z \end{aligned} \tag{65.8}$$

It is clear from Fig. 65-1 that the first set of equations (65.8) corresponds to the rotation of the axes through a small angle ϵ about the positive z axis. In addition, from the second set of equations (65.8), we see that the momenta p_x, p_y transform under an infinitesimal rotation about the z axis in the same way that the coordinates x, y transform.

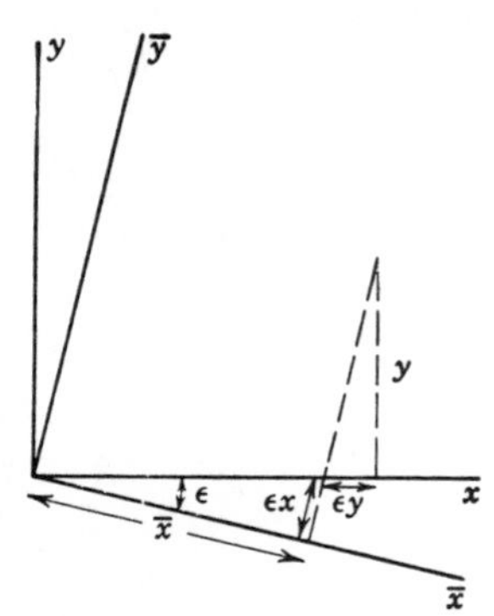

Fig. 65-1. Rotation about negative z axis through a small angle ϵ.

The transformation theory of dynamics has been developed starting from the more elementary theory based directly on Newton's laws of motion. The language of the transformation theory provides a very simple and general characterization of classical dynamic systems from which the forms of the equations of motion can be deduced.

We assume that the state of our system at a time t is specified by giving the values of f coordinates and f momenta at this time. It is a matter of indifference what set of coordinates and momenta is used, all sets connected by contact transformation being considered equivalent for this purpose. (The numerical values given to the coordinates and momenta do, of course, depend on the coordinates and momenta chosen.) The contact transformations involved here are time-independent because we are considering the state at the fixed time t.

If $q_i(t_1)$, $p_i(t_1)$ describe the state of the system at time t_1 and if $q_i(t_2)$, $p_i(t_2)$ do the same at time t_2, we assume that these two sets of variables are related by a contact transformation. Let the generator of this transformation be ϕ. Thus

$$p_i(t_2)\,dq_i(t_2) - p_i(t_1)\,dq_i(t_1) = d\phi \tag{65.9}$$

We investigate the consequences of this description of a dynamic system in terms of contact transformations.

The time interval $t_2 - t_1$ may be broken up into n equal intervals Δt such that $n\,\Delta t = t_2 - t_1$, and n may be made arbitrarily large. Thus we may treat Δt as small and neglect powers of Δt greater than the first. The second assumption above says that the state at time $t + \Delta t$ is related to the state at time t by contact transformation. We take this transformation to be infinitesimal. It is generated by

$$\begin{aligned}\phi(q(t+\Delta t), q(t); t+\Delta t, t) &= \phi(q(t) + \dot{q}\,\Delta t, q(t); t+\Delta t, t) \\ &= L(q(t), \dot{q}(t), t)\,\Delta t\end{aligned} \tag{65.10}$$

In the limit $n \to \infty$, then,

$$\phi(q(t_2), q(t_1); t_2, t_1) = \int_{t_1}^{t_2} L(q(t), \dot{q}(t), t)\,dt \tag{65.11}$$

because the generator of the resultant of several contact transformations is the sum of the generators of the separate contact transformations.

Equation (65.11) is similar in form to (57.1) and we may calculate the variation of ϕ as we did that of Φ in (57.6):

$$\Delta\phi = \int_{t_1}^{t_2} \left(\frac{\partial L}{\partial q_i} - \frac{d}{dt}\frac{\partial L}{\partial \dot{q}_i}\right)\delta q_i + \left[\left(-\frac{\partial L}{\partial \dot{q}_i}\dot{q}_i + L\right)\Delta t + \frac{\partial L}{\partial \dot{q}_i}\Delta q_i\right]_t^{t_2} \tag{65.12}$$

When $\Delta t_1 = \Delta t_2 = 0$, (65.12) should reproduce (65.9). It does this only when the coefficients of the δq_i vanish, i.e., when Lagrange's equations are satisfied, and when

$$p_i(t) = \frac{\partial L(q(t), \dot{q}(t), t)}{\partial \dot{q}_i(t)} \tag{65.13}$$

Equations (65.13) are a necessary adjunct to Lagrange's equations because the latter are second-order equations in the coordinates alone whereas the state is described by the coordinates and momenta. Equations (65.13) relate the velocities $\dot{q}_i$ to the q_i and p_i which define the state.

The expression for $\Delta\phi$ may be written

$$\Delta\phi = [p_i\,\Delta q_i - H\,\Delta t]_{t_1}^{t_2} \tag{65.14}$$

where $H = p_i\dot{q}_i - L$ as usual, and again Hamilton's equations follow from this definition and from Lagrange's equations as in Sec. 57.

The *form* of the equations of motion can thus be deduced from rather general statements about the transformations in phase space which give the state of a system at one time in terms of the state at another time. Nothing can be concluded about the form of the Lagrangian or Hamiltonian needed to describe a specific system except that (65.13) must be soluble for the velocities. The assignment of a specific Lagrangian to a system can be justified only by comparing the solution of the resulting equations of motion with the observed motion of the system.

66. POISSON BRACKETS

At the beginning of the last section it was pointed out that the functional form of a dynamical variable depends upon the set of coordinates and momenta used to describe the state of the system. We now determine what change in functional form is caused by an ICT.

Let the dynamical variable in question be $Y(q, p, t)$ and let $X(q, p, t)$ be the generator of the ICT so that

$$\delta q_i = \epsilon\frac{\partial X}{\partial p_i}, \qquad \delta p_i = -\epsilon\frac{\partial X}{\partial q_i} \tag{66.1}$$

Because the contact transformation is infinitesimal, the change in the functional form will also be small. This change is defined by

$$\bar{Y}(q, p, t) = Y(q, p, t) + \delta Y(q, p, t) \tag{66.2}$$

where the identity of the arguments is to be noted.

The ICT in question is that described in (65.2). Expanding $\bar{Y}(q, p, t)$ in a Taylor series, keeping only the linear terms and using (66.2), we see that

$$\delta Y(q, p, t) = \frac{\partial Y(q, p, t)}{\partial q_i}\,\delta q_i + \frac{\partial Y(q, p, t)}{\partial p_i}\,\delta p_i \tag{66.3}$$

Now using (66.1), we obtain

$$\delta Y(q, p, t) = \epsilon\left[\frac{\partial Y(q, p, t)}{\partial q_i}\frac{\partial X(q, p, t)}{\partial p_i} - \frac{\partial X(q, p, t)}{\partial q_i}\frac{\partial Y(q, p, t)}{\partial p_i}\right] \tag{66.4}$$

The quantity in parentheses in (66.4) is called the *Poisson bracket* of the two dynamical variables $Y(q, p, t)$ and $X(q, p, t)$. It is usually denoted by (Y, X). Thus

$$\delta Y = \epsilon(Y, X) \tag{66.5}$$

The invariance of the Poisson bracket of two dynamical variables under ICT is readily shown. Let the ICT be generated by $Z(q, p, t)$. From (65.1) it follows that

$$\frac{\partial \bar{f}}{\partial \bar{q}_i} = \frac{\partial f}{\partial q_j}\frac{\partial q_j}{\partial \bar{q}_i} + \frac{\partial f}{\partial p_j}\frac{\partial p_j}{\partial \bar{q}_i}$$

and from (65.2) and (65.4) that

$$\frac{\partial q_j}{\partial \bar{q}_i} = \delta_{ij} + \epsilon\frac{\partial^2 Z}{\partial p_i\,\partial q_j}, \qquad \frac{\partial p_j}{\partial \bar{q}_i} = -\epsilon\frac{\partial^2 Z}{\partial q_i\,\partial q_j}$$

Similar expressions exist for the derivatives with respect to $\bar{p}_i$. With f replaced by X and then by Y and neglecting terms of order ϵ^2, it then appears that

$$\frac{\partial \bar{Y}}{\partial \bar{q}_i}\frac{\partial \bar{X}}{\partial \bar{p}_i} - \frac{\partial \bar{X}}{\partial \bar{q}_i}\frac{\partial \bar{Y}}{\partial \bar{p}_i} + \frac{\partial \bar{Y}}{\partial q_j}\frac{\partial \bar{X}}{\partial p_j} = \frac{\partial \bar{X}}{\partial q_j}\frac{\partial \bar{Y}}{\partial p_j} \tag{66.6}$$

The omission of the arguments in (66.5) is thus justified and *the Poisson bracket of two dynamical variables is a dynamical variable with a value independent of the variables used to describe the state of the system.*

Poisson brackets have the following properties:

$$\begin{aligned} (X, Y) &= -(Y, X), \text{ so that } (X, X) = 0 \\ (X, Y + Z) &= (X, Y) + (X, Z) \\ (X, YZ) &= Y(X, Z) + Z(X, Y) \\ ((X, Y), Z) + ((Z, X), Y) &+ ((Y, Z), X) = 0 \end{aligned} \tag{66.7}$$

The last equation is known as *Jacobi's identity*. Some consequences of it are developed in Sec. 67.

From the definition of Poisson brackets it follows that

$$
\begin{aligned}
(q_i, q_j) &= 0 \\
(p_i, p_j) &= 0 \qquad (66.8) \\
(q_i, p_j) &= \delta_{ij}
\end{aligned}
$$

and also that

$$
\begin{aligned}
(q_i, X) &= \frac{\partial X(q, p)}{\partial p_i} \\
(p_i, X) &= -\frac{\partial X(q, p)}{\partial q_i} \qquad (66.9)
\end{aligned}
$$

If X is an algebraic function of some set of q's and p's, it is possible to find its derivative with respect to any q or p by purely algebraic means using (66.7), (66.8), and (66.9).

Nowhere in this section has the time been varied. The δY that has been found is that due to the change in the canonical coordinates and momenta produced by an ICT, these new coordinates and momenta being considered a new description of the same state. In Example 1 of Sec. 65 it was shown that the ICT generated by the Hamiltonian of the system could be regarded as a transformation from the coordinates and momenta at time t to those at time $t + dt$ in the *same* coordinate system in phase space. Now (66.2) no longer holds because (a) q_i, p_i do not describe the same state as $\bar{q}_i$, $\bar{p}_i$, and (b) the two times are not equal. We can regard the transition from $\bar{q}_i$, $\bar{p}_i$, t to q_i, p_i, $t + dt$ as taking place in two steps. In the first step the canonical coordinates and momenta are changed to their new values; in the second step the time is changed. Our analysis holds for the first step and we must add the effect of the second. If $dY(q, p, t)$ is the entire change in the value of Y, then

$$
dY(q, p, t) = \delta Y(q, p, t) + \frac{\partial Y(q, p, t)}{\partial t} dt
$$

or

$$
dY = [(Y, H) + \dot{Y}]\, dt \qquad (66.10)
$$

Here the symbol $\dot{Y}$ stands for that function of q, p, t which has the value of the partial time derivative of the value of Y.

Equation (66.10) allows the equations of motion to be written in an elegant form. Let Y be one of the coordinates, i.e., $Y = q_i$. Then clearly $\dot{Y} = 0$ (naturally this means not that $\dot{q}_i = 0$ but that

$\partial q_i/\partial t = 0$) and we have

$$\frac{dq_i}{dt} = (q_i, H) \tag{66.11}$$

Similarly, with $Y(q, p, t) = p_i$, we obtain

$$\frac{dp_i}{dt} = (p_i, H) \tag{66.12}$$

As an example, let H be the Hamiltonian of a charged particle in an electromagnetic field,

$$\begin{aligned} H &= \frac{1}{2m}\left(p_i - \frac{e}{c}A_i\right)\left(p_i - \frac{e}{c}A_i\right) + e\phi \\ &= \frac{1}{2m}P_iP_i + e\phi \end{aligned} \tag{66.13}$$

where P_i is the kinetic momentum introduced in Sec. 55. The Poisson bracket of any two functions of the coordinates only vanishes, so

$$(x_j, H) = \frac{1}{m}\left(p_j - \frac{e}{c}A_j\right) = \dot{x}_j \tag{66.14}$$

in accordance with (66.11). Similarly,

$$\begin{aligned} (p_j, H) &= -\frac{e}{cm}\left(p_i - \frac{e}{c}A_i\right)(p_j, A_i) + e(p_j, \phi) \\ &= \frac{e}{c}\dot{x}_i\,\partial_jA_i - e\,\partial_j\phi \\ &= \dot{p}_j \end{aligned} \tag{66.15}$$

agreeing with (66.12)

The Poisson bracket of two kinetic momenta is

$$\begin{aligned} (P_i, P_j) &= m^2(v_i, v_j) = -\frac{e}{c}[(A_i, p_j) + (p_i, A_j)] \\ &= \frac{e}{c}[\partial_iA_j - \partial_jA_i] \end{aligned} \tag{66.16}$$

which does vanish if A is a constant (or is a gradient) so that the magnetic field is zero. This may be written as

$$(v_x, v_y) = \frac{e}{m^2c}B_z, \text{ etc.} \tag{66.17}$$

From (66.10) we can also draw the following conclusion. If Y is a dynamical variable which does not depend explicitly on the time, so that $\dot{Y} = 0$, then Y is a constant of the motion if and only if its Poisson bracket with the Hamiltonian vanishes. We may write (66.5) in this case with $X = H$ as

$$\delta H = -(Y, H) \tag{66.18}$$

by use of the first of equations (66.7). Thus we may say that $Y(q, p)$ *is a constant of the motion if and only if the form of the Hamiltonian is invariant under the ICT generated by Y.*

This statement provides a way of connecting the symmetry of a system directly to integrals of the motion of the system. Here we apply this to Examples 2 and 3 of Sec. 65. In Sec. 67 we shall use this method to derive some less obvious results.

Example 2. Let $X(q, p, t) = -p_x = -\sum_\rho p_{x,\rho}$ so that (65.7)

$$x = \bar{x} - \epsilon$$

$$y = \bar{y}, \qquad z = \bar{z}, \qquad \mathbf{p} = \bar{\mathbf{p}}$$

for each particle ρ. If now H is of the form

$$H = T(p_\rho) + V(r_{\rho\sigma})$$

with

$$r_{\rho\sigma}{}^2 = (x_\rho - x_\sigma)^2 + (y_\rho - y_\sigma)^2 + (z_\rho - z_\sigma)^2$$

the system possesses a potential energy which depends only on the distances $r_{\rho\sigma}$ between the various particles. But these distances may also be written

$$r_{\rho\sigma}{}^2 = (\bar{x}_\rho - \bar{x}_\sigma)^2 + (\bar{y}_\rho - \bar{y}_\sigma)^2 + (\bar{z}_\rho - \bar{z}_\sigma)^2$$

so that $r_{\rho\sigma}$, hence V, is the same function of the barred variables as it is of the unbarred. Since p_ρ is invariant under the transformation, we have

$$\bar{H}(q, p) = H(q, p)$$

the Hamiltonian being the same function of $\bar{q}$, $\bar{p}$, as it is of q, p. Thus $\delta H = 0$, and p_x, the total momentum in the x direction, is a constant of the motion. A similar result applies, of course, for p_y, p_z. Thus, if a system of particles is acted upon only by internal forces which depend on the separations of the particles, the total momentum of the system is constant. This result is, of course, quite elementary, and it has been established in Sec. 13.

In general, any momentum p_k is a constant if H is invariant under the ICT generated by p_k:

$$\delta q_j = \epsilon \delta_{jk}$$
$$\delta p_j = 0$$

i.e., under infinitesimal change of the coordinate q_k. Clearly, if H does not involve q_k explicitly it is so invariant, and the generator of the transformation, i.e., the momentum p_k conjugate to the cyclic coordinate q_k, is, as has been shown previously, a constant of the motion.

Example 3. With $X(q, p, t) = -l_z$, the new coordinates and momenta are given by (65.8). Hence

$$r_{\rho\sigma}^2 = [(\bar{x}_\rho + \epsilon y_\rho) - (\bar{x}_\sigma + \epsilon y_\sigma)]^2 + [(\bar{y}_\rho - \epsilon x_\rho) - (\bar{y}_\sigma - \epsilon x_\sigma)]^2 + (\bar{z}_\rho - \bar{z}_\sigma)^2$$
$$= (\bar{x}_\rho - \bar{x}_\sigma)^2 + (\bar{y}_\rho - \bar{y}_\sigma)^2 + (\bar{z}_\rho - \bar{z}_\sigma)^2$$

to terms of first order in ϵ. Hence, as before, $r_{\rho\sigma}$, and thus any function $V(r_{\rho\sigma})$, is the same function of the $\bar{x}_\rho, \bar{x}_\sigma, \cdots$, as it is of $x_\rho, x_\sigma, \cdots$ Similarly,

$$p_\rho^2 = p_{x,\rho}^2 + p_{y,\rho}^2 + p_{z,\rho}^2$$
$$= \bar{p}_{x,\rho}^2 + \bar{p}_{y,\rho}^2 + \bar{p}_{z,\rho}^2$$

to terms of order ϵ, so that any kinetic energy function $T(p_\rho)$ is the same function of the new variables as it is of the old. Hence for any Hamiltonian of the form

$$H = T(p_\rho) + V(r_{\rho\sigma})$$

again $\delta H = 0$, so that the angular momentum about any axis is a constant of the motion.

For the case of a single particle referred to rectangular cartesian coordinates, the definitions (66.4) and (66.5) of the Poisson bracket of two dynamical variables become

$$(66.19) \quad (X, Y) \equiv \frac{\partial X(r, p, t)}{\partial \mathbf{r}} \cdot \frac{\partial Y(r, p, t)}{\partial \mathbf{p}} - \frac{\partial Y(r, p, t)}{\partial \mathbf{r}} \cdot \frac{\partial X(r, p, t)}{\partial \mathbf{p}}$$

The Poisson bracket of two components of orbital angular momentum $\boldsymbol{l} = \mathbf{r} \times \mathbf{p}$ follows from this definition as

$$(l_y, l_z) = l_x$$

or in general

$$(66.20) \qquad (l_i, l_j) = \epsilon_{ijk} l_k$$

Similarly,

$$(p_i, l_j) = \epsilon_{ijk} p_k \tag{66.21}$$

may be derived from (66.19) or from (66.9), and if l^2 is the square of the total orbital angular momentum

$$\begin{aligned} (l_x, l^2) &= (l_y, l^2) = (l_z, l^2) = 0 \\ (\mathbf{p}, l^2) &= 2\mathbf{l} \times \mathbf{p} \end{aligned} \tag{66.22}$$

From (66.8), two dynamical variables X and Y can be simultaneously introduced as coordinates or momenta only if their Poisson bracket vanishes or if it is unity. In the latter case X and Y form a canonically conjugate pair. Thus from (66.20) no two components of the angular momentum of a particle or a rigid body can be taken as canonical momenta, for in general the Poisson bracket of two angular momenta does not vanish.

One component of angular momentum may be used as a canonical momentum, and indeed in the discussion of central motion [Sec. 64(a)] the z component of angular momentum $l_z = p_\phi$ was so used, along with its canonically conjugate coordinate, the azimuthal angle ϕ. The momentum p_θ introduced in that connection is another canonical momentum, but it is not one of the other components l_x, l_y.

Consider a set of dynamical variables $Y_j(q, p, t)$ which constitute the components of a vector in configuration space. The index j refers to the coordinate curve along with q_j is measured. The vectorial character of the $Y_j(q, p, t)$ specifies the way in which the components transform under a coordinate transformation. They must transform in the same way that the coordinate differentials transform.* A coordinate transformation is a special kind of contact transformation, and we have established how dynamical variables transform under ICT. It is interesting to combine these two approaches to the transformation of vectors.

An infinitesimal coordinate transformation is generated by a dynamical variable $X(q, p)$ linear in the momenta. Then, according to (65.4),

$$\bar{q}_i = q_i - \epsilon \frac{\partial X(q, p)}{\partial p_i}$$

* We are using only contravariant components here. See Appendix II for details of vector transformations.

so that

$$d\bar{q}_i = \left(\delta_{ij} - \epsilon \frac{\partial^2 X(q, p)}{\partial p_i \, \partial q_j}\right) dq_j \tag{66.23}$$

If $Y_j(\bar{q}, \bar{p}, t)$ are the components in the barred system of the vector whose components in the unbarred system are $Y_j(q, p, t)$, then

$$\begin{aligned} Y_i(\bar{q}, \bar{p}, t) &= \left(\delta_{ij} - \epsilon \frac{\partial^2 X(q, p)}{\partial p_i \, \partial q_j}\right) Y_j(q, p, t) \\ &= Y_i(q, p, t) - \epsilon \frac{\partial^2 X(q, p)}{\partial p_i \, \partial q_j} Y_j(q, p, t) \end{aligned} \tag{66.24}$$

This is the geometric approach.

The ICT generated by $X(q, p)$ causes a change in the functional form of any component $Y_j(q, p)$. According to (66.1), (66.2), and (66.5)

$$Y_i(\bar{q}, \bar{p}, t) = Y_i(q, p, t) - \epsilon(Y_i, X) \tag{66.25}$$

Now if the old and new coordinates are of the same kind and if the old and new components depend on their arguments in the same way so that their physical meanings are similar, then

$$Y_i(\bar{q}, \bar{p}, t) = Y_i(\bar{q}, \bar{p}, t) \tag{66.26}$$

Under these circumstances and only then we have, by comparing (66.24) and (66.25),

$$(Y_i, X) = Y_j \frac{\partial^2 X}{\partial q_j \, \partial p_i} \tag{66.27}$$

This result is especially simple when the system consists of a single particle referred to cartesian coordinates and $X(q, p)$ is taken to be a component of angular momentum:

$$X(q, p) = \epsilon_{kmn} x_m p_n = l_k \tag{66.28}$$

Then

$$\frac{\partial^2 X}{\partial q_j \, \partial p_i} = \epsilon_{kji} \tag{66.29}$$

which is a constant. Then, if the Y_i are the components of a vector satisfying (66.26),

$$(Y_i, l_k) = Y_j \epsilon_{jik} \tag{66.30}$$

For example,

$$(l_y, l_z) = l_x, \qquad (l_x, l_z) = -l_y$$

since the angular momentum does satisfy (66.26).

This theorem is more useful when applied to a system of particles. Let the Y_i be the three-dimensional cartesian components of a vector which depends additively on vectors defined for each particle of the system, and let X be the total k component of angular momentum. Then we again obtain (66.30). Examples of such vectors are the total linear momentum of the system and the magnetic moment of the system due to the motion of charged particles. A vector which does not satisfy (66.26) is, for example, the magnetic field **B** impressed on the system by external currents. This does satisfy (66.26) in the special case where **B** is constant and l_k is the component of $\boldsymbol{l}$ in the direction of **B**.

67. JACOBI'S IDENTITY

If the total rate of change dX/dt of a variable X along the path of a dynamic system is zero, then X is called a constant of the motion. If now $X(q_j, p_j)$ is such a constant, it follows that

$$(X, H) = 0 \tag{67.1}$$

i.e., the Poisson bracket of any constant of the motion with the Hamiltonian vanishes. From Jacobi's identity (66.7), with $Z = H$:

$$((X, Y), H) + ((Y, H), X) + ((H, X), Y) = 0$$

It is therefore possible from any two constants of the motion X, Y to find a third constant (X, Y), for, since $(X, H) = 0$, $(Y, H) = 0$, it follows that

$$((X, Y), H) = 0 \tag{67.2}$$

Thus the Poisson bracket of two constants of the motion is itself a constant of the motion.

In many instances the constant so found is quite trivial; for instance, given that $\dot{p}_k = 0$, $\dot{p}_l = 0$ for particular values of k and l, (67.2) merely shows, from the fundamental bracket expressions (66.8), that zero is also a constant of the motion.

Sometimes, however, the third constant of the motion derived from Jacobi's identity is not immediately obvious. Thus, from (66.17), when a charged particle moves in an electromagnetic field, the Poisson bracket of the velocities in two perpendicular directions is proportional to the magnetic field strength in the third orthogonal direction. We notice that this relation couples the properties of the particle, on the one hand, with properties of the field, on the other, the coupling depending on the magnitude of the electric charge. If, then, v_x and v_y are constants of the motion, so is $B_z(x, y, z, t)$.

This result may be seen from the equations of motion, for

$$m\dot{v}_x = 0 = e\left[E_x + \frac{1}{c}(v_yB_z - v_zB_y)\right]$$

$$m\dot{v}_y = 0 = e\left[E_y + \frac{1}{c}(v_zB_x - v_xB_z)\right]$$

Hence

$$\partial_y E_x + \frac{1}{c} v_y \,\partial_y B_z - \frac{1}{c} v_z \,\partial_y B_y = \partial_x E_y + \frac{1}{c} v_z \,\partial_x B_x - \frac{1}{c} v_x \,\partial_x B_z$$

so that, using Maxwell's equations,

$$\partial_x E_y - \partial_y E_x + \frac{1}{c}\frac{\partial B_z}{\partial t} = 0, \qquad \nabla \cdot \mathbf{B} = 0$$

we have

$$\frac{\partial B_z}{\partial t} + v_x \,\partial_x B_z + v_y \,\partial_y B_z + v_z \,\partial_z B_z = 0$$

Hence, since B_z is a function of x, y, z, t,

$$\frac{dB_z}{dt} = 0$$

and B_z is constant along the path of the particle. Similar results hold, of course, for cyclic permutation of x, y, z.

From the Poisson bracket relations (66.19) and (66.20) involving the components of the orbital angular momentum of a particle, application of Jacobi's identity leads to the following results:

1. If p_y, l_z are constants of the motion, so is p_x. (See Fig. 67-1.)
2. If l_y, l_z are constants of the motion, so is l_x.

In result 1, the force on the particle must have no component in the y direction and its line of action must intersect the z axis. Unless the particle is moving in the x-z plane, the force must therefore be parallel to the z axis. Thus it has no component in the x direction, so p_x is a constant of the motion. If the particle is constrained to move in the x-z plane, the system has one less degree of freedom. In this case the Poisson bracket relations of Sec. 66 are no longer applicable, since then $l_x \equiv l_z \equiv 0$, $p_y \equiv 0$, $y \equiv 0$, $l_y = zp_x - xp_z$. Then

$$(p_x, l_y) = p_z$$

$$(p_z, l_y) = -p_x$$

as before, but we can say nothing about Poisson bracket expressions of the form (p_y, l_z), etc. In this case, if p_x and l_y are constants, so is p_z, for then the force has no component in the x direction, and it must pass through the origin. Hence, unless the particle is constrained to move along the z axis, the force on it can have no component in the z direction, and p_z is a constant. If the particle is constrained to move along the z axis, we have a further reduction of the number of degrees of freedom and none of the above relations are applicable.

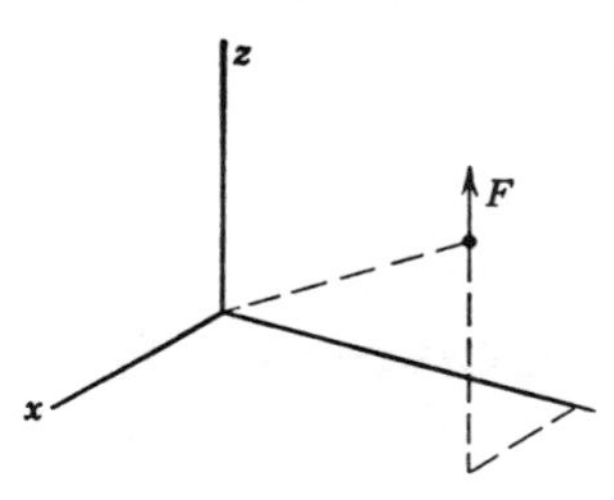

Fig. 67-1. If p_y, l_z are constants of the motion, so is p_z.

In result 2, l_y and l_z are constants, so that the force on the particle must intersect the y and z axes. Thus, unless the particle is in the $y - z$ plane the force on it must also intersect the x axis, so that l_x is also a constant of the motion. If the particle is constrained to move in the $y - z$ plane, it has one less degree of freedom as before, and the Poisson bracket relation is no longer applicable.

68. POISSON BRACKETS IN QUANTUM MECHANICS

In the previous section, and indeed throughout this book, we have mentioned special cases of problems in which, for example, a coordinate and the corresponding momentum were both zero. In classical mechanics we do not inquire into the consistency of such conditions; this would require a closer examination of the physical processes involved when an observation on a system is made. In quantum mechanics such questions have been partially answered, and there it emerges that such special cases have no meaning, for it is physically impossible to measure both x and p_x simultaneously with arbitrary precision and, in particular, to specify that for a given system both of these variables are zero.

The chief interest of Poisson brackets lies in the fact that, if in classical mechanics the Poisson bracket of two dynamical variables vanishes, then according to quantum mechanics these variables may be simultaneously observed. Thus, for example, a coordinate q_k and any momentum p_l ($l \neq k$) other than the one conjugate to q_k may be observed at the same time with arbitrary precision. If, on the other hand, the Poisson bracket of two variables does not vanish (e.g., v_x and v_y in the presence of a field, or l_x and l_y), these two variables may not be observed simultaneously. In quantum mechanics, dynamical variables are described by operators, which may be represented by matrices. The quantum Poisson bracket of two operators X and Y is then defined by

$$(X, Y) = -\frac{i}{\hbar}(XY - YX) \tag{68.1}$$

Hence, if the matrices X, Y commute, the Poisson bracket of X and Y vanishes. The condition that quantum mechanics should include classical mechanics as a valid approximation for all but atomic phenomena is that the form of any Poisson bracket relation in classical theory should remain unchanged in quantum theory when the definition (68.1) is adopted instead of (66.4), (66.5). Hence, if the Poisson bracket of two variables in classical mechanics is zero, the operators representing these variables in quantum theory commute. This is interpreted physically as implying that these variables may be simultaneously observed.

EXERCISES

1. Verify Jacobi's identity [last of equations (66.7)].

2. Show that $(x_i, l_j) = \epsilon_{ijk}x_k$, either from the definition (66.5) or more simply from (66.9).

3. A particle is free to slide under gravity in a fixed, smooth, curved tube which lies in a vertical plane. Are equations (66.20) and (66.21) applicable to its motion (*a*) when the origin lies in that plane; (*b*) when it lies outside of the plane of the tube?

4. When applied to the coordinates and momenta of a particle, the quantum Poisson bracket definition (68.1) becomes

$$(x_j, p_k) = -\frac{i}{\hbar}(x_j p_k - p_k x_j)$$

Show that, if p_k is the operator $-i\hbar(\partial/\partial x_k)$, this yields

$$(x_j, p_k) = \delta_{jk}$$
$$(x_j, x_k) = 0 = (p_j, p_k)$$

which are formally identical with (66.8).

13 FURTHER DEVELOPMENT OF TRANSFORMATION THEORY

69. NOTATION

The state of a dynamic system is defined by the values of the coordinates and momenta of the system. The coordinates and momenta enter the theory in a very symmetrical way, the difference in sign between the two sets of canonical equations alone distinguishing between them. It is convenient to introduce a notation which reflects this symmetry. Let us denote the variables q_1, p_1, q_2, p_2, $\cdots$, q_f, p_f by z_1, z_2, $\cdots$, z_{2f}. Any dynamical variable can be written as a function of the z's and the time. Greek indices will be used to label the z's and will have the range from 1 to $2f$.

Hamilton's equations may now be written in the form

$$\dot{z}_\alpha = A_{\alpha\beta} \frac{\partial H(z, t)}{\partial z_\beta} \tag{69.1}$$

where $A_{\alpha\beta}$ is the $2f \times 2f$ matrix

$$A = \begin{pmatrix} 0 & 1 & 0 & 0 & \cdots \\ -1 & 0 & 0 & 0 & \\ 0 & 0 & 0 & 1 & \\ 0 & 0 & -1 & 0 & \\ \cdot & & & & \\ \cdot & & & & \\ \cdot & & & & \end{pmatrix} \tag{69.2}$$

A couples the canonically conjugate variables with the characteristic sign difference. From (69.2) we see that

$$A^2 = -1, \qquad A^{\mathsf{T}} = -A \tag{69.3}$$

where, as in Appendix II, the transpose is denoted by the symbol ^{T}. The equations describing any ICT have the form of (69.1)

$$\delta z_\alpha = \epsilon A_{\alpha\beta} \frac{\partial X(z, t)}{\partial z_\beta} \tag{69.4}$$

if $\epsilon X(z, t)$ is the generator of the transformation. We shall use matrix notation and not write out all indices explicitly. Thus (69.4) will be written

$$\delta z = \epsilon A \frac{\partial X(z, t)}{\partial z} \tag{69.5}$$

It is interesting to note that the equations of motion (69.1) can be regarded as Lagrange equations resulting from a Lagrangian in phase space,

$$L(z, \dot{z}, t) = \tfrac{1}{2}\dot{z}Az - H(z, t) \tag{69.6}$$

This Lagrangian is linear in the "velocities" $\dot{z}$. It differs in value from the ordinary Lagrangian only by a total time derivative, for

$$\begin{aligned} L(z, \dot{z}, t) &= \tfrac{1}{2}\dot{q}_i p_i - \tfrac{1}{2}\dot{p}_i q_i - H(z, t) \\ &= p_i \dot{q}_i - H(z, t) - \frac{1}{2}\frac{d}{dt}(p_i q_i) \end{aligned}$$

The Poisson bracket of two dynamical variables becomes simply

$$(X, Y) = \frac{\partial X}{\partial z} A \frac{\partial Y}{\partial z} \tag{69.7}$$

70. INTEGRAL INVARIANTS AND LIOUVILLE'S THEOREM

A contact transformation is a transformation among the z's which leaves the form of the equations of motion invariant. Let the $\bar{z}$'s be the new coordinates, which are functions of the old coordinates and the time. We define the matrix M by

$$M_{\alpha\beta} = \frac{\partial \bar{z}_\alpha}{\partial z_\beta} \tag{70.1}$$

The determinant of M is the Jacobian of the transformation. If this is a contact transformation, the old and new Lagrangians differ by a total time derivative. Thus

$$L(z, \dot{z}, t) - \bar{L}(\bar{z}, \dot{\bar{z}}, t) = \frac{d\chi}{dt}$$

Using (69.6) for L, we obtain

$$\tfrac{1}{2}(\dot{z}Az - \dot{\bar{z}}A\bar{z})\, dt - (H(z, t) - \bar{H}(\bar{z}, t))\, dt = d\chi$$

Now

$$\dot{\bar{z}}_\alpha = \frac{\partial \bar{z}_\alpha}{\partial z_\beta} \dot{z}_\beta + \frac{\partial \bar{z}}{\partial t}$$

or, in matrix notation,

$$\dot{\bar{z}} = M\dot{z} + m$$

so that, using the antisymmetry of A, we have

$$(70.2) \quad -\tfrac{1}{2}(zA - \bar{z}AM)\, dz - (H(z, t) - \bar{H}(\bar{z}, t) - \tfrac{1}{2}\bar{z}Am)\, dt = d\chi$$

For this to be a perfect differential the derivatives with respect to z of the coefficients of dz must form a symmetric matrix; i.e.,

$$\frac{\partial}{\partial z_i}[z_j A_{jk} - \bar{z}_j A_{jl} M_{lk}]$$

is symmetric under interchange of i and k. Thus

$$\begin{aligned} A - M^{\mathsf{T}}AM &= (A - M^{\mathsf{T}}AM)^{\mathsf{T}} \\ &= A^{\mathsf{T}} - M^{\mathsf{T}}A^{\mathsf{T}}M \\ &= -(A - M^{\mathsf{T}}AM) \end{aligned}$$

Thus finally

$$(70.3) \qquad M^{\mathsf{T}}AM = A$$

is the condition for a contact transformation. This condition on M was derived by Poincaré and has been used by Courant and Snyder in discussing high energy accelerators. (See Chapter 17.) Matrices satisfying (70.3) are called *symplectic*. The other cross derivatives in (70.2) involve differentiation with respect to the time and serve to determine $\bar{H}(\bar{z}, t)$.

Taking the determinant of both sides of (70.3), we see that

$$(70.4) \qquad (\det M)^2 = 1, \qquad \det M = \pm 1$$

since $\det A = 1$ and $\det M^{\mathsf{T}} = \det M$. If M describes a contact transformation which can be continuously obtained from the identity,

then

$$\det M = 1 \tag{70.5}$$

However, any contact transformation can be obtained as a succession of infinitesimal transformations. Let the generator of a given contact transformation be $\psi'(q, \bar{p})$. Then

$$q_i\bar{p}_i + \lambda(\psi'(q, \bar{p}) - q_i\bar{p}_i)$$

is the generator of the identity transformation for $\lambda = 0$ and the generator of the given transformation for $\lambda = 1$, and so exhibits the given contact transformation as developing from identity. Finally, then, we may conclude that *the Jacobian determinant of any contact transformation is unity.* This is known as *Liouville's theorem.*

In an ordinary coordinate transformation the sign of the Jacobian is irrelevant, as it can be changed by merely interchanging the labels on two of the new coordinates, an operation which interchanges two columns of the determinant. This cannot be done in M, however, because the rows and columns in M are paired, each coordinate being tied to its conjugate momentum by the form of A which is constant. Thus here interchanges of columns must occur in pairs and no sign change results.

We may consider a set of points in phase space at a given time. These points then represent a set of possible initial conditions of the system, or equivalently may be taken as the representative points of a set of similar systems which do not interact with each other. Such a set of systems is called an *ensemble.* It is important in statistical mechanics where the time average of a function of the state of a system is assumed to be equal to the average of that same function over a suitable ensemble. Similar considerations arise in electron and ion optics where the behavior of a beam of particles in an external field is to be studied.

In all problems where a family of possible trajectories of a system is of interest rather than a particular trajectory, the idea of an *integral invariant* is important. Let

$$I_r = \int_{\Omega_r} F(z)\, dV_r \tag{70.6}$$

be an r-dimensional integral over some part of phase space Ω_r. If the value of I_r is unchanged as the points z of Ω_r move along their trajectories, then I_r is an integral invariant of the system. If Ω_r is a closed manifold (e.g., the surface of a hypersphere), the integral invariant is *relative;* otherwise it is *absolute.*

We consider two integral invariants which are of particular importance. Let $F(z)$ in (70.6) be unity and let us integrate over a finite f-dimensional volume V of phase space. Then

$$I_f = \int_{\Omega_f} dV_f = V \tag{70.7}$$

is an absolute integral invariant. To prove this we only need to use the result (70.5) that the Jacobian of a contact transformation is unity. Then if new coordinates $\bar{z}_\alpha$ obtained from the original z_α by contact transformation are introduced, $d\bar{V}_f = dV_f$. Further, the new limits of integration are just the old limits expressed in the new coordinates, so that the value of $I_f = V$ is invariant under contact transformation. The coordinates at any other time are obtained from the z_α by contact transformation, and so V is constant in time.

This result can be derived in a more illuminating way. We may consider the phase points of an ensemble distributed in phase space with a density $\rho(z)$. The velocity $\dot{z}$ of a phase point is given by (69.1). The number of phase points being constant, the density ρ must satisfy an equation of continuity,

$$\begin{aligned} 0 &= \frac{\partial \rho}{\partial t} + \frac{\partial}{\partial z_\alpha}(\rho \dot{z}_\alpha) \\ &= \frac{\partial \rho}{\partial t} + \frac{\partial \rho}{\partial z_\alpha}\dot{z}_\alpha + \rho \frac{\partial \dot{z}_\alpha}{\partial z_\alpha} \end{aligned} \tag{70.8}$$

From (69.1) it follows that

$$\frac{\partial \dot{z}_\alpha}{\partial z_\alpha} = 0 \tag{70.9}$$

because $A_{\alpha\beta}$ is antisymmetric and $\partial^2 H/(\partial z_\alpha\, \partial z_\beta)$ is symmetric. Hence

$$\frac{d\rho}{dt} = \frac{\partial \rho}{\partial t} + \frac{\partial \rho}{\partial z_\alpha}\dot{z}_\alpha = 0 \tag{70.10}$$

and the density of phase points does not change along the trajectory, establishing Liouville's theorem once again.

As the second integral invariant we take

$$I_1 = \oint p_j\, dq_j \tag{70.11}$$

where the integral is over any closed path in phase space. The invariance of this under arbitrary contact transformation follows immediately from the definition (58.6). For the path given in (70.11) there is no time integration, so that

$$\oint p_i \, dq_i - \oint \bar{p}_i \, d\bar{q}_i = 0 \tag{70.12}$$

Again because the coordinates and momenta at one time are connected with those at any other time by contact transformation, the invariance of I_1 follows. I_1 is a relative integral invariant. It is equivalent to an absolute invariant

$$I_2 = \int dp_j \, dq_j$$

over region Ω_2 bounded by the curve defining (70.11). The action integrals introduced in (63.4) are special cases of I_1 where the path of integration involves only one coordinate and its conjugate momentum, and consists of a period of this motion.

71. LAGRANGE BRACKETS

The Lagrange bracket of two dynamical variables is defined as follows:

$$[X, Y] = \frac{\partial q_j}{\partial X}\frac{\partial p_j}{\partial Y} - \frac{\partial q_j}{\partial Y}\frac{\partial p_j}{\partial X} \tag{71.1}$$

The partial differentiation with respect to X and Y implies that $2f - 2$ other dynamical variables have been defined and are to be kept constant during the differentiation. In the notation of the last section this may be written

$$[X, Y] = \frac{\partial z}{\partial X} A \frac{\partial z}{\partial Y} \tag{71.2}$$

The Lagrange bracket of two coordinates or momenta is

$$\begin{aligned}[z_\alpha, z_\beta] &= \frac{\partial z_\gamma}{\partial z_\alpha} A_{\gamma\delta} \frac{\partial z_\delta}{\partial z_\beta} \\ &= \delta_{\gamma\alpha} A_{\gamma\delta} \delta_{\delta\beta} \\ &= A_{\alpha\beta}\end{aligned}$$

or in detail

$$\begin{aligned}[q_i, q_j] &= 0 = [p_i, p_j] \\ [q_i, p_j] &= \delta_{ij}\end{aligned} \tag{71.3}$$

It is relatively easy to show directly that the Lagrange bracket is invariant under contact transformation. Alternatively we may proceed as follows:

If the X_α are a set of dynamical variables, then

$$
\begin{aligned}
(X_\alpha, X_\gamma)[X_\gamma, X_\beta] &= \frac{\partial X_\alpha}{\partial z} A \frac{\partial X_\gamma}{\partial z} \cdot \frac{\partial z}{\partial X_\gamma} A \frac{\partial z}{\partial X_\beta} \\
&= \frac{\partial X_\alpha}{\partial z} AIA \frac{\partial z}{\partial X_\beta} \\
&= -\frac{\partial X_\alpha}{\partial z} \frac{\partial z}{\partial X_\beta} \\
&= -\delta_{\alpha\beta}
\end{aligned}
$$

Thus the matrix of Lagrange brackets is, except for sign, the reciprocal of the matrix of the Poisson brackets. The invariance of one implies the invariance of the other. We saw in Sec. 66 that Poisson brackets are invariant under contact transformation, and so the invariance of the Lagrange bracket is established.

72. CHANGE OF INDEPENDENT VARIABLE

Up to this point we have treated the time t on quite a different basis from the space coordinates q_j, but it is possible to formulate classical dynamic theory without doing this. It has, of course, been natural to develop classical mechanics in a manner which treated the time variable quite differently from the other independent variables, and to give special attention to Hamiltonians which are independent of the time, because of the obvious distinction between time and space variables in our common experience. Nevertheless, the formulation given in this section is not of mere academic interest, for it leads to the generalization of the mechanics so far developed to the special theory of relativity (Chapter 16) in which it is seen that the distinction between space and time variables in fundamental physical theory is not so sharp as our experience might suggest.

Let us consider then the coordinates q_j and the time t as functions of some other parameter τ which labels the path of the system in configuration space, and let us denote differentiation with respect to τ by a prime. The variable τ could, for instance, be the distance along the path from some fixed point on it, but we do not specify τ beyond stating that there is a 1:1 correspondence between the set of values of τ and the set of points on the path of the system.

The functional Φ, whose variation vanishes according to Hamilton's principle, may be written

$$\Phi = \int_{t_1}^{t} (p_j \dot{q}_j - H)\, dt = \int_{\tau_1}^{\tau_2} (p_j q_j' - Ht')\, d\tau \tag{72.1}$$

This definition is symmetrical in the coordinates and the time except for the sign of the last term and the fact that H is given as a function of the q's, p's, and t. If we introduce the momentum h conjugate to t by

$$h + H(q, p, t) = 0 \tag{72.2}$$

then (72.1) becomes

$$\Phi = \int_{\tau_1}^{\tau_2} (p_j q_j' + ht')\, d\tau \tag{72.3}$$

When Φ as given by (72.3) is varied to obtain the equations of motion, the variation must be consistent with (72.2).

Equation (72.3) itself is entirely symmetrical in the $f + 1$ quantities p_j, h and in the $f + 1$ quantities q_j, t. The time is reduced to the role of a coordinate, the negative of the energy to that of its conjugate momentum. The only asymmetry lies in (72.2). If, now, (72.2) is solved for a momentum, p_1 say, so that the equation reads

$$p_1 + P(q_1 \cdots q_f, t, p_2 \cdots p_f, h) = 0 \tag{72.4}$$

p_1 can be eliminated from (72.3). This yields

$$\Phi = \int_{\tau_1}^{\tau_2} \left(\sum_{j=2}^{f} p_j q_j' + ht' - Pq_1' \right) d\tau \tag{72.5}$$

and there is no constraint on the variation applied to obtain the equations of motion. Now let $\tau = q_1$. Hamilton's principle applied to (72.5) will lead to the equations of motion in canonical form, but with q_1 as the independent variable and with P as the Hamiltonian.

$$\begin{aligned} \frac{dq_k}{dq_1} &= \frac{\partial P}{\partial p_k}, & \frac{\partial p_k}{\partial q_1} &= -\frac{\partial P}{\partial q_k} \\ \frac{dt}{dq_1} &= \frac{\partial P}{\partial h}, & \frac{dh}{dq_1} &= -\frac{\partial P}{\partial t} \end{aligned} \qquad (k = 2, 3, \cdots, f) \tag{72.6}$$

For this device to be useful, P must be a single-valued function. Otherwise the requirement that there be a 1 : 1 correspondence between values of $q_1 = \tau$ and points on the path in configuration space is not satisfied. The advantage to be gained by changing the independent variable is sometimes important. In conservative systems ($\partial H/\partial t = 0$) it permits us to reduce the number of degrees of freedom by one, for $\partial H/\partial t = 0$ implies $\partial P/\partial t = 0$, so that the first line of (72.6) is a set of equations for the trajectory without regard to when the system occupies a particular point on the trajectory. The second line of (72.6) furnishes this latter information. In the special theory of relativity it is advantageous to be able to treat the coordinates of a particle and the time on the same footing. This can be done by using the proper time as an independent variable.

The symmetry of (72.3) is emphasized if t, h are denoted by q_{f+1}, p_{f+1} respectively. Allowing μ to run from 1 to $f + 1$, we might write

$$\Phi = \int_{\tau_1}^{\tau_2} p_\mu q_\mu' \, d\tau$$

This is misleading because the constraint (72.2) must be maintained, but this can be incorporated in Φ with a Lagrange multiplier $\lambda(\tau)$, so we write

$$\Phi = \int_{\tau_1}^{\tau_2} (p_\mu q_\mu' + \lambda(H + h)) \, d\tau \tag{72.7}$$

$$= \int_{t_1}^{t_2} (p_j \dot{q}_j - H) \, dt$$

It is clear that these two functionals are identical if and only if

$$\lambda = -t' \tag{72.8}$$

We may define a Lagrangian $\mathcal{L}$ by

$$\mathcal{L} = Lt' = p_\mu q_\mu' - t'(H + h) \tag{72.9}$$

Lagrange's equations are then

$$\frac{d}{d\tau} \frac{\partial \mathcal{L}}{\partial q_\mu'} - \frac{\partial \mathcal{L}}{\partial q_\mu} = 0 \tag{72.10}$$

A Hamiltonian $\mathcal{H}$ may be defined by

$$\mathcal{H} = t'(H + h) \tag{72.11}$$

so that Φ may be written

$$\Phi = \int_{t_1}^{t_2} (p_\mu q_\mu' - \mathcal{H}) \, d\tau \tag{72.12}$$

From this follow the equations of motion in canonical form:

$$q_\mu' = \frac{\partial \mathcal{H}}{\partial p_\mu}, \qquad p_\mu' = -\frac{\partial \mathcal{H}}{\partial q_\mu} \tag{72.13}$$

73. EXTENDED CONTACT TRANSFORMATIONS

The developments of the last section have shown that it is possible to write the Lagrangian and Hamiltonian forms of the equations of motion with an arbitrary independent variable τ and with $f + 1$ coordinates q_μ, $f + 1$ momenta p_μ, subject to the constraint

$$p_{f+1} + H(q_1 \cdots q_f, p_1 \cdots p_f, q_{f+1}) = 0 \tag{73.1}$$

This constraint is satisfied identically by solutions of the equations of motion if the initial conditions satisfy it. It is natural, therefore, to introduce a contact transformation by

$$p_\mu \, dq_\mu - \bar{p}_\mu \, d\bar{q}_\mu = d\phi \qquad (\mu = 1, 2, \cdots, f + 1) \tag{73.2}$$

The only difference between this and (58.6) is that the time variable is no longer unchanged. Such a transformation is called an *extended* contact transformation.

Extended contact transformations suggest a generalization of Poisson brackets. Let us denote such a generalized bracket with braces, and define it by

$$\{X, Y\} = \frac{\partial X}{\partial q_\mu}\frac{\partial Y}{\partial p_\mu} - \frac{\partial Y}{\partial q_\mu}\frac{\partial X}{\partial p_\mu} \tag{73.3}$$

The generalized Poisson bracket of a dynamical variable X with the Hamiltonian $\mathcal{H}$ is given by

$$\begin{aligned} \{X, \mathcal{H}\} &= \frac{\partial X}{\partial q_\mu}\frac{dq_\mu}{d\tau} + \frac{\partial X}{\partial p_\mu}\frac{dp_\mu}{d\tau} \\ &= \frac{dX}{d\tau} \end{aligned} \tag{73.4}$$

Here there cannot arise a term $\partial X/\partial t$ because a dynamical variable may be a function only of the coordinates, momenta, and time.

Integral invariants may be defined in the generalized $(2f + 2)$-dimensional phase space, but as the motion is subject to the constraint (72.2) the representative point is restricted to a $(2f + 1)$-dimensional subspace of the q_μ, p_μ space. In this $2f + 1$ space, with q_j, p_j, t as coordinates, a point represents the state and time of the system. We define generalized absolute and relative integral invariants in this $2f + 1$ space in the same way as they were defined for $2f$ space in Sec. 70. We note, however, that an integral I_r over an r-dimensional domain of this space does not necessarily involve an integration over points associated with the same value of the time.

By analogy with (70.12), it is clear that

$$\oint p_\mu \, dq_\mu = \oint (p_j \, dq_j - H \, dt) \tag{73.5}$$

is a generalized relative integral invariant. If any closed curve is drawn in $2f + 1$ space, and dynamic paths of the system are drawn through each point of this curve, the expression (73.5) has the same value for any closed curve drawn on and around this tube of paths. If the closed curve is drawn in the $2f$-dimensional subspace t = const, the result reduces to the one obtained earlier.

We shall conclude by examining several applications of these transformations which also transform the variable t.

Example 1. Let

$$H = \frac{p^2}{2m} + \frac{1}{2} kq^2$$

the Hamiltonian for a simple harmonic oscillator in one dimension. We have seen that we may choose our units of mass and time so that $H = \frac{1}{2}(\bar{p}^2 + \bar{q}^2)$. This change of units may be represented by a contact transformation generated by

$$\psi = -\lambda_1 p\bar{q} - \lambda_2 h\bar{t}$$

where λ_1 and λ_2 are suitably chosen constants. For from equations analogous to (73.2) [cf. (59.4)], we have

$$\bar{p} = -\frac{\partial \psi}{\partial \bar{q}} = \lambda_1 p, \qquad q = -\frac{\partial \psi}{\partial p} = \lambda_1 \bar{q}$$

$$\bar{h} = -\frac{\partial \psi}{\partial \bar{t}} = \lambda_2 h, \qquad t = -\frac{\partial \psi}{\partial h} = \lambda_2 \bar{t}$$

so that the equation $H + h = 0$ becomes

$$\frac{1}{2}\left(\frac{\lambda_2 \bar{p}^2}{m\lambda_1^2} + k\lambda_1^2 \lambda_2 \bar{q}^2\right) + \bar{h} = 0$$

When this is compared with the transformed equation $\bar{H} + \bar{h} = 0$, we have, for $\lambda_1 = (mk)^{-1/4}$, $\lambda_2 = (m/k)^{1/2}$,

$$\bar{H} = \tfrac{1}{2}(\bar{p}^2 + \bar{q}^2)$$

Example 2. If the Hamiltonian does not involve t explicitly, then in the above formulation t is an ignorable coordinate, and its conjugate momentum h is a constant of the motion. Let us consider for simplicity such a Hamiltonian for a particle with one degree of freedom

$$H = H(q, p)$$

and introduce the contact transformation defined by

$$\begin{aligned} \psi' = \psi'(q_\mu, \bar{p}_\mu) &= \psi'(q, t, \bar{p}, \bar{h}) \\ &= q\bar{h} + t\bar{p} \end{aligned}$$

Then

$$p = \frac{\partial \psi'}{\partial q} = \bar{h}, \qquad h = \frac{\partial \psi'}{\partial t} = \bar{p}$$

$$\bar{q} = \frac{\partial \psi'}{\partial \bar{p}} = t, \qquad \bar{t} = \frac{\partial \psi'}{\partial \bar{h}} = q$$

This transformation therefore interchanges the roles of position and time and of energy and momentum. The equation

$$H(q, p) + h = 0$$

may be written

$$H(\bar{t}, \bar{h}) + \bar{p} = 0$$

and, if this equation is solved for $\bar{h}$, we write its solution thus:

$$-\bar{h} = \bar{H}(\bar{p}, \bar{t})$$

Hence $\bar{H}$ is the transformed Hamiltonian, and therefore $\bar{q}$ is an ignorable coordinate. The existence of the energy integral for systems whose Hamiltonians do not contain the time explicitly can thus formally be expressed as the existence of a momentum integral h, h being conjugate to the ignorable coordinate t. Thus, from (72.6),

$$\begin{aligned} \frac{\partial \bar{H}}{\partial \bar{q}} &= \frac{d\bar{p}}{d\bar{t}} = 0 \\ \frac{\partial \bar{H}}{\partial \bar{p}} &= \frac{d\bar{q}}{d\bar{t}} = -\frac{\partial \bar{H}}{\partial E} \end{aligned} \qquad (\bar{p} = h = \text{const} = -E)$$

where $H = E$.

Thus,

$$\bar{q} = q_0 - \int \frac{\partial \bar{H}}{\partial E}\, d\bar{t}$$

$$t = t_0 + \int \frac{\partial p}{\partial E}\, dq = t_0 + \frac{\partial S}{\partial E}$$

giving the time at which any point on the orbit is reached in the form of the equation (61.12) derived from Hamilton-Jacobi theory.

74. PERTURBATION THEORY

It was pointed out in the very beginning of this book that the systems dealt with in classical mechanics are highly idealized, many complicating features being ignored in the expectation that they are unimportant. It is frequently necessary, however, to take some complications into account, to improve the description of the system, with the result that there is no known way to obtain an exact solution of the less idealized problem. If the added features produce small effects, it may be possible to obtain the necessary solutions to a sufficient accuracy by approximation methods. The systematic methods of obtaining approximate solutions are the subject of perturbation theory. Classical perturbation theory reaches its highest development in celestial mechanics. Here the highly idealized model of the solar system consists of a set of noninteracting planets moving in the gravitational field of the sun. Complicating features to be introduced are the interactions of the planets among themselves, the existence of moons, etc. Our purpose here is to outline some methods that may be used in such problems and to give some elementary examples.

The idealized system is described by a Lagrangian $L_0(q, \dot{q}, t)$. The equations of motion derived from this Lagrangian are assumed to be solvable for arbitrary initial conditions. The real system is described by a Lagrangian $L(q, \dot{q}, t)$; the same coordinates appear in L_0 and L. We write

$$L = L_0 + U \tag{74.1}$$

and call U the perturbation. U is usually a function of the coordinates and the time only, but we do not restrict it in this way.

The system may be described in terms of a Hamiltonian rather than a Lagrangian. With $p_j = \partial L_0/\partial \dot{q}_j$, the unperturbed Hamiltonian is formed from L_0 in the usual way. If the perturbation U depends on the velocities, the perturbed momenta p_j' depend on the coordinates and velocities differently from the unperturbed momenta:

$$\begin{aligned} p_j' &= \frac{\partial L_0}{\partial \dot{q}_j} + \frac{\partial U}{\partial \dot{q}_j} \\ &= p_j + \delta p_j \end{aligned} \tag{74.2}$$

The perturbed Hamiltonian is then

$$H(q, p', t) = p_j' \dot{q}_j - L_0 - U$$

The perturbing Hamiltonian, V, is defined by

$$H(q, p, t) = H_0(q, p, t) + V(q, p, t) \tag{74.3}$$

There are two principal ways in which perturbation theory may be formulated. If the unperturbed system is a separable multiply-periodic system, we may inquire about the change in the periods of the motion produced by a perturbation. The quantum theoretical analog of this is the question of the amount by which energy levels are shifted by a perturbation. For a general system whose state of unperturbed motion is described by a set of initial conditions or constants of integration of the unperturbed Hamilton-Jacobi equation, we may inquire about the rate at which this previously constant state is made to change by a perturbation. The quantum analog of this is the calculation of transition probabilities produced by a perturbation. Following the nomenclature of quantum mechanics, we call these stationary state theory and time-dependent perturbation theory respectively.

Perturbation theory involves an expansion in a power series, the terms in the series becoming, it is hoped, rapidly smaller. In order to identify the various orders of magnitude it is useful to introduce a parameter λ into the Lagrangian, or Hamiltonian, writing it

$$L = L_0 + \lambda U, \qquad H = H_0 + \lambda V$$

$\lambda = 0$ gives the unperturbed motion; $\lambda = 1$ gives the perturbed motion. All quantities are expanded in a power series in λ, and the highest power of λ retained is called the order of the perturbation calculation.

75. STATIONARY STATE PERTURBATION THEORY

The unperturbed system is described by action and angle variables J_k^0, w_k^0, so that the Hamiltonian H_0 depends on the J_k^0 alone. The V, however, is a function of the angle variables as well, so that

$$H = H_0(J^0) + \lambda V(J^0, w^0) \tag{75.1}$$

is dependent on both sets of variables. A contact transformation which introduces new action variables J_k is needed, H being a function of the J_k only. Because V is small, the contact transformation is near identity and is most conveniently described by a generating function $S'(w^0, J)$. We assume

$$S'(w^0, J) = w_k{}^0 J_k + \lambda S_1(w^0, J) + \lambda^2 S_2(w^0, J) + \cdots \tag{75.2}$$

which reduces to the generator of the identity transformation for $\lambda = 0$.

According to (75.2),

$$J_k{}^0 = \frac{\partial S'}{\partial w_k{}^0} = J_k + \lambda \frac{\partial S_1}{\partial w_k{}^0} + \lambda^2 \frac{\partial S_2}{\partial w_k{}^0} + \cdots \tag{75.3}$$

This value for $J_k{}^0$ may be inserted in (75.1) and a power series expansion in λ may be made:

$$\begin{aligned} H(J) &= H_0(J) + \lambda\left(\frac{\partial H_0}{\partial J_k}\frac{\partial S_1}{\partial w_k{}^0} + V\right) + \lambda^2\left(\frac{\partial^2 H_0}{\partial J_k\,\partial J_l}\frac{\partial S_1}{\partial w_k{}^0}\frac{\partial S_1}{\partial w_l{}^0}\right. \\ &\qquad \left. + \frac{\partial H_0}{\partial J_k}\frac{\partial S_2}{\partial w_k{}^0} + \frac{\partial V}{\partial J_k}\frac{\partial S_1}{\partial w_k{}^0}\right) + \cdots \\ &= H_0(J) + \lambda H_1(J) + \lambda^2 H_2(J) + \cdots \end{aligned} \tag{75.4}$$

The new action variables J_k are to be so defined that all time-independent dynamical variables are periodic functions of the new angle variables w_k as they are of the old angle variables $w_k{}^0$. (The $w_k{}^0$ are not linear functions of the time in the presence of the perturbation, but the dependence of the dynamical variables on the $J_k{}^0$, $w_k{}^0$ is not affected by this.) Thus the new and old angle variables must differ by a periodic function with unit period. From (75.2)

$$w_k = w_k{}^0 + \lambda \frac{\partial S_1}{\partial J_k} + \lambda^2 \frac{\partial S_2}{\partial J_k} + \cdots \tag{75.5}$$

$\partial S_1/\partial J_k$, $\partial S_2/\partial J_k$, $\cdots$ are thus periodic functions of the $w_k{}^0$ with unit period, and they can be chosen to have vanishing mean values, as this merely changes w_k by an additive constant. Thus S_1, S_2 may be chosen to be periodic functions of the $w_k{}^0$ with unit period.

Coefficients of powers of λ in (75.4) may be equated separately. This yields a set of equations

$$H_1(J) = \frac{\partial H_0}{\partial J_k}\frac{\partial S_1}{\partial w_k{}^0} + V(J, w^0) \tag{75.6}$$

$$H_2(J) = \frac{\partial^2 H_0}{\partial J_k\,\partial J_l}\frac{\partial S_1}{\partial w_k{}^0}\frac{\partial S_1}{\partial w_l{}^0} + \frac{\partial H_0}{\partial J_k}\frac{\partial S_2}{\partial w_k{}^0} + \frac{\partial V}{\partial J_k}\frac{\partial S_1}{\partial w_k{}^0} \tag{75.7}$$

etc. In (75.6) the first term on the right has zero mean value as a function of the $w_k{}^0$ and the left side is independent of these variables.

Hence averaging (75.6) over one period of the $w_k{}^0$, we obtain

$$H_1(J) = \langle V(J, w^0) \rangle \tag{75.8}$$

where the average is over one period of the $w_k{}^0$ and is therefore equal to the *time average* over the *unperturbed* trajectory of the system.

We may now split V into its average value and an oscillating part W with zero mean:

$$V = \langle V \rangle + W \tag{75.9}$$

Equation (75.6) now reads

$$\frac{\partial H_0}{\partial J_k} \frac{\partial S_1}{\partial w_k{}^0} + W(J, w^0) = \nu_k{}^0 \frac{\partial S_1}{\partial w_k{}^0} + W(J, w^0) = 0 \tag{75.10}$$

This equation may be solved for S_1, thus giving the first term in the equation for the new action variables.

Looking now at (75.7), we see that the first term on the right is known, as is the last, and that the second term has vanishing mean value. Equation (75.7) therefore may be averaged to give

$$H_2(J) = \frac{\partial^2 H_0}{\partial J_k \, \partial J_l} \left\langle \frac{\partial S_1}{\partial w_k{}^0} \frac{\partial S_1}{\partial w_l{}^0} \right\rangle + \left\langle \frac{\partial W}{\partial J_k} \frac{\partial S_1}{\partial w_k{}^0} \right\rangle \tag{75.11}$$

S_2 is then obtained by solving

$$\nu_k{}^0 \frac{\partial S_2}{\partial w_k{}^0} + \frac{\partial^2 H_0}{\partial J_k \, \partial J_l} \left\{ \frac{\partial S_1}{\partial w_k{}^0} \frac{\partial S_1}{\partial w_l{}^0} \right\} + \left\{ \frac{\partial V}{\partial J_k} \frac{\partial S_1}{\partial w_k{}^0} \right\} = 0 \tag{75.12}$$

obtained from (75.7) and (75.11), the braces meaning that only the oscillating part of the enclosed function is to be taken.

The approximation may be carried further. It clearly becomes more and more difficult to calculate the successive terms. A more fundamental difficulty presents itself. Let S_1, W be expanded in a Fourier series in $w_k{}^0$

$$S_1 = \sum_{\tau_k = -\infty}^{\infty} S(\tau_k) \exp (2\pi i \tau_k w_k{}^0)$$

$$W = \sum_{\tau_k = -\infty}^{\infty} W(J, \tau_k) \exp (2\pi i \tau_k w_k{}^0) \tag{75.13}$$

where the τ_k are a set of integers. Then (75.10) becomes an algebraic equation for the Fourier coefficients

$$\nu_k{}^0\tau_k S(\tau_k) = \frac{i}{2\pi} W(J, \tau_k) \tag{75.14}$$

which is soluble for $S(\tau_k)$ only if the quantity

$$\nu_k{}^0\tau_k \neq 0 \tag{75.15}$$

A system satisfying this condition is termed *nondegenerate*. Perturbation theory can be applied to degenerate systems provided that the coordinates are chosen in such a way that $W(J, \tau_k) = 0$ for those τ's which make (75.15) untrue. Under these circumstances $S(\tau_k)$ is indeterminate and may be chosen to vanish.

As an example of the use of this perturbation theory, consider the problem of a particle of charge $-e$ moving in a Coulomb field and in a uniform magnetic field. The magnetic field is taken as the perturbing influence.

$$H = \frac{1}{2m}\left(\mathbf{p} + \frac{e}{c}\mathbf{A}\right)^2 - \frac{Ze^2}{r}$$

$$H_0 = \frac{1}{2m}\mathbf{p}^2 - \frac{Ze^2}{r}$$

so that

$$V = \frac{e}{mc}\mathbf{p}\cdot\mathbf{A} + \frac{e^2}{2mc^2}\mathbf{A}^2$$

$$\approx \frac{e}{mc}\mathbf{p}\cdot\mathbf{A}$$

Here

$$A_x = -\tfrac{1}{2}By$$

$$A_y = \tfrac{1}{2}Bx$$

$$A_z = 0$$

The first quantity needed is the average value of V over the unperturbed Kepler orbit. This can be calculated in any variables, but must eventually be expressed as a function of the action variables.

$$V = \frac{e}{mc}\mathbf{p}\cdot\mathbf{A} = \frac{eB}{2mc}(xp_y - yp_x)$$

$$= \frac{eB}{2mc}l_z$$

$$= \frac{eB}{4\pi mc}J_\phi$$

In this particular problem the perturbation (to first order in **A**) is not a function of the angle variables at all

$$H_1(J) = \frac{eB}{4\pi mc} J_\phi$$

so that the only frequency to be changed is the frequency about the field direction

$$\nu_\phi = \nu_\phi^{\,0} + \frac{eB}{4\pi mc}$$

which is the exact solution. One could include the $\mathbf{A}^2$ term which was neglected in the above analysis.

As a second example take a special case of the problem discussed in Sec. 64. A particle moves under the influence of an inverse square law force and a constant force in the plane of the orbit, so that $J_\phi = 0$. We have

$$H_0(q, p) = \frac{1}{2\mu}\left(p_r^{\,2} + \frac{p_\theta^{\,2}}{r^2} + \frac{\kappa}{r}\right)$$

$$V(q, p) = -Fr \cos \theta$$

From (64.3)

$$H_0(J) = -\frac{2\pi^2 \mu \kappa^2}{(J_r + J_\theta)^2}$$

$$H_1(J) = \langle -Fr \cos \theta \rangle$$

This average depends on the orientation of the major axis of the ellipse relative to the applied field. For a general orientation we find, using (37.4),

$$\langle Fr \cos \theta \rangle = \frac{F}{\tau} \int_0^\tau r \cos \theta \, dt$$

$$= -\frac{Fl^5}{\tau \kappa^3 \mu^2} \oint \frac{\cos \theta \, d\theta}{[1 - e \cos (\theta - \theta_0)]^3}$$

$$= -\frac{Fl^5}{\tau \kappa^3 \mu^2} \oint \frac{\cos \theta_0 \cos \phi - \sin \theta_0 \sin \phi}{(1 - e \cos \phi)^3} \, d\phi$$

The integral involving $\sin \phi$ vanishes because the numerator is odd and the denominator is even in ϕ. The integral involving $\cos \phi$ is finite. In fact,

$$\oint \frac{\cos\phi\, d\phi}{(1 - e\cos\phi)^3} = \frac{1}{2}\frac{d}{de}\oint \frac{d\phi}{(1 - e\cos\phi)^2}$$
$$= \frac{1}{2}\frac{\kappa^2\mu}{l^3}\frac{d\tau}{de}$$
$$= 3\pi e(1 - e^2)^{-5/2}$$

by use of (37.12) and (37.9). Expressing everything in terms of action variables, using (37.12) and (64.3),

$$H_1(J) = -\frac{3F\cos\theta_0}{8\pi^2\mu\kappa}(J_r + J_\theta)[J_r(J_r + 2J_\theta)]^{1/2}$$

The correction to the Hamiltonian H_0 is not a function of $J_r + J_\theta$ only, so that the perturbed radial and rotational frequencies are not equal. We see that

$$\Delta\nu_r = \frac{\partial H_1}{\partial J_r} = H_1(J)\left(\frac{1}{J_r + J_\theta} + \frac{J_r + J_\theta}{J_r^2 + 2J_rJ_\theta}\right)$$

$$\Delta\nu_\phi = \frac{\partial H_1}{\partial J_r} = H_1(J)\left(\frac{1}{J_r + J_\theta} + \frac{J_r}{J_r^2 + 2J_rJ_\theta}\right)$$

If $\cos\theta_0 > 0$, $H_1(J) < 0$, and the physical situation is as shown in Fig. 75-1. In this situation the perturbed value of ν_θ is greater than the perturbed value of ν_r so that the major axis of the orbit precesses in the direction of increasing θ_0. It is therefore possible to consider

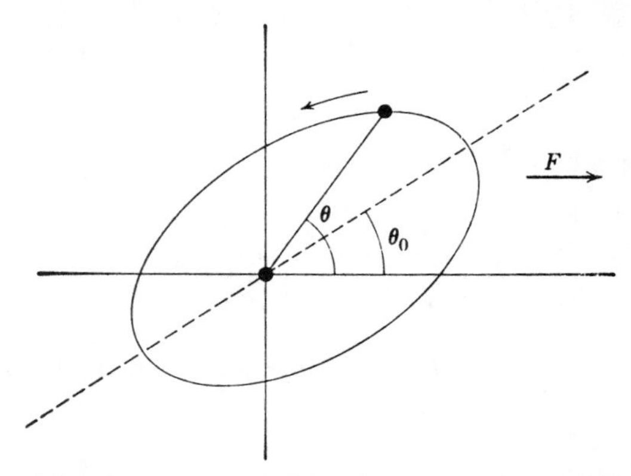

Fig. 75-1. Kepler motion perturbed by a uniform field in the orbit plane. The major axis precesses away from the field direction.

this a periodic system for time intervals over which θ_0 does not change appreciably, but not over long time intervals.

76. TIME-DEPENDENT PERTURBATION THEORY

Let the Hamiltonian of the system be expressed as the sum of two terms as in (74.3). Instead of using action and angle variables which are derived from a time-independent form of the Hamilton-Jacobi equation we use the full, time-dependent theory. The generator of the contact transformation from constant coordinates and momenta, q_0 and p_0, to the variables used to describe the system originally, q and p, is a complete integral of the Hamilton-Jacobi equation:

$$\frac{\partial \phi}{\partial t} + H\left(q, \frac{\partial \phi}{\partial q}, t\right) = 0 \tag{76.1}$$

We make the transformation in two steps. The generator of the first step, $\phi_0(q, \bar{q}, t)$ yields a contact transformation connecting variables $\bar{q}$ with the coordinates q. We choose ϕ_0 so that in the absence of the perturbation the $\bar{q}$ are constants; i.e., ϕ_0 is taken to be a complete integral of the Hamilton-Jacobi equation formed with H_0,

$$\frac{\partial \phi_0}{\partial t} + H_0\left(q, \frac{\partial \phi_0}{\partial q}, t\right) = 0 \tag{76.2}$$

the $\bar{q}$ being the integration constants. The second step is generated by $\phi_1(\bar{q}, q_0, t)$, which yields a contact transformation between the coordinates q_0 which are truly constants and the slowly varying variables $\bar{q}$. Thus the entire generator ϕ is given by

$$\phi(q, q_0, t) = \phi_0(q, \bar{q}, t) + \phi_1(\bar{q}, q_0, t) \tag{76.3}$$

The contact transformation generated by ϕ_0 (a) replaces the q, p by $\bar{q}, \bar{p}$, and (b) replaces the Hamiltonian $H(q, p, t)$ by $H + \partial\phi_0/\partial t = V$, which must be expressed as a function of the barred variables. The equations of motion for the barred variables are then

$$\dot{\bar{q}}_i = \frac{\partial V}{\partial \bar{p}_i}, \qquad \dot{\bar{p}}_i = -\frac{\partial V}{\partial \bar{q}_i} \tag{76.4}$$

Equations (76.4) are exact, but usually not exactly soluble. The usefulness of (76.4) comes about if V is small. The the barred variables are slowly varying functions of the time and these equations can be integrated by successive approximation. Instead of using the canonical set of variables $\bar{q}$, $\bar{p}$, it is sometimes convenient to use $2f$ functions of them, C_α ($\alpha = 1, 2, \cdots, 2f$) which do not constitute a set of canonical variables. Then

$$\begin{aligned} \dot{C}_\alpha &= (C_\alpha, V) \\ &= (C_\alpha, C_\beta) \frac{\partial V}{\partial C_\beta} \end{aligned} \tag{76.5}$$

which is a form first derived by Poisson. Using the fact that the matrix of the Lagrange brackets is the negative reciprocal of that of the Poisson brackets, (76.5) may be rewritten in Lagrange's form:

$$[C_\alpha, C_\beta]\dot{C}_\alpha = \frac{\partial V}{\partial C_\beta} \tag{76.6}$$

Both (76.5) and (76.6) reduce to (76.4) when the C_α form a canonical set of variables.

Still another way to formulate the theory is to consider the contact transformation generated by a function which is a power series in the parameter λ:

$$\phi(q, \bar{q}, t) = \phi^{(0)}(q, \bar{q}, t) + \lambda\phi^{(1)}(q, \bar{q}, t) + \lambda^2\phi^{(2)}(q, \bar{q}, t) + \cdots \tag{76.7}$$

The functions $\phi^{(k)}$ cannot be considered as the generators of a sequence of contact transformations as were the ϕ_0 and ϕ_1 of (76.2) because the arguments here are all the same. Let us assume that the power series is rapidly convergent at $\lambda = 1$, so that $\phi^{(n+1)} \ll \phi^{(n)}$. Then under the transformation generated by ϕ we have

$$\bar{H}\left(\bar{q}, -\frac{\partial \phi}{\partial \bar{q}}, t\right) = H\left(q, \frac{\partial \phi}{\partial q}, t\right) + \frac{\partial \phi}{\partial t} \tag{76.8}$$

We expand the right-hand side in powers of λ:

$$\begin{aligned} H &= H_0 + \lambda V \\ H_0\left(q, \frac{\partial \phi}{\partial q}, t\right) &= H_0\left(q, \frac{\partial \phi_0}{\partial q}, t\right) + \lambda \frac{\partial H_0}{\partial p_i}\frac{\partial \phi_1}{\partial q_i} \\ &\quad + \lambda^2\left(\frac{\partial H_0}{\partial p_i}\frac{\partial \phi_2}{\partial q_i} + \frac{1}{2}\frac{\partial^2 H_0}{\partial p_i \partial p_j}\frac{\partial \phi_1}{\partial q_i}\frac{\partial \phi_1}{\partial q_j}\right) + \cdots \\ \lambda V\left(q, \frac{\partial \phi}{\partial q}, t\right) &= \lambda V\left(q, \frac{\partial \phi_0}{\partial q}, t\right) + \lambda^2 \frac{\partial V}{\partial p_i}\frac{\partial \phi_1}{\partial q_i} + \cdots \end{aligned} \tag{76.9}$$

This yields, since ϕ does not depend explicitly on the momenta,

$$(76.10)\quad \bar{H} = H_0 + \frac{\partial \phi^{(0)}}{\partial t} + \lambda\left[(\phi^{(1)}, H_0) + V + \frac{\partial \phi^{(1)}}{\partial t}\right]$$
$$+ \lambda^2\left[(\phi^{(2)}, H_0) + (\phi^{(1)}, V) + \frac{1}{2}(\phi^{(1)}, (\phi^{(1)}, H_0)) + \frac{\partial \phi^{(2)}}{\partial t}\right] + \cdots$$

If $\bar{H} = 0$, ϕ is the solution of the exact Hamilton-Jacobi equation. We may make $\bar{H}$ as small as we wish by making the coefficients of sufficiently many successive powers of λ vanish. Thus $\bar{H}$ is of order $(\lambda V)^3$ if

$$H_0\left(q, \frac{\partial \phi^{(0)}}{\partial q}, t\right) + \frac{\partial \phi^{(0)}}{\partial t} = 0$$

$$(76.11)\quad (\phi^{(1)}, H_0) + \frac{\partial \phi^{(1)}}{\partial t} = -V$$

$$(\phi^{(2)}, H_0) + \frac{\partial \phi^{(2)}}{\partial t} = -(\phi^{(1)}, V) - \frac{1}{2}(\phi^{(1)}, (\phi^{(1)}, H))$$

Equations (76.11) are a set of differential equations for the $\phi^{(k)}$. They can be solved successively because the right side of each equation depends on known quantities only once the preceding equations have been solved. Thus if equations (76.11) are solved, $\bar{H}$ is zero to second order, and $(\phi^{(0)} + \phi^{(1)} + \phi^{(2)})$ is the second approximation to the solution of the problem. It is usually impractical to carry the calculation out to terms higher than the second.

Instead of applying the contact transformation generated by $(\phi^{(0)} + \phi^{(1)} + \phi^{(2)} + \cdots + \phi^{(n)})$, we may apply that generated by $(\phi^{(1)} + \phi^{(2)} + \cdots + \phi^{(n)})$, omitting $\phi^{(0)}$. If this is done,

$$\bar{H} = H_0 + 0(\lambda^{n+1}V^{n+1})$$

and we have introduced new variables which, to within terms of order $(n + 1)$, obey equations of motion derived from the unperturbed Hamiltonian.

The first-order term $\phi^{(1)}$ is easily understood. It is the integral of the negative of the perturbation over the unperturbed trajectory. Since

$$\phi = \int_{t_0}^{t} L\, dt$$

if $L = L_0 - V$ (which is true if the V does not depend on the momenta), a first approximation to ϕ is naturally

$$\phi \approx \phi^{(0)} + \phi^{(1)} = \int_{t_0}^{t} (L_0 - V)\, dt$$

taken along the trajectory determined by L_0 alone. The higher order terms are more complicated.

77. QUASI COORDINATES AND QUASI MOMENTA

In discussing the motion of a rigid body, it was found convenient to introduce the components ω_j of the angular velocity. It was shown in Sec. 49 that these quantities are not "velocities," i.e., time derivatives of coordinates, but are nonintegrable linear combinations of velocities. In this section and the next, we wish to develop some properties of such quantities.

Let us consider a dynamic system whose configuration is specified by the f coordinates q_j. We define f independent linear combinations of the velocities

$$\omega_j = \alpha_{jk}\dot{q}_k \tag{77.1}$$

where the α_{jk} are functions of the coordinates. The case of interest is that in which

$$\frac{\partial \alpha_{jk}}{\partial q_l} - \frac{\partial \alpha_{jl}}{\partial q_k} \neq 0 \tag{77.2}$$

for some j, k, l, which means that the ω_j are not total time derivatives of any coordinates. It is customary to write, however,

$$\omega_j\, dt = d\theta_j \tag{77.3}$$

and to call $d\theta_j$ the differentials of *quasi coordinates*. $d\theta_j$ is not a differential, and θ_j is an undefined symbol.

Because the ω_j are independent, (77.1) may be solved for the velocities. We write the solution:

$$\dot{q}_k = \beta_{kl}\omega_l \tag{77.4}$$

The matrix β is the reciprocal of the matrix α:

$$\alpha_{jk}\beta_{kl} = \delta_{jl} \qquad \text{or} \qquad \alpha\beta = 1 \tag{77.5}$$

If $F(q)$ is a function of the coordinates, its differential may be expressed in terms of the $d\theta_j$,

$$dF = \frac{\partial F}{\partial q_k} dq_k = \frac{\partial F}{\partial q_k} \beta_{kl}\, d\theta_l$$

which suggests defining the symbol $\partial F/\partial \theta_l$ by

$$\frac{\partial F}{\partial \theta_l} = \frac{\partial F}{\partial q_k} \beta_{kl} \tag{77.6}$$

Thus, in particular,

$$\frac{\partial q_k}{\partial \theta_l} = \beta_{kl} \tag{77.7}$$

a definition that may be obtained directly from (77.4).

Let the Lagrangian of the system be $L(q, \dot{q}, t)$. The $\dot{q}$'s may be replaced by the ω's, yielding

$$L(q, \dot{q}, t) = \bar{L}(q, \omega, t) \tag{77.8}$$

Writing

$$\sigma_l = \frac{\partial \bar{L}}{\partial \omega_l} \tag{77.9}$$

we have

$$p_k = \frac{\partial L}{\partial \dot{q}_k} = \frac{\partial \bar{L}}{\partial \omega_j} \alpha_{jk} = \sigma_j \alpha_{jk}$$

$$\dot{p}_k = \frac{\partial L}{\partial q_k} = \frac{\partial \bar{L}}{\partial q_k} + \sigma_j \frac{\partial \alpha_{jl}}{\partial q_k} \dot{q}_l$$

so that Lagrange's equations become

$$\frac{d}{dt} (\sigma_j \alpha_{jk}) = \frac{\partial \bar{L}}{\partial q_k} + \sigma_j \frac{\partial \alpha_{jl}}{\partial q_k} \beta_{lm} \omega_m \tag{77.10}$$

which may be written as

$$\frac{d}{dt} \frac{\partial \bar{L}}{\partial \omega_j} - \frac{\partial \bar{L}}{\partial \theta_j} = -\gamma_{jkl} \omega_k \frac{\partial \bar{L}}{\partial \omega_l} \tag{77.11}$$

with

$$\gamma_{jkl} = \left(\frac{\partial \alpha_{lm}}{\partial q_n} - \frac{\partial \alpha_{ln}}{\partial q_m} \right) \beta_{mj} \beta_{nk} \tag{77.12}$$

If the θ_j exist, $\gamma_{jkl} = 0$ and equations (77.11) are just the usual Lagrange's equations in the new variables.

The σ_l introduced in (77.9) resemble momenta. They do not constitute a set of momenta in the usual sense unless the θ_j exist. To see this, it is sufficient to evaluate the Poisson bracket of two σ's,

$$(\sigma_i, \sigma_j) = -\sigma_l \gamma_{lij} \tag{77.13}$$

which vanishes only when the γ_{lij} vanish. It is possible to define some *quasi momenta* which act as though they were canonically conjugate to the quasi coordinates. Let

$$d\pi_j = dp_k \beta_{kj} \tag{77.14}$$

so that

$$d\theta_j \, d\pi_j = dq_k \, dp_k$$

Extending the definition (77.7) in an obvious way, we have

$$\frac{\partial \pi_j}{\partial p_k} = \beta_{kj}, \qquad \frac{\partial \pi_j}{\partial q_k} = 0$$

along with

$$\frac{\partial \theta_i}{\partial q_k} = \alpha_{ik}, \qquad \frac{\partial \theta_i}{\partial p_k} = 0$$

so that with these definitions

$$(\theta_i, \pi_j) = \alpha_{ik}\beta_{kj} = \delta_{ij}$$

Equations formally resembling Hamilton's equations may also be written. If H is the Hamiltonian corresponding to the Lagrangian L,

$$dH = \frac{\partial H}{\partial q_k} \beta_{kj} \, d\theta_j + \frac{\partial H}{\partial p_k} \alpha_{jk} \, d\pi_j + \frac{\partial H}{\partial t} dt = \frac{\partial H}{\partial \theta_j} d\theta_j + \frac{\partial H}{\partial \pi_j} d\pi_j + \frac{\partial H}{\partial t} dt$$

However, $H = p_j \dot{q}_j - L$ so that

$$dH = dp_j \dot{q}_j - \dot{p}_j \, dq_j - \frac{\partial L}{\partial t} dt = d\pi_j \omega_j - \dot{\pi}_j \, d\theta_j - \frac{\partial L}{\partial t} dt$$

from which it follows that

$$\omega_j = \frac{\partial H}{\partial \pi_j}, \qquad \dot{\pi}_j = -\frac{\partial H}{\partial \theta_j} \tag{77.15}$$

These equations are not very useful because the differentials of the quasi momenta are not especially interesting quantities, although the ω_j are.

The quantities σ_j are related to the momenta p_k as the $d\pi_j$ are to the $d\sigma_k$, and are of interest even though they do not constitute a set of canonical momenta. We may replace the p's in the Hamiltonian with the σ's, just as the $\dot{q}$'s in the Lagrangian were replaced by ω's. Thus

$$H(q, p, t) = \bar{H}(q, \sigma, t) = \sigma_j \omega_j - \bar{L} \tag{77.16}$$

so that

$$d\bar{H} = \frac{\partial \bar{H}}{\partial q_k} dq_k + \frac{\partial \bar{H}}{\partial \sigma_k} d\sigma_k + \frac{\partial \bar{H}}{\partial t} dt = -\frac{\partial \bar{L}}{\partial q_k} dq_k + \omega_k \, d\sigma_k - \frac{\partial \bar{L}}{\partial t} dt$$

With the use of (77.10) and (77.12), this yields

$$\omega_j = \frac{\partial \bar{H}}{\partial \sigma_j} \tag{77.17}$$

$$\dot{\sigma}_j + \gamma_{jkl} \omega_k \sigma_l = -\frac{\partial \bar{H}}{\partial \theta_j}$$

These equations are not in canonical form unless the γ_{jkl} vanish.

The σ's may replace the p's in any dynamical variable:

$$X(q, p) = \bar{X}(q, \sigma) \tag{77.18}$$

The ICT generated by X can be written

$$\delta\theta_j = \epsilon \frac{\partial \bar{X}}{\partial \sigma_j}$$

$$\delta\sigma_j + \gamma_{jkl} \, \delta\theta_k \sigma_l = -\epsilon \frac{\partial \bar{X}}{\partial \theta_j} \tag{77.19}$$

These equations are of the same form as (77.17), as was to be expected since H also generates an ICT.

EXERCISES

1. Verify (77.17) and (77.19) in detail.

2. For the case of a rigid body referred to space axes, comparison of (49.11) and (77.13) shows that $\gamma_{jkl} = -\epsilon_{jkl}$. Obtain the matrix β_{ij} connecting the space components of $\boldsymbol{\omega}$ with the Euler angles and verify this conclusion directly from (77.12).

3. Obtain the matrix β_{ij} connecting the body components of $\boldsymbol{\omega}$ with the Euler angles from (49.12), and show that now $\gamma_{jkl} = \epsilon_{jkl}$.

4. Show that

$$\gamma_{jkl} = \alpha_{lm} \left(\frac{\partial \beta_{mk}}{\partial q_n} \beta_{nj} - \frac{\partial \beta_{mj}}{\partial q_n} \beta_{nk} \right)$$

14 SPECIAL APPLICATIONS

78. NONCENTRAL FORCES

In the definition of a particle given in Sec. 2, the only part of the description of a particle which could depend upon the time was taken to be its position. We now want to relax this restriction and consider a particle with an "intrinsic" angular momentum or *spin*, $\boldsymbol{\sigma}$, independent of the orbital motion of the particle but which may be subjected to torques. In particular, we are interested in a spin of constant magnitude with which is associated a magnetic moment $\boldsymbol{\mu}$:

$$\boldsymbol{\mu} = g\boldsymbol{\sigma} \tag{78.1}$$

This particular system is interesting because it is a reasonable classical analog to the quantum-mechanical electron of the Dirac theory. There the electron has a spin $\hbar/2$ which vanishes in the classical limit so that the analogy is not exact in the sense of the correspondence principle, but nonetheless certain features of interest emerge from a classical model.

The components of $\boldsymbol{\sigma}$ in a fixed cartesian coordinate system are quantities of the kind denoted by σ_j in the last section. They are connected to the structure of the particle and the angular velocities in a way which we shall see below. For simplicity, the particle is assumed to possess spherical symmetry so that the inertia tensor may be taken to be a multiple of the unit tensor and denoted by the number J.

As the Lagrangian for our particle, assumed charged, we may take

$$\bar{L} = \frac{1}{2} m\mathbf{v}^2 + e\frac{\mathbf{v}}{c} \cdot \mathbf{A} - e\phi + \frac{1}{2} J\boldsymbol{\omega}^2 + gJ\boldsymbol{\omega} \cdot \mathbf{B} \tag{78.2}$$

in which the two last terms describe the kinetic energy associated with the spin and the coupling of the spin to an external magnetic field $\mathbf{B}$ respectively. The momentum of the orbital motion is

$$\mathbf{p} = \frac{\partial \bar{L}}{\partial \mathbf{v}} = m\mathbf{v} + \frac{e}{c}\mathbf{A} \tag{78.3}$$

as in (55.3). The spin is given by

$$\boldsymbol{\sigma} = \frac{\partial \bar{L}}{\partial \boldsymbol{\omega}} = J\boldsymbol{\omega} + gJ\mathbf{B} \tag{78.4}$$

Thus

$$\mathbf{v} = \frac{\mathbf{p} - (e/c)\mathbf{A}}{m}, \qquad \boldsymbol{\omega} = \boldsymbol{\sigma}/J - g\mathbf{B} \tag{78.5}$$

The Hamiltonian corresponding to $\bar{L}$ is

$$\begin{aligned} \bar{H} &= \mathbf{p}\cdot\mathbf{v} + \boldsymbol{\sigma}\cdot\boldsymbol{\omega} - \bar{L} \\ &= \frac{1}{2m}\left(\mathbf{p} - \frac{e}{c}\mathbf{A}\right)^2 + e\phi + \frac{1}{2J}(\boldsymbol{\sigma} - gJ\mathbf{B})^2 \end{aligned} \tag{78.6}$$

From this Hamiltonian we can see that the orbital motion is that of a charged particle in an electromagnetic field with an additional force given by

$$-\nabla \frac{1}{2J}(\boldsymbol{\sigma} - gJ\mathbf{B})^2$$

If we restrict our attention to particles with small J and in weak fields so that the angular velocity is very much larger than the Larmor angular velocity, $\omega \gg gB$, then the extra force may be written

$$\nabla(g\boldsymbol{\sigma}\cdot\mathbf{B}) = \nabla(\boldsymbol{\mu}\cdot\mathbf{B}) \tag{78.7}$$

which is the usual expression for the force on a dipole. It vanishes if $\mathbf{B}$ is constant.

The equation of motion for the spin follows from (77.17). $\bar{H}$ contains only the x's, the p's, and the σ's, so $\partial\bar{H}/\partial\theta_j = 0$ and thus

$$\dot{\boldsymbol{\sigma}} - \boldsymbol{\omega} \times \boldsymbol{\sigma} = 0$$

Using (78.5), this becomes

$$\begin{aligned} \dot{\boldsymbol{\sigma}} &= -g\mathbf{B} \times \boldsymbol{\sigma} \\ &= \boldsymbol{\mu} \times \mathbf{B} \end{aligned} \tag{78.8}$$

the familiar expression for the torque on a magnet in a magnetic field.

The vector $\boldsymbol{\sigma}$ appears in $\bar{H}$ combined with the vector **B**. Therefore $\bar{H}$ will have no simple transformation properties under a rotation of the axes to which the positional coordinates and momenta are referred unless at the same time the axes to which $\boldsymbol{\sigma}$ is referred are also rotated. The former axes are rotated by the ICT generated by a component of the orbital angular momentum, say l_z. The coupled rotation is generated by a component of the total angular momentum $\boldsymbol{j} = \boldsymbol{l} + \boldsymbol{\sigma}$, say j_z given by

$$j_z = xp_y - yp_x + \sigma_z \tag{78.9}$$

Using Sec. 66 and (77.19) with $\bar{X} = j_z$, we see that

$$\begin{aligned} \delta x &= -\epsilon y, & \delta y &= \epsilon x, & \delta z &= 0 \\ \delta p_x &= -\epsilon p_y, & \delta p_y &= \epsilon p_x, & \delta p_z &= 0 \\ \delta \theta_x &= 0, & \delta \theta_y &= 0, & \delta \theta_z &= \epsilon \\ \delta \sigma_x &= -\epsilon \sigma_y, & \delta \sigma_y &= \epsilon \sigma_x, & \delta \sigma_z &= 0 \end{aligned} \tag{78.10}$$

which is the coupled rotation of space and spin axes about the negative z direction through the angle ϵ.

As an example, we may consider the case in which B is the magnetic field due to a dipole μ' fixed at the origin. Thus

$$\mathbf{B} = -\frac{\boldsymbol{\mu}'}{r^3} + \frac{3\mathbf{r}(\boldsymbol{\mu}' \cdot \mathbf{r})}{r^5}$$

so that

$$V = -\boldsymbol{\mu} \cdot \mathbf{B} = \frac{\boldsymbol{\mu} \cdot \boldsymbol{\mu}'}{r^3} - \frac{3(\boldsymbol{\mu} \cdot \mathbf{r})(\boldsymbol{\mu}' \cdot \mathbf{r})}{r^5}$$

This corresponds to a vector potential

$$\mathbf{A} = \frac{\boldsymbol{\mu}' \times \mathbf{r}}{r^3}$$

We shall suppose that the direction of the dipole $\boldsymbol{\mu}'$ is fixed, and take this as the z direction. Then

$$\begin{aligned} \bar{H} = \frac{1}{2m}\left[\left(p_x + \frac{e\mu' y}{cr^3}\right)^2 + \left(p_y - \frac{e\mu' x}{cr^3}\right)^2 + p_z{}^2\right] + e\phi + \frac{\sigma^2}{2J} \\ + \frac{\mu'}{r^5}[-3xz\mu_x - 3yz\mu_y + (x^2 + y^2 - 2z^2)\mu_z] \end{aligned}$$

Under the transformation generated by j_z [equation (78.9)] each of these terms is invariant if ϕ is symmetrical about the z axis. Thus

j_z, the sum of the orbital and spin angular momenta about the z axis, is a constant of the motion.

For the special case $\mu = 0$, the motion is that of a charged particle in the field of a fixed magnetic dipole. This is of special interest in the study of cosmic radiation incident upon the magnetic field of the earth or sun.

We may also consider the more general problem in which the position and orientation of neither dipole is fixed. If we neglect any interaction not arising from the dipole moments ($e = 0$) and characterize the two particles by the suffixes 1, 2, the Hamiltonian may be written

$$\bar{H} = \frac{p_1^2}{2m_1} + \frac{p_2^2}{2m_2} + \frac{\sigma_1^2}{2J_1} + \frac{\sigma_2^2}{2J_2} - \frac{g_1 g_2 X_{12}}{r^3} \tag{78.11}$$

where

$$\begin{aligned} X_{12} &= 3(\boldsymbol{\sigma}_1 \cdot \hat{\mathbf{r}})(\boldsymbol{\sigma}_2 \cdot \hat{\mathbf{r}}) - \boldsymbol{\sigma}_1 \cdot \boldsymbol{\sigma}_2 \\ \boldsymbol{\mu}_1 &= g_1 \boldsymbol{\sigma}_1, \qquad \boldsymbol{\mu}_2 = g_2 \boldsymbol{\sigma}_2 \end{aligned} \tag{78.12}$$

and $\hat{\mathbf{r}}$ is a unit vector in the direction of the relative separation of the particles. This Hamiltonian is invariant under the ICT generated by

$$\begin{aligned} X = j_z &= (x_1 p_{1y} - y_1 p_{1x}) + (x_2 p_{2y} - y_2 p_{2x}) + \sigma_{1z} + \sigma_{2z} \\ &= l_z + \sigma_z \end{aligned}$$

i.e., the total angular momentum (orbital and spin) about the z axis. This transformation corresponds to (78.10) for each particle. We leave to the reader to verify that these changes leave $\bar{H}$ invariant.

Under certain conditions the Hamiltonian (78.11) is also invariant under the ICT generated by

$$X = (\boldsymbol{\sigma}_1 + \boldsymbol{\sigma}_2)^2$$

i.e., the square of the total spin angular momentum of the two particles. Under such conditions it follows that X is a constant of the motion. Since $\boldsymbol{\sigma}_1$ and $\boldsymbol{\sigma}_2$ are of constant length, this result is equivalent to stating that the angle between $\boldsymbol{\sigma}_1$ and $\boldsymbol{\sigma}_2$ stays constant throughout the motion. We have

$$\begin{aligned} \delta\boldsymbol{\theta}_1 &= 2\epsilon(\boldsymbol{\sigma}_1 + \boldsymbol{\sigma}_2) = \delta\boldsymbol{\theta}_2 \\ \delta\boldsymbol{\sigma}_1 &= 2\epsilon(\boldsymbol{\sigma}_1 + \boldsymbol{\sigma}_2) \times \boldsymbol{\sigma}_1 \\ &= 2\epsilon(\boldsymbol{\sigma}_2 \times \boldsymbol{\sigma}_1) = -\delta\boldsymbol{\sigma}_2 \end{aligned}$$

so that

$$\delta X_{12} = 6\epsilon[(\boldsymbol{\sigma}_1 \times \boldsymbol{\sigma}_2) \cdot \hat{\mathbf{r}}][(\boldsymbol{\sigma}_1 - \boldsymbol{\sigma}_2) \cdot \hat{\mathbf{r}}]$$

The term $\boldsymbol{\sigma}_1 \cdot \boldsymbol{\sigma}_2$ in X_{12} [equation (78.12)] is clearly invariant under the transformation. X_{12} is invariant if any two of $\boldsymbol{\sigma}_1, \boldsymbol{\sigma}_2, \mathbf{r}$ are parallel, or if the components of $\boldsymbol{\sigma}_1$ and $\boldsymbol{\sigma}_2$ in the direction of $\mathbf{r}$ are equal. In particular, if $\boldsymbol{\sigma}_1$ and $\boldsymbol{\sigma}_2$ are initially parallel or antiparallel, the magnitude of their sum, the total spin angular momentum, remains constant.

These results are of particular interest in the attempt to understand the properties of the deuteron in terms of an interaction between proton and neutron which includes, in addition to central forces, a non-central or "tensor" interaction of the form of X_{12}. Because of the coupling between spin and orbital motion so postulated, the orbital angular momentum of the system is then no longer a constant of the motion; instead, we must use the total angular momentum and magnitude of the spin as a starting point for prescribing the properties of the system.

79. SPIN MOTION

In this section we discuss the equation of motion (78.8) of a spinning particle under the influence of an externally applied magnetic field $\mathbf{B}$:

$$\dot{\boldsymbol{\sigma}} = g(\boldsymbol{\sigma} \times \mathbf{B}) \tag{79.1}$$

This equation describes with remarkable accuracy the precession of the spin of an elementary particle, and is therefore of basic significance in the theory of electron and nuclear magnetic resonance, since classical mechanics is applicable to many situations in which such resonances are measured.

We note first, from (79.1), that

$$\frac{d\sigma^2}{dt} = 2\boldsymbol{\sigma} \cdot \frac{d\boldsymbol{\sigma}}{dt} = 0$$

so that a magnetic field cannot change the magnitude of the spin angular momentum of a particle. The equation is really concerned with the way in which the *direction* of the spin changes for different forms of $\mathbf{B}$.

In many situations it is easier to understand the consequences of (79.1) if we introduce a coordinate system which is rotating about the particle with a constant angular velocity $\boldsymbol{\Omega}$. From (49.22) we may then write

$$\dot{\boldsymbol{\sigma}} = \frac{\partial \boldsymbol{\sigma}}{\partial t} + \boldsymbol{\Omega} \times \boldsymbol{\sigma} \tag{79.2}$$

where $\partial/\partial t$ denotes the rate of change with respect to the time as seen in this rotating system. The vector $\boldsymbol{\sigma}$, however, continues to denote the spin angular momentum relative to the inertial coordinate system with respect to which (79.1) is valid. Hence

$$\frac{\partial \boldsymbol{\sigma}}{\partial t} = g\boldsymbol{\sigma} \times (\mathbf{B} + g^{-1}\boldsymbol{\Omega}) \tag{79.3}$$

i.e., in the rotating system the spin precesses as if the particle were being acted upon by an effective magnetic field

$$\mathbf{B}_{\text{eff}} = \mathbf{B} + g^{-1}\boldsymbol{\Omega}$$

In the special case $\mathbf{B} = $ const, we may choose

$$\boldsymbol{\Omega} = -g\mathbf{B} \tag{79.4}$$

and note that if the coordinate system is rotating at just this rate the spin vector appears in that system to be at rest (cf. Fig. 79-1). This

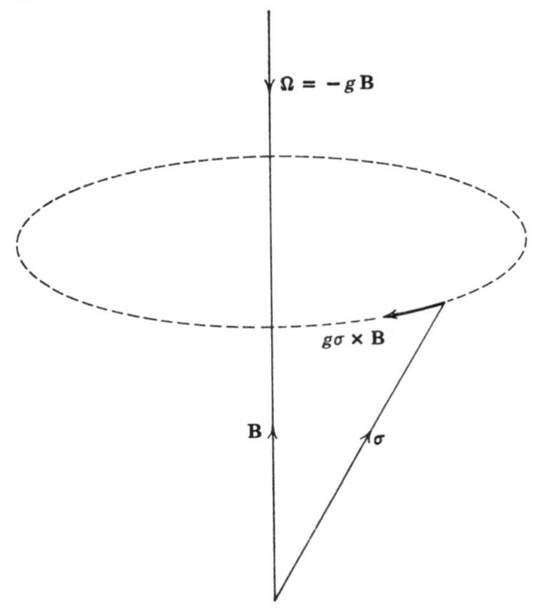

Fig. 79-1. Precession of spin in a uniform magnetic field for g positive.

result may also be seen directly from (79.1).

In many problems concerning magnetic spin resonance it is necessary to study the equation governing the spin motion for the case in which the externally applied field $\mathbf{B}$ consists of a constant field $\mathbf{B}_0$ upon which is superposed a relatively small rotating field $\mathbf{B}_1$, so that (79.3) may be written

$$\frac{\partial \boldsymbol{\sigma}}{\partial t} = g\boldsymbol{\sigma} \times (\mathbf{B}_0 + \mathbf{B}_1 + g^{-1}\boldsymbol{\Omega})$$

Since any convenient constant value may be chosen for $\boldsymbol{\Omega}$, we shall now choose it so that in the rotating system the rotating field $\mathbf{B}_1$ is

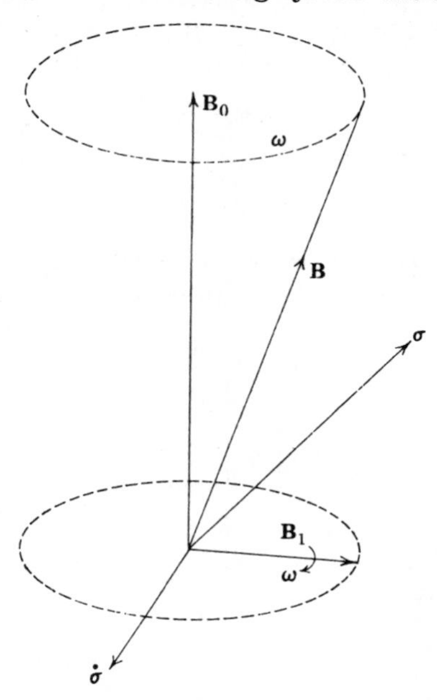

Fig. 79-2. Spin motion in uniform field plus rotating field (inertial system).

independent of the time. If, for example, $\mathbf{B}_1$ is rotating about the axis of $\mathbf{B}_0$ with angular velocity $-\boldsymbol{\omega}$, we simply choose $\boldsymbol{\Omega} = -\boldsymbol{\omega}$, to obtain

$$\text{(79.5)} \qquad \frac{\partial \boldsymbol{\sigma}}{\partial t} = g\boldsymbol{\sigma} \times (\mathbf{B}_0 + \mathbf{B}_1 - g^{-1}\boldsymbol{\omega}) = g\boldsymbol{\sigma} \times \mathbf{B}_{\text{eff}}$$

(cf. Figs. 79-2, 79-3). Since $\mathbf{B}_0$ and $\boldsymbol{\omega}$ are parallel and $\mathbf{B}_1$ is perpendicular to each of them the magnitude of the effective field $\mathbf{B}_{\text{eff}}$ is therefore

$$|\mathbf{B}_{\text{eff}}| = \left[\left(B_0 - \frac{\omega}{g}\right)^2 + B_1^2\right]^{1/2}$$

or

$$\text{(79.6)} \qquad g|\mathbf{B}_{\text{eff}}| = \omega_{\text{eff}} = [(\omega_0 - \omega)^2 + \omega_1^2]^{1/2} \qquad (\omega_0 = gB_0,\ \omega_1 = gB_1)$$

In this rotating system the spin precesses about the *constant* effective field $\mathbf{B}_{\text{eff}}$ with the constant angular frequency ω_{eff}. If the angle ϕ between $\boldsymbol{\sigma}$ and $\mathbf{B}_{\text{eff}}$ is sufficiently large ($>\frac{1}{2}\pi - \theta$) the component of $\boldsymbol{\sigma}$ in the direction of $\mathbf{B}_0$ can therefore become negative during part of the motion, and if $\theta = \frac{1}{2}\pi$ ($\omega = \omega_0$) a spin initially parallel to $\mathbf{B}_0$ will

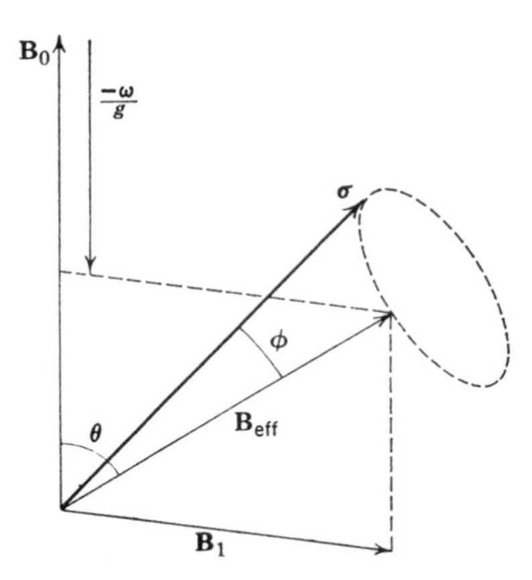

Fig. 79-3. Spin motion in uniform field plus rotating field, as seen from system in which rotating field is at rest.

precess about $\mathbf{B}_1$, becoming successively antiparallel and parallel to $\mathbf{B}_0$ during its motion.

In the same way that the effect of a single uniform field was shown in (79.4) to disappear in an appropriately chosen rotating coordinate system, so also the effect of the constant $\mathbf{B}_{\text{eff}}$ field in our singly rotating coordinate system may be removed by transforming to a doubly rotating system. Such a set of coordinates would rotate around the direction of $\mathbf{B}_{\text{eff}}$ with an angular velocity $-\omega_{\text{eff}}$, the spin direction remaining constant in this system.

80. VARIATIONAL PRINCIPLES IN ROCKET MOTION

We consider a rocket moving in a gravitational field which is described by the potential $U(r, t)$ under the influence of a thrust of arbitrary direction and magnitude. Thus if x_i ($i = 1,2,3$) denotes the position of the rocket at time t, we have [cf. (27.2)]

$$\ddot{x}_i + \partial_i U = a_i \qquad \left(\partial_i = \frac{\partial}{\partial x_i}\right) \tag{80.1}$$

with

$$a_i = \frac{\dot{M}}{M} c_i \tag{80.2}$$

the vector c_i denoting the exhaust velocity relative to the rocket. Thus

$$\frac{\dot{M}}{M} = -\frac{|\ddot{x}_i + \partial_i U|}{c} \tag{80.3}$$

where $c = |\mathbf{c}|$ is the magnitude of the exhaust velocity. We shall assume that c is a constant.

We write

$$|\ddot{x}_i + \partial_i U| = [(\ddot{x}_i + \partial_i U)(\ddot{x}_i + \partial_i U)]^{1/2} = u \tag{80.4}$$

$$a_i = \alpha_i u$$

where α_i is a unit vector in the direction of the exhaust velocity. Hence from (80.3)

$$\frac{M_f}{M_i} = \exp\left(-\frac{1}{c}\int_0^t u(t')\,dt'\right) \tag{80.5}$$

where M_f, M_i are the final and initial masses and t is the time of flight. For a given origin and destination of the rocket then, the ratio M_f/M_i is a maximum if the trajectory is such as to make the integral

$$\int_0^t u(t')\,dt'$$

have a minimum value.

This is a variational problem in which the integrand is a function of $x_i(t')$, t', and the *second* derivatives $\ddot{x}_i(t')$ of the coordinates of the rocket. The resulting Euler-Lagrange equation is

$$\frac{d^2}{dt^2}\frac{\partial u}{\partial \ddot{x}_i} + \frac{\partial u}{\partial x_i} = 0 \tag{80.6}$$

From (80.4),

$$\frac{\partial u}{\partial \ddot{x}_i} = \alpha_i$$

$$\frac{\partial u}{\partial x_i} = \alpha_j \partial_i \partial_j U$$

so that

$$\ddot{\alpha}_i + \alpha_j \partial_i \partial_j U = 0 \tag{80.7}$$

This is a differential equation for the unit vector α_i showing how the direction of the thrust should be programmed to make M_f/M_i a maximum,when compared with neighboring paths with the same end points.

For the special cases of zero gravity and of a constant gravitational field, (80.7) reduces to

$$\ddot{\alpha}_i = 0$$

which, coupled with the fact that α_i is a unit vector, implies that each component of α_i is constant, i.e., the direction of the thrust is constant throughout the motion. Thus, for this optimum programming of the thrust vector the attitude of the rocket may change but the angle between the thrust vector and the major axis of the rocket must be varied by gimbaling or other means so as to preserve a thrust direction constant in space.

Since α_i is a unit vector, i.e., $\alpha_i\alpha_i = 1$, we have $\alpha_i\ddot{\alpha}_i = -\dot{\alpha}_i\dot{\alpha}_i$ so that, from (80.7),

$$\dot{\alpha}_i\dot{\alpha}_i = -\alpha_i\alpha_j\partial_i\partial_j U \tag{80.8}$$

Since for a Kepler field $[U = -(\kappa/r)\ (\kappa > 0)]$ we have

$$\partial_i\partial_j U = \frac{\kappa}{\kappa^3}\left(\delta_{ij} - \frac{3x_ix_j}{r^2}\right) \tag{80.9}$$

it follows from (80.8) that

$$(\dot{\boldsymbol{\alpha}})^2 = \frac{\kappa}{r^3}\left[1 - 3(\boldsymbol{\alpha}\cdot\hat{\mathbf{r}})^2\right] \tag{80.10}$$

where $\hat{\mathbf{r}}$ is a unit vector in the direction of the local upward vertical. Hence

$$\cos^2\psi \equiv (\boldsymbol{\alpha}\cdot\hat{\mathbf{r}})^2 \leq \tfrac{1}{3}$$

or

$$125°16' \geq \psi \geq 54°44'$$

where ψ is the angle between $\hat{\mathbf{r}}$ and the direction of thrust. Thus for chemically propelled rockets, for which the magnitude c of the exhaust velocity is basically limited, the optimum thrust direction for motion in the field of the earth always lies outside of the cone which has an axis along the local vertical and a semiangle 54°44′ (see Fig. 80-1). However, although in principle this result is also valid for a rocket projected almost vertically, the orbit that would arise from following this program would then intersect the earth's surface after take-off!

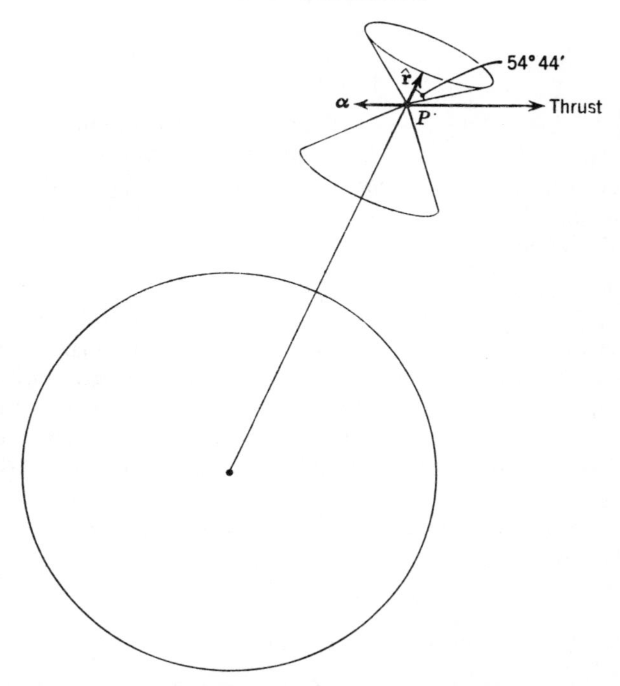

Fig. 80-1. Optimum thrust direction for propelled rocket moving in the field of the earth. The thrust should lie outside of the cone in order that $|\boldsymbol{\alpha} \cdot \hat{\mathbf{r}}| \leq 1/\sqrt{3}$.

81. THE BOLTZMANN AND NAVIER-STOKES EQUATIONS

The theorem of Liouville discussed in Sec. 70 refers to the density of representative points in the $2f$-dimensional phase space of the system described by the Hamiltonian $H(z_\alpha, t)$. Thus if the system under consideration consists of a gas of N interacting point particles, the space to which the theorem applies is the $6N$-dimensional phase space of the whole system, a space usually referred to as Γ space in the theory of statistical mechanics. The necessity for considering this multi-dimensional space arises from the fact that in general the Hamiltonian of the system is a function of all of the coordinates and all of the momenta of the N particles.

In this section we consider a theorem that is valid in the much simpler six-dimensional phase space, called the μ space, of a single particle. In such a space the system (a viscous gas) may be represented by N points, each point corresponding to the position and momentum of one of the particles in the system. As time develops, these N points trace out paths in μ space, and if no particles are created or destroyed we may write down a law expressing the conservation of these points which is analogous to the differential form for the law of conservation of electric charge in ordinary three-dimensional space.

If f is the density of representative points in μ space and if μ_m ($m = 1 \cdot \cdot \cdot 6$) denotes the coordinates in μ space ($\mu_1, \mu_2, \mu_3 \equiv \mathbf{r}, \mu_4, \mu_5, \mu_6 \equiv \mathbf{p}$), we have

$$\frac{\partial f}{\partial t} + \sum_m \frac{\partial}{\partial \mu_m} (f\dot{\mu}_m) = 0 \tag{81.1}$$

$$\frac{\partial f}{\partial t} + \frac{\partial}{\partial x_i} (fv_i) + \frac{\partial}{\partial p_i} (f\dot{p}_i) = 0 \tag{81.2}$$

The similarity of this equation to (70.10) lies in the fact that each is expressing a law of continuity, but here the continuity refers to the number of particles in a system, rather than, as with (70.10), to the number of systems that are under consideration.

Equation (81.2) is one form of the *Boltzmann equation*. In applications of this equation to problems involving a large number of particles it is usually convenient to divide $\dot{p}_i$ into two parts, that due to externally applied fields, which are assumed to be known, and that arising from collisions of the particles with each other and with the walls of the container. For further study of this basic equation of statistical mechanics the reader is referred to the books listed at the end of this chapter.

If (81.2) is integrated over momentum space, the last term becomes an integral of the normal component of the current of particles over the bounding surface of the space. Since no particles enter or leave momentum space, such integrals are zero. Hence defining the spatial density n of particles by

$$n(\mathbf{r}) = \int f \, dp_x \, dp_y \, dp_z \tag{81.3}$$

we have

$$\frac{\partial n}{\partial t} + \frac{\partial}{\partial x_i} (n\bar{v}_i) = 0 \tag{81.4}$$

where

$$n\bar{v}_i = \int fv_i \, dp_x \, dp_y \, dp_z \tag{81.5}$$

so that $\bar{v}_i$ is the average velocity of the particles that are in the neighborhood of the point x_i. Equation (81.4) merely expresses the law of conservation of particles in coordinate space.

Multiplication of (81.2) by p_j and by p^2, followed by integration over momentum space, leads to the equations

$$\frac{\partial(n\bar{p}_j)}{\partial t} + \frac{\partial}{\partial x_i}(\overline{nv_ip_j}) = n\bar{\dot{p}}_i \tag{81.6}$$

$$\frac{\partial}{\partial t}(n\overline{p^2}) + \frac{\partial}{\partial x_i}(\overline{nv_ip^2}) = 2n\overline{\mathbf{p}\cdot\dot{\mathbf{p}}} = n\overline{\frac{d}{dt}p^2} \tag{81.7}$$

The right-hand sides of these equations are obtained by integrating by parts and setting integrals over the surface of momentum space equal to zero. As in (81.5), the bar over a variable denotes its value averaged over momentum space, in the sense defined by

$$n\bar{X} = \int fX\,dp_x\,dp_y\,dp_z \tag{81.8}$$

If there are no magnetic fields present and the particles all have the same mass m, we may write $\mathbf{p} = m\mathbf{v}$, $\dot{\mathbf{p}} = \mathbf{F}$ so that (81.6) and (81.7) become

$$m\frac{\partial}{\partial t}(n\bar{v}_j) + m\frac{\partial}{\partial x_i}(\overline{nv_iv_j}) = mn\bar{\dot{v}}_j = n\bar{F}_j \tag{81.9}$$

$$\frac{1}{2}m\frac{\partial}{\partial t}(n\overline{v^2}) + \frac{m}{2}\frac{\partial}{\partial x_i}(\overline{nv_iv^2}) = n\overline{\mathbf{v}\cdot\mathbf{F}} \tag{81.10}$$

We may imagine the volume of the gas divided up into a large number of cells, each of linear dimensions ϵ, where ϵ is greater than the range of forces between individual molecules. All collisions would then occur between molecule pairs that were in the same cell. Since momentum is conserved during a collision, the net force on all the molecules in a given cell would be unaltered by a collision between two of their number, and in the limit $\epsilon \to 0$ the term $\bar{F}_i$ on the right-hand side of (81.9) would represent the average *external* force per unit volume. Thus, for very short range forces between the particles we may greatly simplify the analysis by neglecting the effects of collisions on the right-hand side of (81.9), whereas for long range forces, such as occur in plasmas and liquids, transfer of momentum between different cells during collisions must be considered.

From (81.4), (81.9) we have, by eliminating $\partial n/\partial t$,

$$\rho\,Dv_i = \rho G_i - \frac{\partial}{\partial x_j}p_{ij} \tag{81.11}$$

where

$$p_{ij} = \rho(\overline{v_iv_j} - \bar{v}_i\bar{v}_j), \qquad G_i = \overline{F_i} \tag{81.12}$$

D denotes differentiation along the path of a particle moving with the average velocity $\bar{v}_k$:

$$D \equiv \frac{\partial}{\partial t} + \bar{v}_k \frac{\partial}{\partial x_k}$$

and $\rho = nm$ is the mass density. The average of the diagonal elements of p_{ij} is called the *mean pressure*

$$\bar{p} = \frac{1}{3} \rho \sum_{i=1}^{3} \left(\overline{v_i v_i} - \bar{v}_i \bar{v}_i \right) \tag{81.13}$$

On the other hand, it is found experimentally that

$$p_{ij} \approx -\eta \left(\frac{\partial \bar{v}_j}{\partial x_i} + \frac{\partial \bar{v}_i}{\partial x_j} \right) \qquad (i \neq j) \tag{81.14}$$

where η is the viscosity of the gas. In general η could be a function of position. To understand (81.14) in a particular case, we note that in the coordinate system moving with the neighboring part of the gas ($\bar{v}_i = 0$) we would have from (81.14) for flow in the z direction

$$-\eta \frac{\partial \bar{v}_z}{\partial x} = \rho \overline{v_x v_z}$$

and since the right-hand side is the tangential force per unit area due to transport of molecules, this reduces to the usual definition of viscosity for a gas. In a liquid, however, long range intermolecular forces may produce a major contribution to the viscosity which is not included here.

We therefore assume that p_{ij} may be written thus:

$$p_{ij} = \left(p - \lambda \frac{\partial \overline{v_k}}{\partial x_k} \right) \delta_{ij} - \mu \left(\frac{\partial \overline{v_j}}{\partial x_i} + \frac{\partial \overline{v_i}}{\partial x_j} \right) \tag{81.15}$$

where λ may be a scalar function of position (called the *dilatation viscosity*) and p also is a scalar. Equation (81.15) reduces to (81.14) for $i \neq j$, and is the most general symmetrical tensor that can be constructed as a linear function of the first derivatives of the $\overline{v_k}$ with respect to the x_j.

Comparison of (81.15) with (81.13) yields

$$\bar{p} = p - \kappa \operatorname{div} \bar{\mathbf{v}}$$

where

$$\kappa = \frac{2\mu + 3\lambda}{3}$$

is the *bulk viscosity*.

For monatomic gases, $\kappa = 0$, a result known as *Stokes' relation*, and in this case p, the thermodynamic pressure, becomes identical with the mean pressure $\bar{p}$.

From (81.11), (81.15), for $\kappa = 0$, $\lambda = -(2\mu/3) = \text{const}$, we have then the *Navier-Stokes equation:*

$$\rho\, D\bar{\mathbf{v}} = \rho\mathbf{G} - \nabla p + \frac{\mu}{3}\nabla(\nabla \cdot \bar{\mathbf{v}}) + \mu\nabla \cdot (\nabla\bar{\mathbf{v}}) \tag{81.16}$$

GENERAL REFERENCES

B. D. Fried, Chapter 4 in *Space Technology*, H. S. Seifert (ed.). Wiley, New York (1959).

R. D. Present, *Kinetic Theory of Gases*. McGraw-Hill, New York (1958), Chapter 2.

R. H. Fowler and E. A. Guggenheim, *Statistical Thermodynamics*. Cambridge University Press (1939), p. 271.

I. I. Rabi, N. F. Ramsey, and J. Schwinger, *Rev. Mod. Phys.*, 26, 167 (1954).

D. J. X. Montgomery, *Cosmic Ray Physics*. Princeton University Press (1949), p. 312.

15 CONTINUOUS MEDIA AND FIELDS

82. THE STRETCHED STRING

In this chapter we want to formulate the problem of describing the motion of a continuous medium and by extension the problem of describing fields such as the electromagnetic field. This does not imply that there is a mechanical model for the electromagnetic field, but only that the mathematical methods suitable for treating continuous mechanical media can be used to describe fields also. We begin with the simplest continuous medium, the stretched uniform string with only longitudinal displacements taken into account.

In Sec. 44 we studied the system consisting of a light string loaded at uniform intervals a with equal masses μ. The longitudinal displacement of the ρth particle is ψ_ρ, and the kinetic and potential energies were found to be

$$T = \frac{1}{2}\sum_{\rho=0}^{n+1} \mu\dot{\psi}_\rho^{\,2}$$

$$(82.1) \qquad V = \frac{1}{2}\sum_{\rho=1}^{n+1} Ka\left(\frac{\psi_{\rho-1} - \psi_\rho}{a}\right)^2$$

Here $\psi_0 = \psi_{n+1} = 0$ if the string has fixed ends, or $\psi_0 = \psi_n$, $\psi_1 = \psi_{n+1}$ if periodic boundary conditions are applied. If n is allowed to approach infinity and a to approach zero so that $na = l$, a constant, T and V become integrals over the length of the string,

$$T = \int_0^l dx\,\sigma\left(\frac{\partial\psi}{\partial t}\right)^2 \tag{82.2}$$
$$V = \int_0^l dx\,K\left(\frac{\partial\psi}{\partial x}\right)^2$$

where ψ is now a function of two continuous variables, x and t, and where $\sigma = \lim \mu/a$ is the linear density of the string.

The equations of motion for the discrete mass problem are obtained from the Lagrangian $T - V = L$ by use of Hamilton's principle. Under periodic boundary conditions these equations were seen in Sec. 44 to have traveling wave solutions of the form

$$\psi_\rho = A\sin(k\rho a - \omega t + \theta) \tag{82.3}$$

where

$$k = \frac{2\pi m}{na} \quad (m \text{ integer}), \qquad \omega^2 = \frac{2K}{\mu a}(1 - \cos ka) \tag{82.4}$$

The equations of motion for the continuous string are obtained in a similar way. Hamilton's principle may be written

$$\Delta\int_{t_1}^{t_2} dt \int_0^l dx\,\frac{1}{2}\left[\sigma\left(\frac{\partial\psi}{\partial t}\right)^2 - K\left(\frac{\partial\psi}{\partial x}\right)^2\right] = 0 \tag{82.5}$$

Here the variation is to consist of a change in the dependence of ψ on x and on t, this change to vanish at $t = t_1, t_2$ and to be the same (perhaps zero) at $x = 0, l$.

We carry out the calculation of the variation for a general Lagrangian L which is the integral of a *Lagrangian density* $\mathfrak{L}(\psi, \psi_x, \psi_t)$ over x. Here $\psi_x = \partial\psi/\partial x, \cdots$. Then

$$L = \int_{x_1}^{x_2} dx\mathfrak{L}, \qquad \Phi = \int_{t_1}^{t_2} dtL \tag{82.6}$$

and

$$\Delta\Phi = \int_{t_1}^{t_2} dt \int_{x_1}^{x_2} dx\left\{\frac{\partial\mathfrak{L}}{\partial\psi}\,\delta\psi + \frac{\partial\mathfrak{L}}{\partial\psi_x}\,\delta\psi_x + \frac{\partial\mathfrak{L}}{\partial\psi_t}\,\delta\psi_t\right\} \tag{82.7}$$
$$= \int_{t_1}^{t_2} dt \int_{x_1}^{x_2} dx\left\{\frac{\partial\mathfrak{L}}{\partial\psi} - \frac{\partial}{\partial x}\frac{\partial\mathfrak{L}}{\partial\psi_x} - \frac{\partial}{\partial t}\frac{\partial\mathfrak{L}}{\partial\psi_t}\right\}\delta\psi$$

The second step involves integration by parts, the integrated terms vanishing by hypothesis. Because $\delta\psi$ is arbitrary except at the boundaries, the vanishing of $\Delta\Phi$ implies that

$$\frac{\partial \mathfrak{L}}{\partial \psi} - \frac{\partial}{\partial x}\frac{\partial \mathfrak{L}}{\partial \psi_x} - \frac{\partial}{\partial t}\frac{\partial \mathfrak{L}}{\partial \psi_t} = 0 \tag{82.8}$$

This partial differential equation is the equation of motion of the medium. It is the limiting form of the n ordinary differential equations describing the motion of the string with discrete masses as $n \rightarrow \infty$.

For the particular Lagrangian density

$$\mathfrak{L} = \frac{1}{2}\left[\sigma\left(\frac{\partial \psi}{\partial t}\right)^2 - K\left(\frac{\partial \psi}{\partial x}\right)^2\right] \tag{82.9}$$

of (82.5), this partial differential equation is

$$K\frac{\partial^2 \psi}{\partial x^2} - \sigma\frac{\partial^2 \psi}{\partial t^2} = 0 \tag{82.10}$$

which is the one-dimensional wave equation. It possesses traveling wave solutions of the form

$$\psi(x, t) = A \sin(kx - \omega t + \theta) \tag{82.11}$$

with

$$k = \frac{2\pi m}{l} \quad (m \text{ integer}), \qquad \omega^2 = \frac{K}{\sigma}k^2 \tag{82.12}$$

as may be verified by direct substitution. These solutions are the limiting forms of (82.3), (82.4) as $a \rightarrow 0$. This is a necessary condition for the reasonableness of our description of the continuous medium.

The motion of the string has associated with it an energy density, an energy flux, and a momentum density. The energy density, U, is the sum of the kinetic energy per unit length and the potential energy per unit length

$$U = \frac{1}{2}\left[\sigma\left(\frac{\partial \psi}{\partial t}\right)^2 + K\left(\frac{\partial \psi}{\partial x}\right)^2\right] \tag{82.13}$$

The energy flux, S, is the rate at which the string to the left of a point x does work on the string to the right of x. This is the force exerted by the left part of the string on the right part times the velocity of the string at x, i.e.,

$$S = \left(-K\frac{\partial \psi}{\partial x}\right)\frac{\partial \psi}{\partial t} \tag{82.14}$$

The momentum density, G, is the product of the density of the string and the velocity. The density is affected by the amount the string has been stretched in the neighborhood of the point x

$$\sigma(x, t) = \sigma\left(1 - \frac{\partial\psi}{\partial x}\right) \tag{82.15}$$

so that

$$G = \sigma\left(1 - \frac{\partial\psi}{\partial x}\right)\frac{\partial\psi}{\partial t} \tag{82.16}$$

The part of G which is linear in ψ, namely $\sigma\,\partial\psi/\partial t$, arises from the equilibrium density of the string, and will average to zero over any long time interval. The other part of G depends on the density change, and if the quantities $\partial\psi/\partial x$ and $\partial\psi/\partial t$ are related as in a traveling wave, it will not have a vanishing average. Thus with a wave of the form (82.11),

$$G = \sigma[1 - Ak\cos(kx - \omega t)][-A\omega\cos(kx - \omega t)]$$

In field theories there is nothing analogous to the equilibrium density of the string. Since this term vanishes either in the mean or completely, we ignore it henceforth and let

$$G = -\sigma\frac{\partial\psi}{\partial x}\frac{\partial\psi}{\partial t} \tag{82.17}$$

G is positive for a wave traveling to the right, negative for one traveling to the left.

The energy flux and the momentum density are proportional to each other

$$S = \frac{K}{\sigma}G = v^2G \tag{82.18}$$

v being the phase velocity of the wave as given by (44.18). The time average of the energy density is related to the time average of the energy flux

$$\bar{S} = \bar{U}v \tag{82.19}$$

which is easily seen for a wave of the form (82.11). It is true for any wave which is a superposition of waves of this form with different ω's, for the time average of the product of two sines with different frequencies vanishes. All the waves in the superposition must be traveling to the right. Waves traveling to the left give rise to a minus sign in (82.19) and a mixture of waves in two directions would result in cancellations in $\bar{S}$ but not in $\bar{U}$ so that (82.19) would not be true. Thus in a standing wave $\bar{S} = 0$ but $\bar{U} \neq 0$.

83. ENERGY-MOMENTUM RELATIONS

In the simple problem of the longitudinally vibrating string the expressions for the energy density, energy flux, and momentum density can be obtained directly from their definitions. They can also be obtained from the Lagrangian density in a way which can be applied to more complicated systems.

To see how this can be done we return to Hamilton's principle for discrete systems and look at the total variation of the functional Φ as given by (57.7):

$$\Delta\Phi = [-H\,\Delta t + p_i\,\Delta q_i]_{t_1}^{t_2}$$

If X is a cartesian coordinate of the center of mass of the system, choose the variation of the coordinates to be

$$\Delta q_i = \frac{\partial q_i}{\partial X}\,\Delta X$$

The momentum P conjugate to X is the component of the total linear momentum of the system in the direction of X

$$P = p_i\,\frac{\partial q_i}{\partial X}$$

and thus for this variation

$$\Delta\Phi = [-H\,\Delta t + P\,\Delta X]_{t_1}^{t_2} \tag{83.1}$$

The energy of the system is the coefficient of $-\Delta t$ in $\Delta\Phi$, and the X component of the linear momentum is the coefficient of ΔX. Δt describes a displacement of the entire system in time, and ΔX describes a rigid displacement of the entire system in space.

If H, and therefore L, does not contain the time explicitly, then Φ is not changed by making a change Δt at both t_1 and t_2, so that with $\Delta X = 0$, H must have the same value at these times and energy is conserved. If L is not changed by the translation ΔX, then if the same translation is made at times t_1 and t_2, Φ is not altered and P must have the same value at these times and linear momentum is conserved. In the continuum case we identify the coefficient of $-\Delta t$ as the energy, and the coefficient of a translation ΔX as the linear momentum. If $\mathcal{L}$ does not depend explicitly on t or on X, then by the same arguments energy and momentum are conserved.

From (82.6), we may calculate the variation $\Delta\Phi$ produced by varying ψ as a function of x and t, by varying the boundaries x_1 and x_2, and by varying the end times t_1 and t_2. We omit all terms which lead to (82.8).

$$\Delta\Phi = \int_{t_1}^{t_2} dt \int_{x_1}^{x_2} dx \left\{ \frac{\partial}{\partial x} \left[\left(\mathfrak{L} - \frac{\partial \mathfrak{L}}{\partial \psi_x} \psi_x \right) \Delta x + \frac{\partial \mathfrak{L}}{\partial \psi_x} (\Delta\psi - \psi_t \, \Delta t) \right] + \frac{\partial}{\partial t} \left[\left(\mathfrak{L} - \frac{\partial \mathfrak{L}}{\partial \psi_t} \psi_t \right) \Delta t + \frac{\partial \mathfrak{L}}{\partial \psi_t} (\Delta\psi - \psi_x \, \Delta x) \right] \right\} \tag{83.2}$$

where we have introduced the total variation

$$\begin{aligned} \Delta\psi &= \delta\psi + \psi_x \, \Delta x + \psi_t \, \Delta t \\ &= \delta\psi + \psi_\mu \, \Delta\xi_\mu \qquad (\mu = x, t) \end{aligned} \tag{83.3}$$

and where the partial integration terms left over from (82.7) have not been integrated. The form of (83.2) suggests the usefulness of the quantity

$$T_{\mu\nu} = \mathfrak{L}\delta_{\mu\nu} - \frac{\partial \mathfrak{L}}{\partial \psi_\mu} \psi_\nu \qquad (\mu, \nu = x, t) \tag{83.4}$$

because then (83.2) becomes, if $\Delta\psi = 0$,

$$\Delta\Phi = \int_{t_1}^{t_2} dt \int_{x_1}^{x_2} \frac{\partial}{\partial \xi_\mu} (T_{\mu\nu} \, \Delta\xi_\nu) \tag{83.5}$$

with $\xi_x = x$, $\xi_t = t$.

Equation (83.5) is valid for any variation of the boundaries if ψ is not changed on the boundaries. In particular, let us choose the $\Delta\xi_\nu$ independent of x and t so that they represent displacements of the entire system in the space and time directions. These displacements are comparable to those occurring in (83.1). Then, if $\Delta\xi_x = 0$,

$$\begin{aligned} \Delta\Phi &= \int_{t_1}^{t_2} dt \int_{x_1}^{x_2} dx \left\{ \frac{\partial T_{xt}}{\partial x} + \frac{\partial T_{tt}}{\partial t} \right\} \Delta t \\ &= \left\{ \int_{t_1}^{t_2} dt [T_{xt}]_{x_1}^{x_2} + \int_{x_1}^{x_2} dx [T_{tt}]_{t_1}^{t_2} \right\} \Delta t \end{aligned} \tag{83.6}$$

The first term vanishes under either periodic or fixed boundary conditions. The second term is to be identified with $-[H]_{t_1}^{t_2}$ in the discrete system. This integral is over the spatial extent of the system and the integrand may be identified with the energy density, except for sign:

$$U = -T_{tt} \tag{83.7}$$

If Δt is chosen to vanish outside the spatial interval $x_2' - x_1'$, which is now taken to be less than the length of the system, $\Delta\Phi$ still has the form (83.6) but the first integral now no longer must vanish. The second integral

$$\int_{x_1}^{x_2} dx[T_{tt}]_{t_1}^{t_2}$$

gives the decrease in the energy within this space interval during the time $t_2 - t_1$. The first integral can now be interpreted as the energy flow at the ends integrated over this time. Thus we are led to consider $-T_{xt}$ as an energy flow:

$$S = -T_{xt} \tag{83.8}$$

If now we put $\Delta\xi_t = \Delta t = 0$, we obtain

$$\Delta\Phi = \int_{t_1}^{t_2} dt \int_{x_1}^{x_2} dx \left\{ \frac{\partial T_{xx}}{\partial x} + \frac{\partial T_{tx}}{\partial t} \right\} \Delta x \tag{83.9}$$

By reasoning similar to that above we conclude that there is a momentum density given by

$$G = T_{tx} \tag{83.10}$$

and a momentum flux (force) given by

$$F = T_{xx} \tag{83.11}$$

The quantities $T_{\mu\nu}$ constitute the energy-momentum tensor of the medium (or of the field). It will be generalized to three-dimensional systems in the next section. The reader may readily verify that for the Lagrangian density $\mathfrak{L}$ of (82.9) these results for U, S, G agree with those obtained earlier. F is not the force exerted by one part of the string on the other part, but is a force on the wave, producing a rate of change of the momentum of the wave.

Conservation of energy and momentum are consequences of the relations

$$\frac{\partial T_{\mu\nu}}{\partial \xi_\mu} = 0 \tag{83.12}$$

which when written out in detail are

$$\begin{aligned} \frac{\partial F}{\partial x} + \frac{\partial G}{\partial t} &= 0 \qquad (\nu = x) \\ \frac{\partial S}{\partial x} + \frac{\partial U}{\partial t} &= 0 \qquad (\nu = t) \end{aligned} \tag{83.13}$$

If either is integrated over such a large range of x that F or S vanishes at the end points, or if periodic boundary conditions are applied and the x integration is over one period, then there results

$$\frac{d}{dt}\int_{x_1}^{x_2} G\,dx = 0, \qquad \frac{d}{dt}\int_{x_1}^{x_2} U\,dx = 0 \tag{83.14}$$

which are the conservation laws for total linear momentum and energy. Equation (83.12) is a consequence of the equation of motion

$$\begin{aligned}\frac{\partial T_{\mu\nu}}{\partial \xi_\mu} &= \frac{\partial}{\partial \xi_\mu}\left(\mathcal{L}\,\delta_{\mu\nu} - \frac{\partial \mathcal{L}}{\partial \psi_\mu}\psi_\nu\right) \\ &= \frac{\partial \mathcal{L}}{\partial \psi}\psi_\nu + \frac{\partial \mathcal{L}}{\partial \psi_\mu}\frac{\partial \psi_\mu}{\partial \xi_\nu} - \frac{\partial \mathcal{L}}{\partial \psi_\mu}\frac{\partial \psi_\nu}{\partial \xi_\mu} - \left(\frac{\partial}{\partial \xi_\mu}\frac{\partial \mathcal{L}}{\partial \psi_\mu}\right)\psi_\nu \\ &= 0\end{aligned} \tag{83.15}$$

because $\partial\psi_\mu/\partial\xi_\nu = \partial^2\psi/\partial\xi_\mu\,\partial\xi_\nu = \partial\psi_\nu/\partial\xi_\mu$.

84. THREE-DIMENSIONAL MEDIA AND FIELDS

The method developed in the last two sections can be generalized immediately to three dimensions. Let the configuration of the system be described by the r functions $\psi^\alpha(x, y, z, t)$. The dynamic properties of the system are described by a Lagrangian density $\mathcal{L}(\psi^\alpha, \nabla\psi^\alpha, \dot\psi^\alpha)$, which we assume not to depend on the coordinates or time explicitly. Hamilton's principle leads to equations of motion

$$\frac{\partial \mathcal{L}}{\partial \psi^\alpha} - \frac{\partial}{\partial x_j}\frac{\partial \mathcal{L}}{\partial \psi_j{}^\alpha} - \frac{\partial}{\partial t}\frac{\partial \mathcal{L}}{\partial \dot\psi^\alpha} = 0 \tag{84.1}$$

There is one such equation for each ψ^α coming from the independent variations $\delta\psi^\alpha$. Here $\psi_j{}^\alpha = \partial\psi^\alpha/\partial x_j$.

Equation (94.1) has a remarkable symmetry between the space coordinates and the time. It is unchanged on multiplying the coordinates and the time by arbitrary numerical factors. This suggests introducing coordinates x_μ in space-time by

$$x_\mu = \{x, y, z, at\} \tag{84.2}$$

with a a parameter to be chosen later. Then (84.1) becomes

$$\frac{\partial \mathcal{L}}{\partial \psi^\alpha} - \frac{\partial}{\partial x_\mu}\frac{\partial \mathcal{L}}{\partial \psi_\mu{}^\alpha} = 0 \tag{84.3}$$

This form is especially useful when formulating a relativistic theory

If $\mathfrak{L}$ is invariant under Lorentz transformation and if $a = ic$, where c is the velocity of light, then (84.3) is a manifestly covariant equation.

The functional ϕ of (82.6) is generalized to

$$\phi = \int d^4x\mathfrak{L}(\psi^\alpha, \psi_\mu{}^\alpha) \tag{84.4}$$
$$= \int_{t_1}^{t_2} dt \int_V d^3x\mathfrak{L}(\psi^\alpha, \nabla\psi^\alpha, \dot{\psi}^\alpha)$$

The variation of ϕ is to be calculated when the ψ^α are varied in their dependence on x_μ, when the end times are varied, and when the spatial boundary is varied. The integration is taken between two times, t_1 and t_2, which is not a relativistically invariant statement but which could be altered so as to become so at the expense of some mathematical complexity. Because our variations are not described relativistically, there is no point in treating the time and the space variables symmetrically. Let $\mathbf{n}$ be the outward normal to the spatial boundary surface S, and let $d\sigma$ be a surface element of that surface. Then

$$\Delta\phi = \int_{t_1}^{t_2} dt \int_V d^3x \left\{\frac{\partial\mathfrak{L}}{\partial\psi^\alpha}\,\delta\psi^\alpha + \frac{\partial\mathfrak{L}}{\partial(\nabla\psi^\alpha)}\cdot\delta\,\nabla\psi^\alpha \right. \tag{84.5}$$
$$\left. + \frac{\partial\mathfrak{L}}{\partial\dot{\psi}^\alpha}\,\delta\dot{\psi}^\alpha\right\} + \left[\int d^3x\mathfrak{L}\right]_{t_1}^{t_2}\Delta t + \int_{t_1}^{t_2} dt \int_S d\sigma\mathfrak{L}\mathbf{n}\cdot\Delta\mathbf{r}$$

The second term in the first integral may be integrated by parts using the relation

$$\nabla\cdot(\lambda\mathbf{v}) = \mathbf{v}\cdot\nabla\lambda + \lambda\nabla\cdot\mathbf{v}$$

and the third term is integrated by parts in the usual way:

$$\Delta\phi = \int_{t_1}^{t_2} dt \int_V d^3x \left\{\frac{\partial\mathfrak{L}}{\partial\psi^\alpha} - \nabla\cdot\frac{\partial\mathfrak{L}}{\partial(\nabla\psi^\alpha)} - \frac{\partial}{\partial t}\frac{\partial\mathfrak{L}}{\partial\dot{\psi}^\alpha}\right\}\delta\psi^\alpha \tag{84.6}$$
$$+ \left[\int_V d^3x\left(\mathfrak{L}\,\Delta t + \frac{\partial\mathfrak{L}}{\partial\dot{\psi}^\alpha}\,\delta\psi^\alpha\right)\right]_{t_1}^{t_2}$$
$$+ \int_{t_1}^{t_2} dt \int_S d\sigma\left(\mathfrak{L}\mathbf{n}\cdot\Delta\mathbf{r} + \frac{\partial\mathfrak{L}}{\partial(\nabla\psi^\alpha)}\cdot\mathbf{n}\,\delta\psi^\alpha\right)$$

ϕ is to vanish if there are no variations on the boundaries, so the coefficient of $\delta\psi^\alpha$ in the first integral of (84.6) must vanish. This gives (84.1).

We now introduce the total variation of ψ^α on the boundaries. This is more complicated in three dimensions than it was in one dimension because the components ψ^α describing the field may undergo linear transformations among themselves if the system is rotated relative to the coordinate system to which the components refer. This variation must be included along with those arising from the change in the values of the coordinates when the system is displaced.

Let the new coordinates of a point differ from the old by Δx_k, where

$$\Delta x_k = a_k + \omega_{kl}x_l, \qquad \omega_{kl} + \omega_{lk} = 0 \tag{84.7}$$

The a_k represent a rigid translation of the system, the ω_{kl} a rigid rotation of the system about the origin. Only the rotation produces a mixing of the field components. Let this mixing be described by

$$\delta_R\psi^\alpha = \tfrac{1}{2}S_{kl}^{\alpha\beta}\omega_{kl}\psi^\beta, \qquad S_{kl}^{\alpha\beta} = -S_{lk}^{\alpha\beta} \tag{84.8}$$

Then the total variation in ψ^α may be written as

$$\Delta\psi^\alpha = \delta\psi^\alpha + \nabla\psi^\alpha \cdot \Delta\mathbf{r} + \dot{\psi}^\alpha\,\Delta t + \delta_R\psi^\alpha \tag{84.9}$$

Equation (84.6) may be rewritten dropping the terms which lead to the equations of motion (84.1). Thus

$$\begin{aligned}\Delta\phi = \Bigg[\int_V d^3x \Bigg\{&\left(\mathfrak{L} - \frac{\partial\mathfrak{L}}{\partial\dot{\psi}^\alpha}\dot{\psi}^\alpha\right)\Delta t \\ &+ \frac{\partial\mathfrak{L}}{\partial\dot{\psi}^\alpha}(\Delta\psi^\alpha - \delta_R\psi^\alpha - \nabla\psi^\alpha\cdot\Delta\mathbf{r})\Bigg\}\Bigg]_{t_1}^{t_2} \\ + \int_{t_1}^{t_2} dt \int_S d\sigma \Bigg\{&\left(\mathfrak{L}n_k - \frac{\partial\mathfrak{L}}{\partial\psi_l{}^\alpha}n_l\psi_k{}^\alpha\right)\Delta x_k \\ &+ \frac{\partial\mathfrak{L}}{\partial\psi_l{}^\alpha}n_l(\Delta\psi^\alpha - \delta_R\psi^\alpha - \dot{\psi}_\alpha\,\Delta t)\Bigg\}\end{aligned} \tag{84.10}$$

The form of this equation suggests introducing the tensor

$$T_{\mu\nu} = \mathfrak{L}\delta_{\mu\nu} - \frac{\partial\mathfrak{L}}{\partial\psi_\mu{}^\alpha}\psi_\nu{}^\alpha \qquad (\mu,\nu = 1, 2, 3, 4) \tag{84.11}$$

in terms of which (84.10) becomes, with $\Delta\psi^\alpha = 0$,

$$\begin{aligned}\Delta\phi = \Bigg[\int_V d^3x \left\{T_{44}\,\Delta t + T_{4k}\,\Delta x_k - \frac{\partial\mathfrak{L}}{\partial\dot{\psi}^\alpha}\delta_R\psi^\alpha\right\}\Bigg]_{t_1}^{t_2} \\ + \int_{t_2}^{t_1} dt \int_S d\sigma \left\{T_{lk}n_l\,\Delta x_k + T_{l4}n_l\,\Delta t - \frac{\partial\mathfrak{L}}{\partial\psi_l{}^\alpha}n_l\,\delta_R\psi^\alpha\right\}\end{aligned} \tag{84.12}$$

If the volume integral is taken over all space, the coefficient of Δt should be the negative of the energy of the field. Thus

$$H = \int d^3x(-T_{44}) \tag{84.13}$$

We may take $-T_{44}$ to be the energy density in the field. Then (84.12) can be interpreted when the volume integral is extended over only a part of the field. Taking $\Delta x_k = 0$, hence $\delta_R \psi^\alpha = 0$,

$$\Delta\phi = \left[\int_V d^3x T_{44}\, \Delta t\right]_{t_1}^{t_2} + \int_{t_1}^{t_2} dt \int_S d\sigma T_{l4} n_l\, \Delta t \tag{84.14}$$

The first term is the decrease in the energy contained in V between times t_1 and t_2, which vanishes if V contains the whole field so that energy is conserved. The surface integral can be interpreted as the energy flux through the surface S during this time interval. Thus we may take $-T_{l4}$ to be the flux of energy in the l direction. These assignments are not unique because the value of an integral does not in general determine the value of the integrand at every point.

If the displacement of the system is a translation, the coefficient of $\Delta x_k = a_k$ should be the total linear momentum in the k direction, P_k. There again $\delta_R \psi^\alpha = 0$ and we have, putting $\Delta t = 0$,

$$\Delta\phi = \left[\int_V d^3x T_{4k}\, \Delta x_k\right]_{t_1}^{t_2} + \int_{t_1}^{t_2} dt \int_S d\sigma T_{lk} n_l\, \Delta x_k \tag{84.15}$$

By an argument similar to that used above, T_{4k} may be taken as the density of the k component of linear momentum and T_{lk} as the l component of the flux of the k component of momentum, i.e., as the lk component of the stress. The tensor $T_{\mu\nu}$ is called the stress-energy tensor of the field.

Finally, let us consider a rigid rotation of the system. Let

$$\omega_{kl} = -\Delta\theta_j \epsilon_{jkl} \tag{84.16}$$

so that $\Delta\theta_j$ is the infinitesimal angle of rotation of the system about the j direction. Then from (84.8)

$$\delta_R \psi^\alpha = -\tfrac{1}{2}\Delta\theta_j \epsilon_{jkl} S_{kl}^{\alpha\beta} \psi^\beta$$

Keeping only variations resulting from the rotation and allowing the volume integral to include the entire field, (84.12) becomes

$$\Delta\phi = \left[\Delta\theta_j \epsilon_{jkl} \int d^3x \left\{ -T_{4k} x_l + \frac{1}{2} \frac{\partial \mathfrak{L}}{\partial \dot{\psi}^\alpha} \psi^\beta S_{kl}^{\alpha\beta} \right\}\right]_{t_1}^{t_2} \tag{84.17}$$

The j component of angular momentum, J_j, is the coefficient of $\Delta\theta_j$. Letting $J_j \epsilon_{jkl} = M_{kl}$, we see that

$$M_{kl} = \int d^3x \left\{ -T_{4k} x_l + T_{4l} x_k + \frac{\partial \mathfrak{L}}{\partial \dot{\psi}^\alpha} \psi^\beta S_{kl}^{\alpha\beta} \right\} \tag{84.18}$$

The first two terms are exactly what one would expect for an angular momentum, being analogous to $x_1 p_2 - x_2 p_1 = l_3$ in the case of a particle. The last term is dependent on the transformation properties of the field variables. In quantum theory this term is referred to as "spin" angular momentum of the field. For a scalar field (only one component) it vanishes.

85. HAMILTONIAN FORM OF FIELD THEORY

In the last three sections we have shown how a field theory can be formulated in terms of a Lagrangian, or a Lagrangian density. The analogy with particle mechanics can be carried further and the equations of motion can be written in Hamiltonian form, the analogs of canonically conjugate variables being defined. This form of the theory can be made relativistic throughout. We do not do this here as it involves rather more development of the formal aspects of relativity theory than is made in this book.

The starting place for this discussion is the end point part of the variation $\Delta\Phi$. In this section we shall always take the volume integrals occurring to be over the entire field so that the surface integrals vanish. From (84.10), then, we write

$$\Delta\Phi = \left[\int d^3x \left\{\left(\mathcal{L} - \frac{\partial\mathcal{L}}{\partial\dot\psi^\alpha}\dot\psi^\alpha\right)\Delta t + \frac{\partial\mathcal{L}}{\partial\dot\psi^\alpha}\Delta\psi^\alpha\right\}\right]_{t_1}^{t_2} \tag{85.1}$$

where the terms coming from the coordinate variations have been dropped. This expression is similar in form to the corresponding one in particle mechanics, (57.6). Denoting the momentum canonically conjugate to $\psi^\alpha(x)$ by $\pi^\alpha(x)$, we see that by analogy with (53.2),

$$\pi^\alpha(x) = \frac{\partial\mathcal{L}}{\partial\dot\psi^\alpha(x)} \tag{85.2}$$

If (85.2) can be solved for the $\dot\psi^\alpha$, then the quantity

$$\mathcal{H} = -\mathcal{L} + \frac{\partial\mathcal{L}}{\partial\dot\psi^\alpha}\dot\psi^\alpha \tag{85.3}$$

may be expressed in terms of $\psi^\alpha(x)$, $\pi^\alpha(x)$, and their gradients. [The gradient of $\pi^\alpha(x)$ does not occur in most systems of interest.] The Hamiltonian is then

$$H[\psi^\alpha, \pi^\alpha] = \int d^3x\mathcal{H} \tag{85.4}$$

H is a functional of the field. H differs from the energy calculated in the last section only in its form; it must be expressed as a functional of these argument functions before it can be called the Hamiltonian, just as in particle mechanics $H(q, p)$ differs from the energy only by this requirement of form.

The equations of motion follow from the variation of H, as in particle mechanics:

$$(85.5)\quad \delta H = \int d^3x \left\{ \frac{\partial \mathcal{H}}{\partial \psi^\alpha} \delta\psi^\alpha + \frac{\partial \mathcal{H}}{\partial(\nabla \psi^\alpha)} \delta \nabla\psi^\alpha + \frac{\partial \mathcal{H}}{\partial \pi^\alpha} \delta\pi^\alpha + \frac{\partial \mathcal{H}}{\partial(\nabla \pi^\alpha)} \delta \nabla\pi^\alpha \right\}$$

$$= \int d^3x \left\{ \left(\frac{\partial \mathcal{H}}{\partial \psi^\alpha} - \nabla \cdot \frac{\partial \mathcal{H}}{\partial(\nabla\psi^\alpha)} \right) \delta\psi^\alpha + \left(\frac{\partial \mathcal{H}}{\partial \pi^\alpha} - \nabla \cdot \frac{\partial \mathcal{H}}{\partial(\nabla\pi^\alpha)} \right) \delta\pi^\alpha \right\}$$

where, as always in this section, surface terms are dropped. Using the above expression for $\mathcal{H}$ and the Lagrangian field equations we obtain

$$(85.6)\quad \delta H = \int d^3x \left\{ - \frac{\partial \mathcal{L}}{\partial \psi^\alpha} \delta\psi^\alpha - \frac{\partial \mathcal{L}}{\partial(\nabla\psi^\alpha)} \delta \nabla\psi^\alpha - \pi^\alpha \delta\dot{\psi}^\alpha + \delta\pi^\alpha \dot{\psi}^\alpha + \pi^\alpha \delta\dot{\psi}^\alpha \right\}$$

$$= \int d^3x \{ -\dot{\pi}^\alpha \delta\psi^\alpha + \dot{\psi}^\alpha \delta\pi^\alpha \}$$

Identifying coefficients of the independent variations $\delta\psi^\alpha$, $\delta\pi^\alpha$, we obtain

$$(85.7)\quad \begin{aligned} \dot{\psi}^\alpha(x) &= \frac{\partial \mathcal{H}}{\partial \pi^\alpha} - \nabla \cdot \frac{\partial \mathcal{H}}{\partial(\nabla\pi^\alpha)} = \frac{\delta H}{\delta\pi^\alpha(x)} \\ \dot{\pi}^\alpha(x) &= -\frac{\partial \mathcal{H}}{\partial \psi^\alpha} + \nabla \cdot \frac{\partial \mathcal{H}}{\partial(\nabla\psi^\alpha)} = -\frac{\delta H}{\delta\psi^\alpha(x)} \end{aligned}$$

The derivative $\delta H/\delta\pi^\alpha(x)$ is called a variational derivative. Equations (85.7) are the field equations in Hamiltonian form.

A dynamical variable X of the field is a functional of the field. We consider only those functionals which can be written in the form

$$(85.8)\quad X[\psi, \pi] = \int d^3x X(\psi^\alpha, \nabla\psi^\alpha, \pi^\alpha, \nabla\pi^\alpha)$$

The Poisson bracket of two such variables is defined by

$$(85.9)\quad (X, Y) = \int d^3x \left\{ \frac{\delta X}{\delta\psi^\alpha(x)} \frac{\delta Y}{\delta\pi^\alpha(x)} - \frac{\delta Y}{\delta\psi^\alpha(x)} \frac{\delta X}{\delta\pi^\alpha(x)} \right\}$$

Let X be the value of a component ψ^α at a point x'. This can be written in the form (85.8), namely

$$X = \int d^3x \, \delta(\mathbf{x} - \mathbf{x}')\psi^\alpha(x)$$

where $\delta(\mathbf{x} - \mathbf{x}')$ is the three-dimensional Dirac delta function. Similarly, let

$$Y = \int d^3x \, \delta(\mathbf{x} - \mathbf{x}'')\pi^\beta(x)$$

With these definitions of X and Y, we see that

$$\begin{aligned}(X, Y) &= \int d^3x \, \delta(\mathbf{x} - \mathbf{x}') \, \delta(\mathbf{x} - \mathbf{x}'') \, \delta_{\alpha\beta} \\ &= \delta_{\alpha\beta} \, \delta(\mathbf{x}' - \mathbf{x}'')\end{aligned}$$

This is usually written in the form

$$(\psi^\alpha(x'), \pi^\beta(x'')) = \delta_{\alpha\beta} \, \delta(\mathbf{x}' - \mathbf{x}'') \tag{85.10}$$

This is the analog of $(q_j, p_k) = \delta_{jk}$.

In its relativistically covariant form this theory can serve as the basis for the quantum theory of fields. It is important to use a covariant form there because of the infinities which appear, and which can be interpreted only on the basis of their formal transformation properties.

EXERCISES

1. A string fixed at both ends is plucked at its middle. Find the fraction of the initial potential energy associated with each of the three lowest frequencies which are excited.

2. Write down the Lagrangian describing the transverse motion of a uniform stretched membrane, using polar coordinates. Hence find the equations of motion of a circular membrane.

3. A system is described by the Lagrangian density

$$\mathcal{L} = \tfrac{1}{2}[\dot{\psi}^2 - (\nabla\psi)^2 - \kappa^2\psi^2]$$

Find the field equations. Evaluate the stress-energy tensor, and find the Hamiltonian of the system.

4. If X is a dynamical variable of a field and H the Hamiltonian, show that

$$\frac{dX}{dt} = (X, H)$$

5. Evaluate the angular momentum tensor M_{kl} for a system whose Lagrangian density is

$$\mathcal{L} = \tfrac{1}{2}[\dot{\psi}_k{}^2 - (\nabla\psi_k)^2 - \kappa^2\psi_k{}^2]$$

where the ψ_k ($k = 1, 2, 3$) transform like the components of a vector under rotations.

16 INTRODUCTION TO SPECIAL RELATIVITY THEORY

86. INTRODUCTION

Throughout this volume it has been stressed that the equations of motion which were developed are limited in their range of validity. In this chapter we investigate the generalization of the preceding theory which it is necessary to introduce in order to describe the behavior of systems which are moving with very large velocities relative to the observer. It is found, for example, that, if a particle is moving with a velocity v, the theory developed so far in this book yields results which for $v \ll c$, the velocity of light, are in accordance with observation, whereas for v of order c the results are completely at variance with the results of experiment. It is therefore necessary to develop a more general theory which, when applied to problems in which $v \ll c$, leads to results experimentally indistinguishable from those obtained earlier, but which for $v \sim c$ yields an essentially different set of mechanical laws. Such a theory is the special theory of relativity. So successful has this been in describing phenomena associated with extremely high velocities that half a century after its inception it is referred to as classical mechanics, the theory which we have outlined earlier being called nonrelativistic classical mechanics.

As has been pointed out, there are other restrictions on the range of validity of the latter, some of which have been resolved by quantum mechanics, but we cannot discuss these here. There also exists, of course, a generalization of relativity theory to include observers who are accelerated relative to each other. This also we omit.

Numerous books and articles on relativity theory have been written (see references at the end of the chapter), so that all that we propose to include here is a brief and rather formal discussion of the theory. Readers unacquainted with relativity theory should consult one of the references cited, for we do not feel that their introduction to the theory should be through the necessarily condensed discussion presented here.

87. SPACE-TIME AND THE LORENTZ TRANSFORMATION

If we consider a particle with position vector $\mathbf{r}$ relative to a cartesian coordinate system O, the force on it is given, according to Newton's nonrelativistic classical dynamics, by

$$\mathbf{F} = m\ddot{\mathbf{r}} \tag{87.1}$$

If we consider another cartesian coordinate system $\bar{O}$, moving with constant velocity $\mathbf{v}$ relative to O, the position vector $\bar{\mathbf{r}}$ of the particle relative to $\bar{O}$ appears to be given by

$$\bar{\mathbf{r}} = \mathbf{r} - \mathbf{v}t \tag{87.2}$$

so that

$$\bar{\mathbf{F}} = m\ddot{\bar{\mathbf{r}}} = m\ddot{\mathbf{r}} = \mathbf{F}$$

Thus the force on the particle appears to be the same relative to the two observers O, $\bar{O}$, and, on account of (87.2), Newton's equation (87.1) retains its form for all observers moving with constant relative velocity.

On the other hand, if we consider Maxwell's equation for the scalar potential ϕ, for example, at points not occupied by electric charges

$$\nabla^2\phi - \frac{1}{c^2}\frac{\partial^2\phi}{\partial t^2} = 0 \tag{87.3}$$

we note that under the transformation (87.2) the form of the equation is completely destroyed (see Exercise 1 at the end of this chapter). In order to reconcile optical theory with the Galilean transformation (87.2), then, it is necessary to suppose that some special coordinate system (the ether) exists with respect to which Maxwell's equations assume the particularly simple form corresponding to (87.3). Furthermore, since c denotes the velocity of light relative to this special system, the velocity of light relative to an observer who is moving with respect to the ether should, by (87.2), be different from c. The earth must be moving relative to the ether, so that this absolute motion of the earth should cause the velocity of light to an observer on the earth to be different in different directions. The experiment of Michelson and Morley, designed to detect this difference, and accurate enough to detect it if it were present, verified instead the hypothesis that the velocity of light is the *same* for all observers moving with constant relative velocity. This would imply that for all such observers the form of (87.3) should be the same, i.e., (87.3) should be covariant with respect to the group of coordinate systems moving with constant relative velocity. We have seen, however, that Maxwell's equations are not covariant with respect to the group of transformations (87.2). Einstein was therefore led to the idea that the intuitively obvious Galilean transformation (87.2) (or with $\mathbf{v}$ in the z direction)

$$\bar{x} = x, \qquad \bar{y} = y, \qquad \bar{z} = z - vt \tag{87.4}$$

requires modification. Instead, we may retain the covariance of Maxwell's equations by supposing that space *and time* coordinates as measured by two observers moving with constant relative velocity v in the z direction may be related thus:

$$\begin{aligned} \bar{x} &= x, & \bar{y} &= y \\ \bar{z} &= \gamma(z - vt), & \bar{t} &= \gamma\left(t - \frac{vz}{c^2}\right) \end{aligned} \tag{87.5}$$

where $\gamma = (1 - \beta^2)^{-1/2}$, $\beta = v/c$. For under this transformation the form of (87.3) is preserved. Moreover, (87.5) may be solved for the unbarred coordinates thus:

$$\begin{aligned} x &= \bar{x}, & y &= \bar{y} \\ z &= \gamma(\bar{z} + v\bar{t}), & t &= \gamma\left(\bar{t} + \frac{v\bar{z}}{c^2}\right) \end{aligned}$$

which may be obtained from (87.5) by interchanging corresponding barred and unbarred coordinates and reversing the sign of v. Like (87.2), this transformation is therefore symmetrical between the two observers, each agreeing that the other is moving in the z direction with velocity of magnitude v. Equation (87.5) is known as the *Lorentz transformation.* It involves the velocity of light explicitly as a fundamental constant, the same for each observer of the group. For $v \ll c$, $vz \ll c^2t$, however, it reduces, as it must do, to the Galilean transformation (87.4). Many experiments have decided in favor of (87.5) rather than (87.4) as the correct transformation equations between inertial systems. But now Newton's equation (87.1) is not covariant with respect to the Lorentz transformation. Are we to say, then, that there is a special coordinate system with respect to which (87.1) holds, but that for an observer moving with constant velocity relative to this special frame (87.1) takes on an unusually complicated form? (See Exercise 4 at the end of this chapter.) We must either do this or modify the fundamental equation of mechanics so that it, too, is covariant with respect to Lorentz transformations. The theory of relativity adopts the latter program, and it is the purpose of this chapter to study such a modification of the theory of dynamics. Clearly, this implies that the velocity of light should appear explicitly in the equations of mechanics.

In addition to the above, the Lorentz transformation differs from the Galilean transformation in that it involves a transformation of time intervals as well as space intervals. It is therefore convenient to introduce the concept of an *event* which is specified by four coordinates x, y, z, t which locate it in space and time. The time between two events may therefore be different for different observers. The Lorentz transformation is then easily shown to leave invariant the quantity

$$ds^2 = c^2\,dt^2 - dr^2$$

where ds is called the *interval* or the *proper time* between the two nearby events, $(\mathbf{r}, t)$ and $(\mathbf{r} + d\mathbf{r}, t + dt)$. Except for the relative sign of $c^2\,dt^2$ and dr^2, the Lorentz transformation therefore bears a strong resemblance to a rotation. The relative sign may formally be changed by introducing an imaginary time coordinate ict. This is a purely formal device and does not make the Lorentz group equivalent to a real rotation group.

We denote the coordinates x, y, z, ict by x_1, x_2, x_3, x_4 respectively, or by x_μ $(\mu = 1, 2, 3, 4)$. Then

$$-ds^2 = dx_\mu\,dx_\mu \tag{87.6}$$

the summation convention applying as before. The introduction of the negative sign in (87.6) is purely a matter of convenience; as we shall see, it has the effect of making the interval along the space-time path of a particle real.

In (47.12) it was shown how a three-dimensional rotation could be represented by a 3×3 orthogonal matrix S with

$$\bar{x} = Sx, \qquad SS^{\mathsf{T}} = 1, \qquad \det S = 1 \tag{87.7}$$

Described in terms of the coordinates x_μ, a Lorentz transformation may be represented by a 4×4 orthogonal matrix S of a special kind. Thus if we take, by analogy with (II.35),

$$S = \begin{pmatrix} \cos\theta & \sin\theta & 0 & 0 \\ -\sin\theta & \cos\theta & 0 & 0 \\ 0 & 0 & 1 & 0 \\ 0 & 0 & 0 & 1 \end{pmatrix} \tag{87.8}$$

it is easily verified that S satisfies (87.7) (hence is called a ***proper*** Lorentz transformation) and that

$$\bar{x}_1 = x_1 \cos\theta + x_2 \sin\theta, \qquad \bar{x}_3 = x_3$$

$$\bar{x}_2 = x_2 \cos\theta - x_1 \sin\theta, \qquad \bar{x}_4 = x_4$$

This is, of course, identically the rotation (II.35), with the additional information that time coordinates are the same for two observers at the same point with axes pointing in different directions. Had we taken $S_{33} = -1$ instead of $+1$, the transformation would yield the above rotation together with a reflection in the 1-2 plane. For such a transformation $\det S = -1$, and the transformation is called *improper*.

Another transformation which is formally similar to (87.8) is given by

$$S = \begin{pmatrix} 1 & 0 & 0 & 0 \\ 0 & 1 & 0 & 0 \\ 0 & 0 & \cosh\theta & i\sinh\theta \\ 0 & 0 & -i\sinh\theta & \cosh\theta \end{pmatrix}$$

so that

$$\bar{x} = x, \qquad \bar{z} = z\cosh\theta - ct\sinh\theta$$

$$\bar{y} = y, \qquad \bar{t} = t\cosh\theta - \frac{z}{c}\sinh\theta$$

If we write $\gamma = \cosh\theta = (1 - \beta^2)^{-1/2}$, $\tanh\theta = v/c = \beta$, this is identically the Lorentz transformation (87.5). Thus

$$S = \begin{pmatrix} 1 & 0 & 0 & 0 \\ 0 & 1 & 0 & 0 \\ 0 & 0 & \gamma & i\beta\gamma \\ 0 & 0 & -i\beta\gamma & \gamma \end{pmatrix} \tag{87.9}$$

represents a transformation without rotation to axes moving with constant velocity v in the positive z direction. If $v < c$, θ is real. In general, a Lorentz transformation between relatively moving axes is represented by an orthogonal matrix whose elements with only one index 4 are purely imaginary, the others being real.

The transformation (87.5) may also be written

$$d\bar{x} = dx, \qquad d\bar{z} = \gamma(dz - v\,dt)$$

$$d\bar{y} = dy, \qquad d\bar{t} = \gamma\left(dt - \frac{v\,dz}{c^2}\right)$$

If we write

$$\bar{u}_z = \frac{d\bar{z}}{d\bar{t}}, \qquad u_z = \frac{dz}{dt}$$

then

$$\bar{u}_z = \frac{u_z - v}{1 - \dfrac{u_z v}{c^2}} \tag{87.10}$$

If barred variables refer to the observations on a particle made by a particular observer $\bar{O}$ and unbarred to those of O on the same particle, it follows that if u_z is the z component of the velocity of the particle relative to O, then $\bar{u}_z$ is the z component of the velocity of the particle relative to $\bar{O}$. These quantities are related by (87.10), which becomes, for $u_z v \ll c^2$,

$$\bar{u}_z = u_z - v \tag{87.11}$$

the familiar law of composition of velocities in nonrelativistic mechanics. The latter law is a good approximation to the Lorentz transformation provided that the z velocity of the moving point relative to O multiplied by the velocity v of $\bar{O}$ relative to O in the z direction is small compared with the square of the velocity of light. For relative velocities which do not satisfy this condition, however, (87.11) is to be replaced by (87.10).

The covariance of any set of equations with respect to a group of transformations can be proved by writing the equations as vector or tensor equations, the vectors or tensors being defined by their transformation properties under the group in question. Thus the Lorentz covariance of Maxwell's equations may be proved by introducing the quantities

$$A_\mu = (A_x, A_y, A_z, i\phi) \tag{87.12}$$

and

$$f_{\mu\nu} = \frac{\partial A_\nu}{\partial x_\mu} - \frac{\partial A_\mu}{\partial x_\nu} \tag{87.13}$$

which are a vector and tensor respectively for Lorentz transformations, $\mathbf{A}$, ϕ being the vector and scalar electromagnetic potentials. The reader may verify that

$$f_{\mu\nu} = \begin{pmatrix} 0 & B_z & -B_y & -iE_x \\ -B_z & 0 & B_x & -iE_y \\ B_y & -B_x & 0 & -iE_z \\ iE_x & iE_y & iE_z & 0 \end{pmatrix} \tag{87.14}$$

since

$$\mathbf{E} = -\nabla\phi - \frac{1}{c}\frac{\partial \mathbf{A}}{\partial t}, \qquad \mathbf{B} = \nabla \times \mathbf{A}$$

Under the Lorentz transformation characterized by S, the electromagnetic field strengths (87.13) transform thus (cf. Appendix I):

$$\bar{f}_{\mu\nu} = S_{\mu\sigma}S_{\nu\tau}f_{\sigma\tau} = S_{\mu\sigma}f_{\sigma\tau}S^{\mathsf{T}}{}_{\tau\nu}$$

This is the condition that the set of quantities $f_{\mu\nu}$ form a tensor with respect to the group of Lorentz transformations. For S given by (87.9), we have

$$\text{(87.15)} \quad \begin{aligned} \bar{f}_{23} &= \bar{B}_x = \gamma(B_x + \beta E_y), & i\bar{f}_{14} &= \bar{E}_x = \gamma(E_x - \beta B_y) \\ \bar{f}_{31} &= \bar{B}_y = \gamma(B_y - \beta E_x), & i\bar{f}_{24} &= \bar{E}_y = \gamma(E_y + \beta B_x) \\ \bar{f}_{12} &= \bar{B}_z = B_z, & i\bar{f}_{34} &= \bar{E}_z = E_z \end{aligned}$$

In general, if the velocity of $\bar{O}$ relative to O is $\mathbf{v}$, the transformation yields

$$\text{(87.16)} \quad \begin{aligned} \bar{\mathbf{E}}_{\parallel} &= \mathbf{E}_{\parallel}, & \bar{\mathbf{E}}_{\perp} &= \gamma\left(\mathbf{E} + \frac{\mathbf{v} \times \mathbf{B}}{c}\right)_{\perp} \\ \bar{\mathbf{B}}_{\parallel} &= \mathbf{B}_{\parallel}, & \bar{\mathbf{B}}_{\perp} &= \gamma\left(\mathbf{B} - \frac{\mathbf{v} \times \mathbf{E}}{c}\right)_{\perp} \end{aligned}$$

where the symbols $\parallel$, $\perp$ denote directions respectively parallel to and perpendicular to $\mathbf{v}$. Thus components of $\mathbf{E}$ and $\mathbf{B}$ in a particular direction are the same for two observers moving with constant relative velocity in that direction. Components of electric and magnetic fields in a perpendicular direction, however, are mixed by the transformation. In particular, if the sources of the fields are at rest relative to $\bar{O}$ ($\bar{\mathbf{B}} = 0$), then

$$\mathbf{E}_{\perp} = \gamma\bar{\mathbf{E}}_{\perp}, \qquad \mathbf{B} = \mathbf{B}_{\perp} = \left(\frac{\mathbf{v} \times \mathbf{E}}{c}\right)_{\perp} = c^{-1}[\mathbf{v} \times (\bar{\mathbf{E}}_{\parallel} + \gamma\bar{\mathbf{E}}_{\perp})]$$

i.e., O observes a magnetic field in the plane perpendicular to the relative velocity.

The covariance of the *mechanical* equations of motion is also assured if they can be written as vector equations in the four-dimensional space-time manifold. We now examine the form these equations assume for the motion of a single particle and investigate their transformation properties.

88. THE MOTION OF A FREE PARTICLE

In nonrelativistic mechanics, the time was an invariant under all transformations of interest and the position was described by a vector, the radius vector $\mathbf{r}$. The velocity was then $d\mathbf{r}/dt$. Relativistically, however, dt is not an invariant but is proportional to a component of the radius vector of an event, with cartesian components x_μ. In the last section, however, we showed that the interval ds between two neighboring events is an invariant, and if two such events lie on the path of a particle in space-time we may define the *four-velocity* u_μ of the particle by

$$u_\mu = \frac{dx_\mu}{ds} = \dot{x}_\mu \tag{88.1}$$

A dot now denotes differentiation with respect to ds, the interval along the path of the particle in space-time.

$$ds = c\,dt(1 - \beta^2)^{1/2}$$

where $\beta = (1/c)|d\mathbf{r}/dt|$ is the speed v of the particle divided by that of light. For $v \ll c$, we note that the infinitesimal interval between two events on the path of the particle is, apart from the constant factor c, simply the time interval between those events.

From (87.6) we see that

$$u_\mu u_\mu = -1 \tag{88.2}$$

so that the four-velocity is a unit vector. The negative sign shows that u_4 never vanishes, but the other three components may. u_μ is therefore called a *time-like* vector. A vector λ_μ with $\lambda_\mu\lambda_\mu > 0$ is called *space-like*, and, if $\lambda_\mu\lambda_\mu = 0$, λ_μ is a *null-vector*.

The nonrelativistic velocity, or three-velocity, is given by

$$v_x = \frac{dx}{dt} = \frac{dx_1}{d(x_4/ic)} = ic\frac{u_1}{u_4}, \quad \text{etc.} \tag{88.3}$$

and therefore has no simple transformation properties under the Lorentz group. Note that $u_4 = ic\,dt/ds = i\gamma$. The nonrelativistic acceleration has no simple transformation properties either.

The three-momentum $\mathbf{p}$ can be generalized to a four-vector. If this is done, the equations of motion can be expressed in a covariant form which reduces to the nonrelativistic form for small velocities. Let us write

$$p_\mu = (p_x, p_y, p_z, iE/c) \tag{88.4}$$

For small velocities the three space components of p_μ are to be the nonrelativistic momentum mv_x, mv_y, mv_z. In the system in which the particle is at rest, only p_4 does not vanish. If we now transform to a system moving in the $-z$ direction, the particle will appear from that frame to be moving in the $+z$ direction.

$$\bar{p}_\mu = S_{\mu\nu}p_\nu \tag{88.5}$$

or

$$\begin{pmatrix} \bar{p}_1 \\ \bar{p}_2 \\ \bar{p}_3 \\ \bar{p}_4 \end{pmatrix} = \begin{pmatrix} 1 & 0 & 0 & 0 \\ 0 & 1 & 0 & 0 \\ 0 & 0 & \gamma & -i\beta\gamma \\ 0 & 0 & i\beta\gamma & \gamma \end{pmatrix} \begin{pmatrix} 0 \\ 0 \\ 0 \\ iE/c \end{pmatrix} = \begin{pmatrix} 0 \\ 0 \\ \beta\gamma E/c \\ i\gamma E/c \end{pmatrix} \tag{88.6}$$

Now if $\beta \ll 1$, we must have $p_3 \approx mv_z$ and so, neglecting terms in $(v/c)^2$, we obtain

$$mv_z = \frac{E}{c^2} v_z$$

Therefore the quantity E must be given by

$$E = mc^2 \tag{88.7}$$

In the moving coordinate system the value of $\bar{p}_4$ is given by

$$\bar{p}_4 = \frac{i\gamma}{c} mc^2 = \frac{i}{c}\left(mc^2 + \frac{1}{2} mv^2 + \cdots\right)$$

The second term of this expansion, which is the first nonconstant term, is the nonrelativistic kinetic energy of the particle. The first term is independent of the velocity and is called the *rest energy* of the particle. Thus p_4 is, to within the constant i/c, the total energy of the particle. The arbitrary additive constant in the nonrelativistic expression for the energy has a definite value mc^2 in relativistic mechanics. Thus the momentum and total energy of a particle together form a four-vector.

For large velocities the components of the momentum are

$$p_x = \gamma m v_x, \quad \text{etc.,} \qquad p_4 = i\gamma mc \tag{88.8}$$

Defining the mass $\tilde{m}$ of the moving particle as the ratio $p_x/v_x = p_y/v_y = p_z/v_z$, we obtain

$$\tilde{m} = \gamma m, \qquad E = \tilde{m}c^2 \tag{88.9}$$

The mass increases with the velocity, becoming infinite as $v \to c$.

The machinery for writing down the equations of motion exists already in the formulation of Sec. 72. There Hamilton's principle was stated with an arbitrary parameter as the variable of integration. We choose this parameter to be s, the invariant interval. We must further assume an invariant Lagrangian $\mathcal{L}$. Then the momenta

$$p_\mu = \frac{\partial \mathcal{L}}{\partial \dot{x}_\mu}$$

where the dot signifies d/ds, are the components of a four-vector. Hamilton's principle may be written, in terms of the generalized configuration space (cf. Sec. 88),

$$\Delta \int p_\mu \, dq_\mu = 0$$

so that, for a single particle,

$$\Delta \int p_\mu \, dx_\mu = \Delta \int (\mathbf{p} \cdot d\mathbf{r} - E \, dt) = 0$$

If we postulate

$$\mathcal{L} = \tfrac{1}{2} mcu_\mu u_\mu = \tfrac{1}{2} mc\dot{x}_\mu \dot{x}_\mu \tag{88.10}$$

then

$$p_\mu = mcu_\mu \tag{88.11}$$

which agrees with (88.8).

The equations of motion are

$$\dot{p}_\mu = \frac{\partial \mathcal{L}}{\partial x_\mu} = 0 \tag{88.12}$$

in this case. Thus the free particle has constant momentum and energy.

89. CHARGED PARTICLE IN AN ELECTROMAGNETIC FIELD

The motion of a particle in a potential field has a meaning in relativity only for a limited class of potentials. The Lagrangian describing the motion must be an invariant under Lorentz transformations, and a potential V which depends on the distances between particles is not so invariant. The combination of scalar and vector potentials which enters into the Lagrangian for a charged particle in an electromagnetic field can, however, be brought into the desired form. We recall from (35.8) that the combination which enters is $e[(\mathbf{v}/c) \cdot \mathbf{A} - \phi]$, which can be replaced by the form $eu_\mu A_\mu$, which is an invariant and which for small velocities approaches the old value. Thus we postulate for our Lagrangian

$$\mathcal{L} = \frac{1}{2} mcu_\mu u_\mu + \frac{e}{c} u_\mu A_\mu \tag{89.1}$$

The momenta conjugate to the coordinates x_μ are

$$p_\mu = \frac{\partial \mathcal{L}}{\partial u_\mu} = mcu_\mu + \frac{e}{c} A_\mu \tag{89.2}$$

The equations of motion are

$$\dot{p}_\mu = \frac{\partial \mathcal{L}}{\partial x_\mu} = \frac{e}{c} u_\nu \, \partial_\mu A_\nu \tag{89.3}$$

Elimination of p_μ between (89.2) and (89.3) yields

$$\begin{aligned} mc\dot{u}_\mu &= \frac{e}{c} u_\nu(\partial_\mu A_\nu - \partial_\nu A_\mu) \\ &= \frac{e}{c} f_{\mu\nu} u_\nu \end{aligned} \tag{89.4}$$

These may be written in three-vector notation as

$$\begin{aligned} \frac{d}{dt}(\tilde{m}\mathbf{v}) &= e\mathbf{E} + \frac{e}{c}\mathbf{v} \times \mathbf{B} = \mathbf{F} \\ \frac{d}{dt}(\tilde{m}c^2) &= e\mathbf{v} \cdot \mathbf{E} \end{aligned} \tag{89.5}$$

Thus our Lagrangian describes the Lorentz force on the charged particle and the work done by an electric field in moving a charge.

The time may be restored to its privileged position by introducing instead the Lagrangian

$$L = -mc^2\left(1 - \frac{v^2}{c^2}\right)^{1/2} + \frac{e}{c}\mathbf{A} \cdot \mathbf{v} - e\phi \tag{89.6}$$

so that

$$\mathbf{p} = \frac{\partial L}{\partial \mathbf{v}} = \tilde{m}\mathbf{v} + \frac{e}{c}\mathbf{A}$$

$$\frac{d\mathbf{p}}{dt} = \frac{\partial L}{\partial \mathbf{x}} = \frac{e}{c}\nabla(\mathbf{A} \cdot \mathbf{v}) - e\,\nabla\phi$$

which leads directly to the first of equations (89.5). The second of these equations may be obtained from the first by taking the scalar product of each side with $\mathbf{v}$.

In discussing the motion of a single particle we may define, by analogy with Sec. 13, an angular momentum tensor

$$L_{\mu\nu} = x_\mu p_\nu - x_\nu p_\mu \tag{89.7}$$

the space components of which describe the angular momentum about the origin:

$$(L_{23}, L_{31}, L_{12}) = (\mathbf{r} \times \mathbf{p})$$

If an external magnetic field is acting on the particle, we note that the angular momentum so defined is not the same as $\mathbf{r} \times \tilde{m}\mathbf{v}$, but rather we may write

$$L_{\mu\nu} = M_{\mu\nu} + F_{\mu\nu} \tag{89.8}$$

where

$$M_{\mu\nu} = mc(x_\mu u_\nu - x_\nu u_\mu) \tag{89.9}$$

$$F_{\mu\nu} = \frac{e}{c}(x_\mu A_\nu - x_\nu A_\mu) \tag{89.10}$$

The space components of (89.9) denote the angular momentum of the particle about the origin, and they vanish when the particle is at rest or when its velocity is parallel to the radius vector. The corresponding components of (89.10), on the other hand, are functions only of the *position* of the particle, and they denote the angular momentum resident in the electromagnetic field due to the presence of the particle of charge e in the external magnetic field described by the vector potential $\mathbf{A}$. The electric field due to the charge combines with the external magnetic field to produce a Poynting vector which in general leads to a contribution $(e/c)\mathbf{r} \times \mathbf{A}$ to the angular momentum.

90. HAMILTONIAN FORMULATION OF THE EQUATIONS OF MOTION

In general, we may pass to the Hamiltonian formulation of the equations of motion of a particle by defining

$$\mathcal{H} = p_\mu \dot{x}_\mu - \mathcal{L} \tag{90.1}$$

so that

$$\dot{p}_\mu = -\frac{\partial \mathcal{H}}{\partial x_\mu}, \qquad \dot{x}_\mu = \frac{\partial \mathcal{H}}{\partial p_\mu} \tag{90.2}$$

As a particular example, the Hamiltonian for a particle moving in an electromagnetic field may be written, by (89.1) and (90.1), thus:

$$\mathcal{H} = \frac{1}{2mc}\left(p_\mu - \frac{e}{c}A_\mu\right)\left(p_\mu - \frac{e}{c}A_\mu\right)$$

$$= \frac{1}{2mc^3}\left[-(E - e\phi)^2 + c^2\left(\mathbf{p} - \frac{e}{c}\mathbf{A}\right)^2\right]$$

The equations of motion (90.2) yield

$$p_\mu - \frac{e}{c}A_\mu = mc\dot{x}_\mu$$

i.e.,

$$\mathbf{p} = \tilde{m}\mathbf{v} + \frac{e}{c}\mathbf{A}$$

$$E = \tilde{m}c^2 + e\phi$$

and

$$\dot{p}_\mu = \frac{e}{c}\dot{x}_\nu\,\partial_\mu A_\nu$$

i.e.,

$$\ddot{x}_\mu = \frac{e}{mc^2}f_{\mu\nu}\dot{x}_\nu$$

as before (89.4).

Another method of writing the Hamiltonian that is often employed is

$$\mathsf{H} = c\left[m^2c^2 + \left(\mathbf{p} - \frac{e}{c}\mathbf{A}\right)^2\right]^{\frac{1}{2}} + e\phi \tag{90.3}$$

with the Hamiltonian equations

$$\mathbf{v} = \frac{\partial \mathsf{H}}{\partial \mathbf{p}} = \frac{c^2}{\mathsf{H} - e\phi}\left(\mathbf{p} - \frac{e}{c}\mathbf{A}\right)$$

$$\dot{p}_i = -\frac{\partial \mathsf{H}}{\partial x_i} = \frac{ce}{\mathsf{H} - e\phi}\left(p_j - \frac{e}{c}A_j\right)\partial_i A_j - e\,\partial_i\phi \qquad (i, j = 1, 2, 3)$$

Substituting the first of these equations back in the Hamiltonian, we have

$$\mathsf{H} - e\phi = \gamma mc^2 = \tilde{m}c^2$$

so that the momentum equations follow as before. This Hamiltonian may be derived from the Lagrangian (89.6).

The Hamilton-Jacobi equation for the motion of a particle is

$$\mathcal{H}\left(x_\mu, \frac{\partial \Phi}{\partial x_\mu}\right) = \text{const} \tag{90.4}$$

If its complete solution is written in the form $\Phi(x_\mu, \alpha_\mu)$, the equations of motion are

$$p_\mu = \frac{\partial \Phi}{\partial x_\mu}, \qquad \beta_\mu = \frac{\partial \Phi}{\partial \alpha_\mu}$$

where the α_μ, β_μ are a set of eight constants.

In the above example of a particle in an electromagnetic field, $\mathcal{H}$ is equal in value to $-\frac{1}{2}mc$, so that the constant which appears on the right-hand side of (90.4) is proportional to the *rest* energy. The $H - J$ equation is then

$$\left(\frac{\partial \Phi}{\partial x_\mu} - \frac{e}{c} A_\mu\right)\left(\frac{\partial \Phi}{\partial x_\mu} - \frac{e}{c} A_\mu\right) = -m^2c^2$$

i.e.,

$$\left(\frac{\partial \Phi}{\partial t} + e\phi\right)^2 - c^2\left(\nabla\Phi - \frac{e}{c}\mathbf{A}\right)^2 = m^2c^4 \tag{90.5}$$

Unlike the nonrelativistic equation, this is not linear in the term $\partial\Phi/\partial t$. In general it is not separable, and the motion is more easily determined directly from the Lagrangian equations. For the special case of a free particle, $\phi = 0$, $\mathbf{A} = 0$, and thus $p_\mu = \text{const}$. The complete solution of (90.5) is then

$$\Phi = xp_x + yp_y + zp_z - Et$$

with the four constants p_x, p_y, p_z, E thus introduced connected with the constant m in the $H - J$ equation by

$$E^2 - c^2(p_x{}^2 + p_y{}^2 + p_z{}^2) = m^2c^4 \tag{90.6}$$

The rest energy mc^2 thus plays the role here that E played in nonrelativistic theory, and E is now just another momentum coordinate. The relation (90.6) is analogous to the relation (61.11) with the constants α_i replaced here by the constants p_μ. The equations of motion

$$\frac{\partial \Phi}{\partial \alpha_i} = \beta_i$$

now become

$$\frac{\partial \Phi}{\partial p_\mu} = \beta_\mu \tag{90.7}$$

with

$$\Phi = p_\mu x_\mu$$

and, from (90.6),

$$p_\mu p_\mu = -m^2c^2 \tag{90.8}$$

In evaluating $\partial\Phi/\partial p_\mu$ in (90.7) it is, of course, necessary to take account of (90.8). Thus we may write

$$\Phi = xp_x + yp_y + zp_z - t[m^2c^4 + c^2(p_x{}^2 + p_y{}^2 + p_z{}^2)]^{1/2}$$

so that, for example,

$$\beta_1 = x_0 = \frac{\partial \Phi}{\partial p_x} = x - tc^2p_xE^{-1}$$

i.e.,

$$p_x = \frac{E}{c^2t}(x - x_0) = \frac{\tilde{m}}{t}(x - x_0)$$

as the integrated expression for the momentum in the x direction as a function of position and time. The result is, of course, quite trivial; we have presented it here to illustrate how the $H - J$ equation may be extended to relativistic problems. The contact transformation generated by Φ is to new coordinates which are the old (constant) momenta and new momenta which are the *initial values* of the original coordinates.

We note that because of the nonlinearity of (90.4) in $\partial\Phi/\partial t$ the solution (90.5) is quite ambiguous with regard to the sign of E as well as of p_x, p_y, p_z. We have chosen (90.5) to conform to the nonrelativistic treatment, but we could have equally well reversed the sign of E throughout. In classical mechanics we may say that no significance is to be attached to the case $E < 0$ for a free particle, and that, because of the continuity of the motion, if E is positive initially it remains so throughout the motion. In quantum mechanics it is necessary to examine the solutions with negative E, because there transitions between different energy levels occur, even transitions from states for which $E > 0$ to states for which $E < 0$ and vice versa. When applied to the electron, such transitions have been interpreted by Dirac as corresponding physically to the annihilation and creation of a particle-antiparticle pair.

91. TRANSFORMATION THEORY AND THE LORENTZ GROUP

The Hamiltonian equations of motion (90.2)

$$\dot{p}_\mu = -\frac{\partial \mathcal{H}}{\partial x_\mu}, \qquad \dot{x}_\mu = \frac{\partial \mathcal{H}}{\partial p_\mu} \tag{91.1}$$

are invariant under the generalized contact transformation

$$p_\mu \, dx_\mu - \bar{p}_\mu \, d\bar{x}_\mu = d\phi(x_\mu, \bar{x}_\mu) \tag{91.2}$$

so that

$$p_\mu = \frac{\partial \phi}{\partial x_\mu}, \qquad \bar{p}_\mu = -\frac{\partial \phi}{\partial \bar{x}_\mu} \tag{91.3}$$

Following Sec. 59, we could, of course, write the generator of the transformation as a function of either of p_μ, x_μ and either of $\bar{p}_\mu$, $\bar{x}_\mu$:

$$\begin{aligned} \psi &= \psi(p_\mu, \bar{x}_\mu) = \phi - p_\mu x_\mu \\ \psi' &= \psi'(x_\mu, \bar{p}_\mu) = \phi + \bar{p}_\mu \bar{x}_\mu \\ \phi' &= \phi'(p_\mu, \bar{p}_\mu) = \phi + \bar{p}_\mu \bar{x}_\mu - p_\mu x_\mu \end{aligned}$$

$$\begin{aligned} x_\mu &= -\frac{\partial \psi}{\partial p_\mu}, \qquad \bar{p}_\mu = -\frac{\partial \psi}{\partial \bar{x}_\mu} \\ p_\mu &= \frac{\partial \psi'}{\partial x_\mu}, \qquad \bar{x}_\mu = \frac{\partial \psi'}{\partial \bar{p}_\mu} \\ x_\mu &= -\frac{\partial \phi'}{\partial p_\mu}, \qquad \bar{x}_\mu = \frac{\partial \phi'}{\partial \bar{p}_\mu} \end{aligned} \tag{91.4}$$

In fact, formally the relations are identical with those which occur in the nonrelativistic theory of time-independent contact transformations. Thus $\mathcal{H}$ itself is invariant under such transformations.

To paraphrase Sec. 59, any generating function ψ' which is linear in the $\bar{p}_\mu$

$$\psi' = f_\mu \bar{p}_\mu + g \tag{91.5}$$

with

$$f_\mu = f_\mu(x_\nu), \qquad g = g(x_\nu)$$

yields a coordinate transformation

$$\bar{x}_\mu = f_\mu, \qquad p_\mu = \bar{p}_\nu \frac{\partial f_\nu}{\partial x_\mu} + \frac{\partial g}{\partial x_\mu} \tag{91.6}$$

If the f_μ are linear inhomogeneous functions of the x_ν, then

$$\bar{x}_\mu = S_{\mu\nu}x_\nu + b_\mu \tag{91.7}$$

where the $S_{\mu\nu}$, b_ν are constants. If the constants $S_{\mu\nu}$ are such that

$$S_{\mu\nu}S_{\mu\sigma} = \delta_{\nu\sigma} \tag{91.8}$$

i.e., $SS^{\mathsf{T}} = 1$, then, as indicated in (87.7), equation (91.7) describes a rotation and translation of the space-time axes, without any other change. The group of transformations generated in this way from the inhomogeneous generating functions f_μ is called the *inhomogeneous Lorentz group*, and the transformation (91.7) is called an *inhomogeneous Lorentz transformation*. For $b_\mu = 0$, corresponding to a pure rotation of the axes, the transformation is called a *homogeneous Lorentz transformation*.

The generating function of the transformation is

$$\psi' = \bar{p}_\mu S_{\mu\nu}x_\nu + b_\mu\bar{p}_\mu + g(x_\nu) \tag{91.9}$$

Thus

$$\begin{aligned} p_\mu &= \frac{\partial\psi'}{\partial x_\mu} = \bar{p}_\nu S_{\nu\mu} + \frac{\partial g}{\partial x_\mu} \\ \bar{x}_\mu &= \frac{\partial\psi'}{\partial\bar{p}_\mu} = S_{\mu\nu}x_\nu + b_\mu \end{aligned} \tag{91.10}$$

Of the two terms on the right-hand sides of these equations, the first is intimately connected with the homogeneous part of the Lorentz transformation of the coordinates, and the second is independent of any coordinate change. Addition of a term of the type $\partial g/\partial x_\mu$ to the momentum p_μ is therefore always possible without any corresponding coordinate transformation. This ambiguity in the definition of the momentum has already been noted in the nonrelativistic theory. We note that for a particle in an electromagnetic field the four-velocity

$$u_\mu = \frac{1}{mc}\left(p_\mu - \frac{e}{c}A_\mu\right)$$

is invariant under the transformation

$$\begin{aligned} p_\mu &= \bar{p}_\mu + \frac{\partial g}{\partial x_\mu} \\ A_\mu &= \bar{A}_\mu + \frac{c}{e}\frac{\partial g}{\partial x_\mu} \end{aligned} \tag{91.11}$$

The electromagnetic field strengths

$$f_{\mu\nu} = \frac{\partial A_\nu}{\partial x_\mu} - \frac{\partial A_\mu}{\partial x_\nu}$$

are then also invariant under the transformation. Such a transformation of the potentials is called a *gauge transformation*. We see from the above that to every gauge transformation there corresponds a contact transformation generated by a function of the coordinates such that if both are applied simultaneously the four-velocity is invariant.

92. THOMAS PRECESSION

The problem of a charged particles possessing an intrinsic angular momentum or spin $\boldsymbol{\sigma}$ and with a magnetic moment $\boldsymbol{\mu}$ moving in an electromagnetic field is of interest in atomic theory. The problem arises in connection with the coupling between the spin and the orbital motion of an electron in an atom. The classical discussion given here is useful in setting up the quantum-mechanical form of the problem. The special theory of relativity has a large effect even for the slowly moving orbital electrons.

We consider only problems in which the orbital motion of the particle is unaffected by the magnetic moment. This means that $e\mathbf{E}$, where $\mathbf{E}$ is the electric field intensity, is large compared with $(\boldsymbol{\mu} \cdot \nabla)\mathbf{B}$, $\mathbf{B}$ being the magnetic intensity at the position of the particle. This condition is not relativistically invariant, and so we are restricted to frames of reference moving slowly with respect to each other and to the particle. As we are interested in the case of closed orbits, there is an inertial frame with respect to which the particle has zero average velocity, and, if the speed of the particle in this frame is small compared with that of light, the above restrictions can be maintained.

Let N be the inertial frame in which the average velocity of the particle vanishes, the particle having coordinates x_μ in this frame, and let P be the inertial frame in which the particle is instantaneously at rest. In P the particle has coordinates $\bar{x}_\mu$. We choose the axes in N and P so that P is moving in the 3 direction in N, and so that the spatial axes of the two systems are parallel. Then the $\bar{x}_\mu$ are derived from the x_μ by a Lorentz transformation S.

$$\bar{x}_\mu = S_{\mu\nu} x_\nu \tag{92.1}$$

$$S = \begin{pmatrix} 1 & 0 & 0 & 0 \\ 0 & 1 & 0 & 0 \\ 0 & 0 & \gamma & i\beta\gamma \\ 0 & 0 & -i\beta\gamma & \gamma \end{pmatrix} \qquad [\beta = v/c,\ \gamma = (1 - \beta^2)^{-1/2}] \tag{92.2}$$

Here v is the speed of the particle in N. Its speed in P is zero.

Looked at in frame P, the particle has a magnetic moment $\bar{\boldsymbol{\mu}}$ with which is associated a vector potential $\bar{\mathbf{A}}$ given by

$$\bar{\mathbf{A}} = \frac{\bar{\boldsymbol{\mu}} \times \bar{\mathbf{r}}}{\bar{r}^3} \tag{92.3}$$

The fourth component of the potential four-vector $i\bar{\phi}$ is zero. Looked at from N, this particle has associated with it a potential $(\mathbf{A}, i\phi)$ given by

$$A_\mu = S^{\mathsf{T}}{}_{\mu\nu}\bar{A}_\nu \tag{92.4}$$

We evaluate A_μ when $v/c \ll 1$ and its square is negligible. Then $\gamma = 1$ and we obtain

$$\mathbf{A} = \bar{\mathbf{A}}, \qquad \phi = \beta\bar{A}_3 \tag{92.5}$$

Recalling that β represents a velocity in the 3 direction, we may write

$$\begin{aligned} \phi &= \frac{v_3}{c}\,\frac{\bar{\mu}_1\bar{x}_2 - \bar{\mu}_2\bar{x}_1}{\bar{r}^3} \\ &= \left(\frac{\mathbf{v}}{c} \times \boldsymbol{\mu}\right) \cdot \frac{\mathbf{r}}{r^3} \end{aligned} \tag{92.6}$$

since $\mathbf{r} = \bar{\mathbf{r}}$ within our approximation. This is the potential of an electric dipole $\boldsymbol{\epsilon}$, where

$$\boldsymbol{\epsilon} = \frac{\mathbf{v}}{c} \times \boldsymbol{\mu} \tag{92.7}$$

Thus a moving magnetic dipole has associated with it an electric dipole whose moment is given by (92.7).

The magnetic and electric dipole moments can be combined into an electromagnetic moment represented by a skew-symmetric tensor or six-vector, $\mu_{\mu\nu}$. We have, since $\boldsymbol{\epsilon}$ vanishes in the rest system P,

$$\bar{\mu}_{\mu\nu} = \begin{pmatrix} 0 & \mu_3 & -\mu_2 & 0 \\ -\mu_3 & 0 & \mu_1 & 0 \\ \mu_2 & -\mu_1 & 0 & 0 \\ 0 & 0 & 0 & 0 \end{pmatrix} \tag{92.8}$$

In the system N the tensor μ is obtained from $\bar{\mu}$ by the transformation

$$\mu = S^{\mathsf{T}}\bar{\mu}S \tag{92.9}$$

Thus

$$\mu_{\mu\nu} = \begin{pmatrix} 0 & \mu_3 & -\mu_2 & i\epsilon_1 \\ -\mu_3 & 0 & \mu_1 & i\epsilon_2 \\ \mu_2 & -\mu_1 & 0 & i\epsilon_3 \\ -i\epsilon_1 & -i\epsilon_2 & -i\epsilon_3 & 0 \end{pmatrix} \tag{92.10}$$

where the components of the vector $\boldsymbol{\epsilon}$ are given by (92.7).

The angular momentum $\bar{\boldsymbol{\sigma}}$ in the frame P is an axial vector like $\boldsymbol{\mu}$ and is therefore also represented by a skew-symmetric tensor $\bar{\sigma}_{\mu\nu}$ of form similar to that of (92.8). In the frame N this becomes

$$\sigma_{\mu\nu} = \begin{pmatrix} 0 & \sigma_3 & -\sigma_2 & i\tau_1 \\ -\sigma_3 & 0 & \sigma_1 & i\tau_2 \\ \sigma_2 & -\sigma_1 & 0 & i\tau_3 \\ -i\tau_1 & -i\tau_2 & -i\tau_3 & 0 \end{pmatrix} \tag{92.11}$$

The vector $\boldsymbol{\tau}$ is given by

$$\boldsymbol{\tau} = \frac{\mathbf{v}}{c} \times \boldsymbol{\sigma} \tag{92.12}$$

In terms of the four-velocity [equation (88.1)], equations (92.7) and (92.12) may be written in the following forms, which reveal the covariance of these relations:

$$\begin{aligned} \mu_{\mu\nu}u_\nu &= 0 \\ \sigma_{\mu\nu}u_\nu &= 0 \end{aligned} \tag{92.13}$$

We must now write down the relativistic analogues of equations (78.1), (78.8), namely:

$$\boldsymbol{\mu} = g\boldsymbol{\sigma}, \qquad \dot{\boldsymbol{\sigma}} = \boldsymbol{\mu} \times \mathbf{B} \tag{92.14}$$

The most natural generalization of the former is

$$\mu_{\mu\nu} = g\sigma_{\mu\nu} \tag{92.15}$$

for this is a relation between two tensors which reduces to the correct equation in the rest system of the particle. The generalization of the equation which describes the spin motion, however, is more complicated. By general arguments based primarily on the law of conservation of angular momentum, it has been shown* that the equa-

* H. J. Bhabha and H. C. Corben, *Proc. Roy. Soc. (London), A*, 178, 302 (1941).

tion which is consistent with (92.13) and reduces to (92.14) in the non-relativistic limit is

$$\dot{\sigma}_{\mu\nu} + \dot{S}_{\mu\nu} = \frac{g}{c}(f_{\mu\lambda}\sigma_{\lambda\nu} - \sigma_{\mu\lambda}f_{\lambda\nu}) \tag{92.16}$$

where

$$\dot{S}_{\mu\nu} = -(\sigma_{\mu\lambda}u_\nu - \sigma_{\nu\lambda}u_\mu)\left(\dot{u}_\lambda - \frac{g}{c}f_{\lambda\rho}u_\rho\right) \tag{92.17}$$

Indeed, if (92.13) holds, we have from (92.16)

$$\dot{\sigma}_{\mu\nu}u_\nu + \dot{S}_{\mu\nu}u_\nu = -\frac{g}{c}\sigma_{\mu\lambda}f_{\lambda\nu}u_\nu$$

and from (92.17)

$$\dot{S}_{\mu\nu}u_\nu = \sigma_{\mu\lambda}\left(\dot{u}_\lambda - \frac{g}{c}f_{\lambda\rho}u_\rho\right)$$

Hence, subtracting,

$$\frac{d}{ds}(\sigma_{\mu\nu}u_\nu) = 0$$

i.e., (92.13) continues to hold throughout the motion. The value $g = e/mc$ for the ratio of the magnetic moment to the spin is correct for electrons and μ mesons (within about 0.1%). It is not correct for protons and neutrons. Additional effects in these cases are ascribed to the meson field associated with the nuclear particles. The amount by which the magnetic moment of a particle differs from $(e/mc)\boldsymbol{\sigma}$ is usually referred to as its *anomalous magnetic moment.*

For the special case $g = e/mc$, it follows from (92.17) that $\dot{S}_{\mu\nu} = 0$ if the particle obeys the Lorentz force equation (89.4). Equation (92.16) for the spin motion then assumes the specially simple form

$$\begin{aligned} \dot{\boldsymbol{\sigma}} &= \frac{e}{mc}(\boldsymbol{\sigma} \times \mathbf{B} + \boldsymbol{\tau} \times \mathbf{E}) \\ \dot{\boldsymbol{\tau}} &= \frac{e}{mc}(-\boldsymbol{\sigma} \times \mathbf{E} + \boldsymbol{\tau} \times \mathbf{B}) \end{aligned} \qquad \left(g = \frac{e}{mc}\right) \tag{92.18}$$

However, even for this special case, the term $\dot{S}_{\mu\nu}$ in (92.16) does not vanish identically, because the magnetic and electric dipole moments of the particle may interact with inhomogeneous magnetic and electric fields to produce a small force on the particle which is not included in the Lorentz force equation (89.4). It might be expected that the appropriate generalization of (89.4) would be

$$\dot{u}_\mu = \frac{e}{mc} f_{\mu\nu} u_\nu + \frac{g}{2mc^2} \sigma_{kl} \partial_\mu f_{kl} \tag{92.19}$$

since the extra term so introduced is relativistically covariant and reduces to the correct nonrelativistic limit [equation (78.7)]. As it stands, however, (92.19) leads immediately to the conclusion that $\sigma_{kl}\dot{f}_{kl} = 0$, or in the nonrelativistic limit $\boldsymbol{\sigma} \cdot (d\mathbf{B}/dt) = 0$. There is no experimental or theoretical reason for such a condition so that extra terms must be added to (92.19). It may be shown that the simplest self-consistent form which the equation of motion can assume is

$$\begin{aligned} \dot{u}_\mu + \frac{1}{mc}\frac{d}{ds}(\sigma_{\mu\nu}\dot{u}_\nu) \\ = \frac{e}{mc} f_{\mu\nu} u_\nu + \frac{g}{2mc^2}\left[\sigma_{kl}\partial_\mu f_{kl} + \frac{d}{ds}(u_\mu \sigma_{kl} f_{kl} + 2\sigma_{\mu k} f_{kl} u_l)\right] \end{aligned} \tag{92.20}$$

However, except in the presence of extremely strong fields these extra terms are negligible.

Equation (92.16) for the spin motion possesses an integral

$$\sigma_{\mu\nu}\sigma_{\mu\nu} = \text{const}$$

i.e.,

$$\sigma^2 - \left(\frac{\mathbf{v}}{c} \times \boldsymbol{\sigma}\right)^2 = \sigma_0{}^2 \tag{92.21}$$

where σ_0 is the magnitude of the spin vector in the system in which the particle is at rest. For a longitudinally polarized particle ($\mathbf{v} \parallel \boldsymbol{\sigma}$) it follows that $\sigma = \sigma_0$, whereas if the spin of the particle is normal to its velocity we have

$$\sigma = \gamma\sigma_0 \qquad [\gamma = (1 - \beta^2)^{-1/2}]$$

Although we may refer to $u_{\mu\nu}$ as the magnetic and electric moment tensor, it must be remembered that the expression

$$V = -\boldsymbol{\mu} \cdot \mathbf{B} - \boldsymbol{\epsilon} \cdot \mathbf{E}$$

can be written in the form

$$V = -\mu_{\mu\nu} f_{\mu\nu} \tag{92.22}$$

hence is an invariant quantity. It cannot therefore denote the *energy* of interaction of the dipole moments with the external field, and indeed we note that (92.19) may be written thus:

$$\frac{d}{dt}(m\mathbf{v}\gamma) = e\left(\mathbf{E} + \frac{\mathbf{v} \times \mathbf{B}}{c}\right) + \frac{g}{\gamma}\nabla(\boldsymbol{\sigma} \cdot \mathbf{B} + \boldsymbol{\tau} \cdot \mathbf{E}) \tag{92.23}$$

The effective magnetic and electric moments are therefore $g\boldsymbol{\sigma}/\gamma$ and $g\boldsymbol{\tau}/\gamma$. For a longitudinally polarized particle the magnitudes of these become $g\sigma_0/\gamma$ and 0, tending to zero as the energy increases. For $v \perp \sigma$ the magnitudes of the effective moments are $\mu_{\text{eff}} = g\sigma_0$ and $\epsilon_{\text{eff}} = (v/c)\sigma_0$.

We now investigate the spin motion described by the first of equations (92.18) when the particle is in a bound orbit. The equation is not in the form required to describe precessional motion, namely

$$\dot{\boldsymbol{\sigma}} = \boldsymbol{\omega} \times \boldsymbol{\sigma}$$

On the average, however, if the change in the spin vector is small in one period of the orbital motion, we show that (92.18) has this form. We write

$$(\mathbf{v} \times \boldsymbol{\sigma}) \times \mathbf{E} = \tfrac{1}{2}\boldsymbol{\sigma} \times (\mathbf{E} \times \mathbf{v}) + (\mathbf{E} \cdot \mathbf{v})\boldsymbol{\sigma} - \frac{m}{2e}\frac{d}{dt}(\boldsymbol{\sigma} \cdot \mathbf{v})\mathbf{v} + \frac{m}{2e}(\dot{\boldsymbol{\sigma}} \cdot \mathbf{v})\mathbf{v} \tag{92.24}$$

where we have used the nonrelativistic equation of motion and have assumed the magnetic field to be weak (cf. Sec. 40). We average both sides of this equation over a revolution of the particle in its orbit, assuming that $\dot{\boldsymbol{\sigma}}$ is so small that $\boldsymbol{\sigma}$ can be treated as a constant for this length of time:

$$\langle(\mathbf{v} \times \boldsymbol{\sigma}) \times \mathbf{E}\rangle = \tfrac{1}{2}\langle\boldsymbol{\sigma} \times (\mathbf{E} \times \mathbf{v})\rangle \tag{92.25}$$

The remaining terms have vanishing averages. $\langle\mathbf{E} \cdot \mathbf{v}\rangle = 0$ because in a complete revolution the force binding the particle into a periodic orbit does no work. $\langle(d/dt)(\boldsymbol{\sigma} \cdot \mathbf{v})\mathbf{v}\rangle = 0$ because $(\boldsymbol{\sigma} \cdot \mathbf{v})\mathbf{v}$ is, for the constant $\boldsymbol{\sigma}$ assumed here, a periodic function of the time. Thus on the average (92.18) becomes

$$\langle\dot{\boldsymbol{\sigma}}\rangle = \frac{e}{mc}\left\langle\boldsymbol{\sigma} \times \left[\mathbf{B} + \frac{1}{2}\left(\mathbf{E} \times \frac{\mathbf{v}}{c}\right)\right]\right\rangle \tag{92.26}$$

The average rate of precession of the spin is thus given by

$$\boldsymbol{\omega} = -\frac{e}{mc}\left(\mathbf{B} + \frac{1}{2}\mathbf{E} \times \frac{\mathbf{v}}{c}\right) \tag{92.27}$$

From straightforward relativistic considerations we should have expected the second term of (92.27) to contain $\mathbf{E} \times \mathbf{v}/c$ without the factor $\frac{1}{2}$, since $\mathbf{E} \times \mathbf{v}/c$ is the magnetic field measured by an observer moving with velocity $\mathbf{v}$ through an electric field $\mathbf{E}$. Therefore we write

$$\boldsymbol{\omega} = -\frac{e}{mc}\left(\mathbf{B} + \mathbf{E} \times \frac{\mathbf{v}}{c}\right) + \boldsymbol{\omega}_T \tag{92.28}$$

where

$$\boldsymbol{\omega}_T = \frac{e}{2mc^2}\,\mathbf{E} \times \mathbf{v} = \frac{1}{2c^2}\,\dot{\mathbf{v}} \times \mathbf{v} \tag{92.29}$$

to within terms quadratic in v/c. $\boldsymbol{\omega}_T$ is called the *Thomas precession;* we proceed to show that it has a purely kinematic origin.

The reference frame N introduced earlier may be considered to be the laboratory frame. Since the particle of interest is a charged particle in a field, it is accelerated, and so the frame P which is the rest frame of the particle at one instant is no longer the rest frame after a time δt. The new rest frame we denote by P'. P' is moving with velocity $\delta\mathbf{v}$ relative to N. We seek the transformation from N to P'. Since $\delta\mathbf{v}$ is infinitesimal relative to v, and v is taken to be infinitesimal relative to c, and since products of v and δv are of concern, terms up to and including $(v/c)^3$ must be treated correctly. This means that $\gamma - 1$ cannot be neglected. Thus the transformation matrix S carrying N into P must be written

$$S = \begin{pmatrix} 1 & 0 & 0 & 0 \\ 0 & 1 & 0 & 0 \\ 0 & 0 & \gamma & i\beta\gamma \\ 0 & 0 & -i\beta\gamma & \gamma \end{pmatrix}$$

and the matrix S_1 carrying P into P' may be written

$$S_1 = \begin{pmatrix} 1 & 0 & 0 & 0 \\ 0 & 1 & 0 & i\beta_2 \\ 0 & 0 & 1 & i\beta_3 \\ 0 & -i\beta_2 & -i\beta_3 & 1 \end{pmatrix}$$

where β_2, β_3 are the components of $\delta\mathbf{v}/c$, which is taken to lie in the 2-3 plane. Thus

$$S_1S = \begin{pmatrix} 1 & 0 & 0 & 0 \\ 0 & 1 & \beta\beta_2 & i\beta_2 \\ 0 & 0 & \gamma & i\beta\gamma \\ 0 & -i\beta_2 & -i\beta\gamma & \gamma \end{pmatrix} \tag{92.30}$$

where in each matrix element only the leading term in v/c has been retained. The matrix (92.30) shows that a rotation has taken place as well as a translation to moving axes, since there is a nonvanishing 23 element in the matrix. Rotations take place both before and after the transformation to moving axes.

To investigate these rotations we write S_1S as the product of rotations and a transformation to moving axes:

$$S_1S = R_PSR_N \tag{92.31}$$

Thus

(92.32)

$$S_1S = \begin{pmatrix} 1 & 0 & 0 & 0 \\ 0 & 1 & \delta\theta_P & 0 \\ 0 & -\delta\theta_P & 1 & 0 \\ 0 & 0 & 0 & 1 \end{pmatrix} \begin{pmatrix} 1 & 0 & 0 & 0 \\ 0 & 1 & 0 & 0 \\ 0 & 0 & \gamma & i\beta\gamma \\ 0 & 0 & -i\beta\gamma & \gamma \end{pmatrix} \begin{pmatrix} 1 & 0 & 0 & 0 \\ 0 & 1 & \delta\theta_N & 0 \\ 0 & -\delta\theta_N & 1 & 0 \\ 0 & 0 & 0 & 1 \end{pmatrix}$$

$$= \begin{pmatrix} 1 & 0 & 0 & 0 \\ 0 & 1 & \gamma\,\delta\theta_P + \delta\theta_N & i\beta\,\delta\theta_P \\ 0 & -\delta\theta_P - \gamma\,\delta\theta_N & \gamma & i\beta\gamma \\ 0 & i\beta\gamma\,\delta\theta_N & -i\beta\gamma & \gamma \end{pmatrix}$$

Comparison of (92.32) with (92.30) shows that the angles $\delta\theta_P$ and $\delta\theta_N$ must be given by

$$\delta\theta_P = \frac{\beta_2}{\beta}, \qquad \delta\theta_N = -\frac{\beta_2}{\beta\gamma} \tag{92.33}$$

$\delta\theta_P$ and $\delta\theta_N$ are themselves of the first order in v/c. It is therefore permissible to add them to obtain the total angle of rotation to this approximation:

$$\delta\theta = \delta\theta_P + \delta\theta_N = \frac{\beta_2}{\beta}\left(1 - \frac{1}{\gamma}\right) \tag{92.34}$$

$$= \frac{\beta_2}{\beta}\,[1 - (1 - \beta^2)^{1/2}] = \frac{v_3\,\delta v_2}{2c^2}$$

In general, then, there will be an angular velocity $\boldsymbol{\omega}_T$ given by

$$\boldsymbol{\omega}_T = \frac{1}{2c^2}\,\dot{\mathbf{v}} \times \mathbf{v} \tag{92.35}$$

in agreement with (92.29).

A vector characterizing a particle which appears constant to an observer in the instantaneous rest frame P appears to an observer in another frame N to be precessing with the Thomas frequency $\boldsymbol{\omega}_T$. Thus the effect of the torque applied to the spin angular momentum vector $\boldsymbol{\sigma}$ must be estimated by the difference between the actual precession and Thomas precession rather than by the total precession alone.*

GENERAL REFERENCES

M. Abraham and R. Becker, *Theorie der Elektrizität.* Teubner, Leipzig and Berlin (1933).

W. Pauli, *Enzykl, d. Math. Wiss.*, Bd. V. 2, Teil S 543.

R. C. Tolman, *Relativity, Thermodynamics, Cosmology.* Oxford University Press (1934).

A. S. Eddington, *The Mathematical Theory of Relativity.* Cambridge University Press (1930).

P. G. Bergmann, *An Introduction to the Theory of Relativity.* Prentice-Hall, New York (1942).

H. P. Robertson, *Rev. Mod. Phys.*, 21, 378 (1949).

C. Møller, *The Theory of Relativity.* Oxford, London (1952).

EXERCISES

1. Show that under the Galilean transformation

$$\bar{x} = x, \qquad \bar{y} = y, \qquad \bar{z} = z - vt, \qquad \bar{t} = t$$

the equation

$$\nabla^2\phi - \frac{1}{c^2}\frac{\partial^2\phi}{\partial t^2} = 0$$

becomes

$$\frac{\partial^2\phi}{\partial\bar{x}^2} + \frac{\partial^2\phi}{\partial\bar{y}^2} + \left(1 - \frac{v^2}{c^2}\right)\frac{\partial^2\phi}{\partial\bar{z}^2} + \frac{2v}{c^2}\frac{\partial^2\phi}{\partial\bar{t}\,\partial\bar{z}} - \frac{1}{c^2}\frac{\partial^2\phi}{\partial\bar{t}^2} = 0$$

2. Show that under the Lorentz transformation

$$\bar{x} = x, \qquad \bar{y} = y, \qquad \bar{z} = \gamma(z - vt)$$

$$\bar{t} = \gamma\left(t - \frac{vz}{c^2}\right) \qquad \left[\gamma = \left(1 - \frac{v^2}{c^2}\right)^{-1/2}\right]$$

the equation

$$\nabla^2\phi - \frac{1}{c^2}\frac{\partial^2\phi}{\partial t^2} = 0$$

is invariant in form.

* For alternative discussions of this theorem see: L. H. Thomas, *Phil. Mag.*, 3, 1 (1926); J. Frenkel, *Z. Physik*, 37, 243 (1926); S. Dancoff and D. Inglis, *Phys. Rev.*, 50, 784 (1936); H. Kramers, *Physica*, 1, 825 (1934).

3. Show that the Lorentz transformation may be written vectorially thus:

$$\bar{\mathbf{r}} = \gamma[(\mathbf{n}\cdot\mathbf{r})\mathbf{n} - \mathbf{v}t] + [\mathbf{r} - (\mathbf{n}\cdot\mathbf{r})\mathbf{n}]$$

$$\bar{t} = \gamma\left(t - \frac{\mathbf{v}\cdot\mathbf{r}}{c^2}\right)$$

where $\mathbf{n}$ is a unit vector in the direction of $\mathbf{v}$.

4. Find what Newton's equation $\mathbf{F} = m\ddot{\mathbf{r}}$ becomes under the Lorentz transformation of Exercise 2.

5. Show that according to (89.4), (92.18), for a particle of gyromagnetic ratio $g = e/mc$ moving in a magnetostatic field, the magnitudes of $\mathbf{v}$, $\boldsymbol{\sigma}$, and the angle between $\mathbf{v}$ and $\boldsymbol{\sigma}$ are each separately constants of the motion.

6. When a particle of charge e is placed in the equatorial plane at a distance r from a fixed magnetic dipole of strength μ, the extra angular momentum resident in the electromagnetic field is $(e/cr)\boldsymbol{\mu}$. Show this result:

(*a*) By evaluating the space components of (89.10).

(*b*) By computing the torque on the particle as it is brought in from infinity along a radius.

(*c*) By computing

$$\frac{c}{4\pi}\int \{\mathbf{r}\times(\mathbf{E}\times\mathbf{B})\}\,dv$$

for the final configuration.

7. Prove that the transformation specified by

$$S_{ij} = -(2\mu_i\mu_j + \delta_{ij}) \qquad (\mu_i\mu_i = -1,\ \mu_i = \text{const})$$

is orthogonal and improper. Show that the Lorentz transformation which this describes relates the observations of an inertial observer O to those which O would make on reflection in a point which is moving relative to O with the uniform four-velocity μ_i. Examine the special case in which the point about which reflections are made is at rest relative to O.

8. Compute the rate of precession of the spin of a particle with an anomalous magnetic moment in a uniform field $\mathbf{B}$.

9. Discuss the motion of a free particle according to Eq. (92.20).

17 THE ORBITS OF PARTICLES IN HIGH ENERGY ACCELERATORS

93. INTRODUCTION

The acceleration of charged particles to very high energies has been a problem of interest for many years. The study of the orbits in accelerators has made great advances recently and promises to make more. It is impossible to give a detailed account here of all that has been done but the general formulation of the problem is an example of the application of classical mechanics where the sophisticated methods developed in the last few chapters are indispensable.

We concern ourselves only with accelerators in which the particles to be accelerated are confined in space by an external magnetic field and in which the energy gain per turn is small. This latter restriction means that before the energy of the particle changes appreciably the particle will have made many revolutions in the machine. Thus the acceleration is an adiabatic process and the orbit problem can be studied at a fixed energy.

In this chapter we discuss the general problem of equilibrium orbits, then the betatron oscillations about these orbits in the linear approximation for both weak and strong focusing accelerators, and finally the acceleration process and the associated synchrotron oscillations. We do not discuss the very important and interesting problems connected with the effects of imperfections on the orbits, nor do we enter into the design details of actual accelerators.*

* For more complete discussion and references see E. D. Courant and H. S. Snyder, *Ann. Phys.*, 3, 1 (1958).

94. EQUILIBRIUM ORBITS

We are concerned with the trajectories of charged particles in time-independent magnetic fields. A trajectory which returns to its starting point with its initial velocity after one revolution is called an *equilibrium orbit.* (Neither a trajectory which merely intersects itself nor one which closes smoothly after more than one turn is classed as such an orbit.) A particle accelerator must provide such orbits and others which stay in the neighborhood of such orbits.

Because even relativistically the Lorentz force is perpendicular to the velocity, the kinetic energy of the particle is constant and there is no variation in the relativistic mass $\tilde{m} = m(1 - v^2/c^2)^{-1/2}$. We may therefore deduce the correct equations of motion from the Lagrangian

$$L = \tfrac{1}{2}\tilde{m}v^2 + \epsilon \mathbf{v} \cdot \mathbf{A}, \qquad \epsilon = e/c \tag{94.1}$$

in which $\tilde{m}$ is to be treated as a constant. This Lagrangian will not suffice when energy changes are allowed for. According to (94.1) the momenta $\mathbf{p}$ conjugate to the coordinates $\mathbf{r}$ are given by

$$\mathbf{p} = \tilde{m}\mathbf{v} + \epsilon\mathbf{A} \tag{94.2}$$

A trajectory between two points P_1 and P_2 is determined by the principle of stationary action to be a curve connecting these two points for which the quantity

$$\Sigma = \int_{t_1}^{t_2} \mathbf{p} \cdot \mathbf{v}\, dt$$

has a stationary value when comparison is made with curves with the same end points and traversed with the same energy. Here the energy is the kinetic energy T of the particle so that a common energy implies a common velocity. If τ is the time interval required to go from P_1 to P_2 along a given curve, we have

$$\Sigma = 2T\tau + \epsilon \int_{P_1}^{P_2} \mathbf{A} \cdot d\mathbf{r} \tag{94.3}$$

Now let the end points P_1 and P_2 coincide at P so that we are considering a closed curve Γ. This curve will be a possible trajectory if it can be chosen so that it gives Σ a stationary value. It will in general not be an equilibrium orbit because it need not have a continuous tangent at P. We have

$$\begin{aligned} \Sigma &= 2T\tau + \epsilon \oint \mathbf{A} \cdot d\mathbf{r} \\ &= 2T\tau + \epsilon \int \mathbf{B} \cdot \mathbf{n}\, dS \end{aligned} \tag{94.4}$$

The surface integral is evaluated over a surface bounded by Γ and is equal to $-\epsilon\Phi$, Φ being the magnetic flux through Γ.

Because the period of revolution τ appears explicitly in Σ and because the speed of the particle is not to be varied, the circumference C of the curve Γ is an important parameter. We write

$$\Delta\Sigma = \Sigma_C \, \Delta C + \Delta_s \Sigma \tag{94.5}$$

where ΔC is the variation in the circumference and Δ_s means variation due to a change of shape of Γ without change of circumference. Now $\Delta C = v \, \Delta\tau$ so that by (94.4)

$$\Delta_s \Sigma = -\epsilon \, \Delta\Phi \tag{94.6}$$

A necessary condition for the existence of a closed trajectory through P is the existence of a closed curve Γ through P which encloses a stationary amount of flux for a given circumference. This will occur whenever the magnetic field is bounded, as it always is in practice.

Now with $\Delta_s \Sigma = 0$ we obtain from (94.4)

$$\begin{aligned} \Delta\Sigma &= \left(\frac{2T}{v} - \epsilon \frac{d\Phi}{dC}\right) \Delta C \\ &= \left(p - \epsilon \frac{d\Phi}{dC}\right) \Delta C \end{aligned} \tag{94.7}$$

where p is the momentum $\tilde{m}v$ of the particle. $d\Phi/dC$ depends only on the distribution of magnetic field in space so that for any magnetic field leading to a stationary value of the flux for a given circumference there is a momentum p for which Γ is a trajectory.

The above variations have all been made with a fixed point P on the curve. We may vary P, keeping the circumference constant and the shape such as to enclose a stationary amount of flux until the enclosed flux is stationary with respect to this variation also. If the magnetic field is continuous, this will occur when Γ is a smooth curve everywhere along its length. This curve will then be an equilibrium orbit for some momentum.

All actual accelerators in existence are designed to have a plane of symmetry to which the magnetic field is perpendicular. A particle whose initial position and velocity lie in this plane will remain there. Equilibrium orbits will exist in this plane; these are the equilibrium orbits of primary importance.

A trajectory which remains in the neighborhood of an equilibrium orbit for the same momentum is said to represent *betatron oscillations* about that orbit. These oscillations may be in the symmetry plane (radial oscillations) or perpendicular to it (vertical oscillations). Only if the oscillations obey linear equations of motion are these usually separable.

A given magnetic field will allow for equilibrium orbits with a range of circumferences, hence with a range of momenta. During the acceleration cycle of the machine it is necessary to have equilibrium orbits for particles of all momenta from that of injection to the final momentum. If the machine can accommodate only a narrow range of momenta, it is necessary to make the field increase with time during the cycle. This increase is slow enough in practice to be treated adiabatically. It is possible to design an accelerator which does contain equilibrium orbits for particles of all momenta of interest at one field setting. The cyclotron and the fixed field alternating gradient (FFAG) synchrotron accomplish this. The equilibrium orbits in the FFAG synchrotron become very complicated as the included momentum range increases. It is usually necessary to determine them by numerical integration of the equations of motion for various initial conditions. (It is not advantageous to make the variations described above numerically because this involves considering curves which are not trajectories of any kind. Numerical integration of the equations of motion leads to possible trajectories only, though in general not to equilibrium orbits. These are found by successive approximation.)

The momentum appropriate to a given equilibrium orbit is

$$p = \epsilon \frac{d\Phi}{dC} \tag{94.8}$$

The *momentum compaction* α is defined by

$$\alpha = \frac{C}{p}\frac{dp}{dC} \tag{94.9}$$

where C is the circumference of the equilibrium orbit of a particle of momentum p. If the equilibrium orbits are circles so that B is constant along the orbit,

$$p = \frac{\epsilon CB}{2\pi}$$

so that in this special case

$$\alpha - 1 = \frac{C}{B}\frac{dB}{dC} = \frac{R}{B}\frac{dB}{dR} \tag{94.10}$$

where R is the radius of the circular orbit.

The *field index* n is defined for a general field by

$$n = -\frac{\rho}{B}\frac{dB}{d\rho} \tag{94.11}$$

where ρ is the radius of curvature of the equilibrium orbit. n is generally a function of position. If n is independent of ρ, then

$$B = B_0(\rho/\rho_0)^{-n}$$

When the equilibrium orbits are circles, $\rho = R$ and $\alpha = 1 - n$.

To find the rotation frequency of a particle in an equilibrium orbit it is necessary to use the relativistic relation between velocity and momentum:

$$v = \frac{pc}{(p^2 + m^2c^2)^{1/2}}$$

Then

$$\frac{1}{\tau} = \frac{v}{C} = \frac{1}{C}\frac{pc}{(p^2 + m^2c^2)^{1/2}} \tag{94.12}$$

and

$$\frac{d}{dp}\left(\frac{1}{\tau}\right) = \frac{1}{\tau}\frac{(\alpha - 1)m^2c^2 - p^2}{\alpha pE^2/c^2} = -\frac{\eta}{p\tau} \tag{94.13}$$

where E is the total energy of the particle including the rest energy. If $\alpha = 1$ (as it nearly is in the low energy cyclotron) the rotation frequency is nearly constant for small p but is decreasing. For $\alpha < 1$ the frequency decreases more rapidly with increasing p. For $\alpha > 1$ the frequency increases with increasing p up to the point where $p^2 = (\alpha - 1)m^2c^2$, after which it decreases again. The energy E_t at which this reversal takes place is called the *transition energy*. It plays an important role in the acceleration process. By making α a function of p the transition energy can be made arbitrarily high.

95. BETATRON OSCILLATIONS

In the last section we saw that there exist closed orbits in a stationary magnetic field which are not equilibrium orbits because the velocity vector does not return to its original direction. These orbits are a special case of orbits executing betatron oscillations about an equilibrium orbit. In general, with both radial and vertical betatron oscillations present the orbits need not ever close exactly. If they remain in the neighborhood of an equilibrium orbit, the betatron oscillations are stable.

To discuss these betatron oscillations we introduce the coordinate system illustrated in Fig. 95-1. P is a point not too far from the equilibrium orbit Γ. P_0 is the point of Γ nearest P. (This is unique only when P is less than the radius of curvature from Γ. Hence the restric-

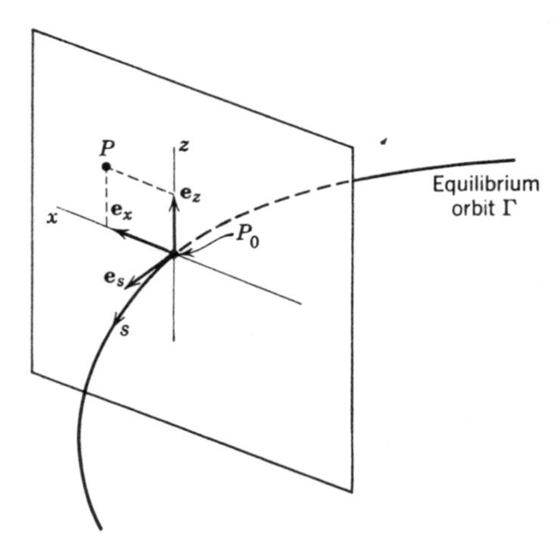

Fig. 95-1. Coordinates used in discussion of betatron oscillations.

tion "not too far.") If $\mathbf{e}_s$ is the unit vector tangent to Γ at P_0, $\mathbf{e}_x$ the principal normal, and $\mathbf{e}_z$ the binormal, then $\mathbf{e}_x \times \mathbf{e}_s = \mathbf{e}_z$ and the radius vector $\mathbf{r}$ to P may be written as

$$\mathbf{r} = \mathbf{r}_0(s) + z\mathbf{e}_z(s) + x\mathbf{e}_x(s) \tag{95.1}$$

where $\mathbf{r}_0(s)$ is the radius vector of P_0 from an arbitrary origin. s is the distance of P_0 from an arbitrary point on Γ measured along Γ. z, x, s form a right-handed coordinate system.

The Frenet formulas for any smooth curve are

$$\begin{aligned} \frac{d\mathbf{e}_s}{ds} &= -\kappa\mathbf{e}_x \\ \frac{d\mathbf{e}_x}{ds} &= \kappa\mathbf{e}_s + \tau\mathbf{e}_z \\ \frac{d\mathbf{e}_z}{ds} &= -\tau\mathbf{e}_x \end{aligned} \tag{95.2}$$

Here κ is the curvature and τ the torsion of the curve. For plane curves $\tau = 0$.

The momenta conjugate to z, x, s may be found by use of a generating function ψ' as in (59.12), as we are dealing with a point transformation

$$\psi' = \mathbf{p} \cdot \mathbf{r}(z, x, s)$$

Thus

$$p_z = \frac{\partial \psi'}{\partial z} = \mathbf{p} \cdot \mathbf{e}_z$$

$$p_x = \frac{\partial \psi'}{\partial x} = \mathbf{p} \cdot \mathbf{e}_x \tag{95.3}$$

$$p_s = \frac{\partial \psi'}{\partial s} = \mathbf{p} \cdot [(1 + \kappa x)\mathbf{e}_s + \tau(x\mathbf{e}_z - z\mathbf{e}_x)]$$

The curvilinear nature of the coordinates shows up only in p_s. From now on we consider only plane curves, i.e., equilibrium orbits in an accelerator with a plane of symmetry, so that $\tau = 0$.

The components of the vector potential are obtained from $\mathbf{A}$ by a set of equations like (95.3) with $\mathbf{A}$ replacing $\mathbf{p}$. The Hamiltonian becomes

$$H = \frac{(\mathbf{p} - e\mathbf{A})^2}{2\tilde{m}} \tag{95.4}$$

$$= \frac{1}{2\tilde{m}} \left[\left(\frac{p_s - \epsilon A_s}{1 + \kappa x} \right)^2 + (p_z - \epsilon A_z)^2 + (p_x - \epsilon A_x)^2 \right]$$

Because we are more interested in the geometry of the orbits than in the time dependence of the motion we change the independent variable from t to s by the method of Sec. 72. H is time-independent so that the momentum h conjugate to H is constant. We solve

$$H + h = 0$$

for p_s and write the solution as

$$G(z, x, p_z, p_x, s) + p_s = 0$$

G is then the new Hamiltonian with s as the independent variable. From (95.4) we obtain

$$G = -(1 + \kappa x)[-2\tilde{m}h - (p_z - \epsilon A_z)^2 - (p_x - \epsilon A_x)^2]^{1/2} - \epsilon A_s \tag{95.5}$$

The sign has been chosen to make p_s positive. The momentum p

used in previous sections is given by

$$p = (-2\tilde{m}h)^{1/2} \tag{95.6}$$

The equations of motion are now, with $' \equiv d/ds$,

$$\begin{aligned} z' &= \frac{\partial G}{\partial p_z}, \qquad p_z' = -\frac{\partial G}{\partial z} \\ x' &= \frac{\partial G}{\partial p_x}, \qquad p_x' = -\frac{\partial G}{\partial x} \end{aligned} \tag{95.7}$$

G is a periodic function of s. The equations refer explicitly to the equilibrium orbit. They are useful only insofar as (1) the equilibrium orbit is known, and (2) the oscillations are of small amplitude so that G may be expanded and only quadratic terms kept. In FFAG synchrotrons neither of these conditions is satisfied. The radial apertures are large and large amplitude betatron oscillations are possible, and, as mentioned before, the equilibrium orbits are very complicated. We do not enter into the discussion of betatron oscillations in FFAG accelerators.

The form of G shows that A_z and A_x are needed only to terms linear in the coordinates, whereas A_s must be expanded to quadratic terms to get the quadratic part of G. We may write

$$\begin{aligned} \epsilon A_z &= -Qx \\ \epsilon A_x &= +Qz \\ \epsilon A_s &= -\kappa px + \tfrac{1}{2}az^2 + bzx + \tfrac{1}{2}cx^2 \end{aligned} \tag{95.8}$$

The coefficients are periodic functions of s. The special forms of A_z, A_x may be obtained from a general linear form for each by a gauge transformation, i.e., by adding a gradient to $\mathbf{A}$. Then

$$\begin{aligned} \epsilon B_{zx} &= \epsilon\left(\frac{\partial A_x}{\partial z} - \frac{\partial A_z}{\partial x}\right) = 2Q \\ \epsilon B_{xs} &= \epsilon\left(\frac{\partial A_s}{\partial x} - \frac{\partial A_x}{\partial s}\right) = -\kappa p + bz + cx - Q'z \\ \epsilon B_{sz} &= \epsilon\left(\frac{\partial A_z}{\partial s} - \frac{\partial A_s}{\partial z}\right) = -Q'x - az - bx \end{aligned} \tag{95.9}$$

are the components of the magnetic field tensor. These must satisfy Maxwell's equations

$$\nabla \times \mathbf{B} = 0$$

The components of B_{mn} are given in curvilinear coordinates. The corresponding vector components are*

$$\epsilon B_z = \frac{1}{1+\kappa x}(-\kappa p + bz + cx - Q'z)$$

$$= -\kappa p + (l - Q')z + (c + \kappa^2 p)x + \cdots$$

(95.10)
$$\epsilon B_x = \frac{-1}{1+\kappa x}[(Q'+b)x + az]$$

$$= -az - (Q'+b)x + \cdots$$

$$\epsilon B_s = 2Q(1+\kappa x) + \cdots$$

The only Maxwell equation to give useful information is $(\nabla \times \mathbf{B})_s = 0$ because nothing has been specified about the dependence of the coefficients on s other than their periodicity. We obtain

$$\epsilon\left(\frac{\partial B_x}{\partial z} - \frac{\partial B_z}{\partial x}\right) = -a - c - \kappa^2 p = 0 \tag{95.11}$$

which can be used to eliminate c. Furthermore,

$$a = -\epsilon\frac{\partial B_x}{\partial z} = -\epsilon\frac{\partial B_z}{\partial x} = \epsilon n\frac{B_z}{\rho} = \frac{np}{\rho^2}$$

using (94.11) and putting $\rho = 1/\kappa$. Thus, inserting (95.8) in (95.5) and using these results we get

$$G = \frac{1}{2p}[(p_z + Qx)^2 + (p - Qz)^2] + \frac{n}{2\rho^2}z^2 + \frac{1-n}{2\rho^2}x^2 - bzx - p \tag{95.12}$$

after expanding the square root.

If G is independent of s so that Q, a, b, and κ are constants, G is simply the Hamiltonian for two coupled harmonic oscillators, which can be separated into two independent oscillators. The effective mass of the oscillators is p. If we now imagine p to increase slowly owing to accelerating fields of some sort, the magnetic field will be increased in

* The covariant components of $\mathbf{B}$ are related to the field tensor B_{mn} by

$$B_i = \tfrac{1}{2} g_{il}\epsilon^{lmn}B_{mn}$$

where $\epsilon^{lmn} = \epsilon(lmn)/\sqrt{g}$. Here $\epsilon(lmn)$ is the antisymmetric symbol and g is the determinant of the metric tensor g_{ij}. See Appendix I.

proportion to keep the radius constant, so that a, b, and Q will be proportional to p. The effective Hamiltonian for one of the oscillators is of the form

$$H = \frac{1}{2p} p_1{}^2 + \frac{1}{2} p\omega^2 x_1{}^2$$

This oscillator has an action integral given by (62.6) which we write as

$$J_1 = \pi p \omega A^2 \tag{95.13}$$

where A is the amplitude of the oscillation. An action integral is an adiabatic invariant so that an adiabatic increase in p leads to a corresponding decrease in A^2. Thus

$$A \sim p^{-1/2} \tag{95.14}$$

This *adiabatic damping* is a very useful effect for it means that the aperture of an accelerator required to accommodate the necessary amplitude of betatron oscillation at injection will be more than ample at the final momentum, and a very well-defined beam can be produced. This result remains valid when the Hamiltonian G is a periodic function of s.

In G the terms involving Q describe a longitudinal magnetic field whereas the term involving b describes a tilt of the symmetry plane of the magnetic field. Both of these are present in an actual accelerator but only because of imperfections, so we shall put them equal to zero. The vertical and radial oscillations are therefore uncoupled.

96. WEAK FOCUSING ACCELERATORS

Weak focusing accelerators are those for which n is not very different from a constant along the orbit. It may be a constant as in a cyclotron or it may differ from a constant primarily because of the presence of field-free regions (straight sections) and the associated fringing field regions. We take the simplest situation, namely $n' = 0$.

The Hamiltonian (95.12) reads

$$G = \frac{1}{2}\left(p_z{}^2 + \frac{n}{\rho^2} z^2\right) + \frac{1}{2}\left(p_x{}^2 + \frac{1-n}{\rho^2} x^2\right) \tag{96.1}$$

the irrelevant term $-p$ having been dropped and all momenta being measured in units of p. The vertical and radial oscillations are harmonic. We see that

$$\omega_z{}^2 = \frac{n}{\rho^2}, \qquad \omega_x{}^2 = \frac{1-n}{\rho^2} \tag{96.2}$$

To have both motions stable, both ω_z^2 and ω_x^2 must be positive so that

$$0 < n < 1 \tag{96.3}$$

The magnetic field must decrease with radius. Equation (94.13) shows that the rotation frequency of the particle in the accelerator decreases with increasing momentum.

If ν_z and ν_x are the numbers of complete vertical and radial betatron oscillations per revolution respectively, we have

$$\nu_z = \sqrt{n}, \qquad \nu_x = \sqrt{1 - n} \tag{96.4}$$

The limiting case $n = 0$ is instructive. Here $\nu_z = 0$ because motion along a uniform field is independent of the field. $\nu_x = 1$, which means that all the orbits in the symmetry plane are circles and are therefore equilibrium orbits. There are thus equilibrium orbits for a given momentum anywhere in the symmetry plane. This is undesirable because any slight field inhomogeneity will cause the orbit to migrate. This can be described by saying that the betatron frequency is resonant with the rotation frequency and that the field inhomogeneity provides a periodic impulse to the particle at the rotation frequency, thus resulting in an indefinitely growing amplitude of the betatron oscillation. This is not the only resonance which can occur in practice.

97. STRONG FOCUSING ACCELERATORS

The necessity to make both the vertical and radial betatron oscillations stable puts very strong limits on the field index n if n is to be a constant along an equilibrium orbit. This then makes the momentum compaction α small so that a large radial aperture is needed to accept an appreciable momentum range. By making n vary along the orbit, however, the stability of both kinds of betatron oscillations can be achieved in many ways and the momentum compaction can be increased, reducing the required radial aperture. We again restrict ourselves to accelerators with a plane of symmetry and to small amplitudes of oscillation. This excludes the FFAG synchrotrons which have been much studied but whose consideration would take us too far.*

Under our assumptions, there is no coupling between the vertical and radial betatron oscillations so that we may confine our attention

* For a discussion of these see Symon, Kerst, Jones, Laslett, and Terwilliger, *Phys. Rev.*, 103, 1837 (1956). Many detailed studies are contained in the unpublished reports of the Midwestern Universities Research Association.

to the vertical oscillations alone. For these the Hamiltonian is

$$G = \tfrac{1}{2}(p_z^2 + n\kappa^2 z^2) \tag{97.1}$$

Both n and κ may be functions of s. The motion may be most easily visualized if κ is regarded as a constant and all the s dependence is put onto n, which does not affect the analysis at all since only the product $n\kappa^2$ appears. The equations of motion are

$$z' = p_z, \qquad p_z' = -n\kappa^2 z \tag{97.2}$$

leading to the second-order differential equation

$$z'' + n(s)\kappa^2 z = 0 \tag{97.3}$$

The field index $n(s)$ is a periodic function of s. In the presence of imperfections the period is C, the circumference of the equilibrium orbit, but in a perfect machine the period is $L = C/N$, where N is the number of identical units of which the accelerator is constructed.

According to Floquet's theorem, a complete set of solutions of a differential equation of the form (97.3) can be written as

$$z_\pm(s) = w(s)e^{\pm i\psi(s)} \tag{97.4}$$

where $w(s)$ is a real function of s with period L. The $z_\pm(s)$ are bounded for large positive and negative s when $\psi(s)$ is real. A complex or imaginary ψ leads to unbounded $z(s)$, hence to unstable orbits. A real solution in the stable case is thus

$$z(s) = Aw(s) \cos [\psi(s) + \delta] \tag{97.5}$$

The functions w and ψ satisfy the equations

$$\begin{aligned} w'' + n\kappa^2 w - w^{-3} &= 0 \\ \psi' &= w^{-2} \end{aligned} \tag{97.6}$$

These imply a suitable normalization for w (which has the dimensions of $l^{1/2}$). The mathematical problem here is the same as that met in discussing the motion of a particle in a periodic, time-independent one-dimensional potential in quantum mechanics. The bounded solutions here are the allowed eigenfunctions there. The well-known band structure of the energy eigenvalues in the quantum problem corresponds to bands of possible betatron frequencies here (cf. also Sec. 26).

A convenient way to study the solutions of this problem is to introduce the 2×2 matrix relating $z(s)$, $z'(s)$ with $z(s_0)$, $z'(s_0)$. Such a matrix exists because (97.3) is linear and of second order. This matrix $M(s, s_0)$ must be sympletic as defined in (70.3). For a system with one degree of freedom this condition is simply $\det M = 1$. Let $X(s)$ be the 2 component vector whose components are $z(s)$, $p_z(s) = z'(s)$. Then

$$X(s) = M(s, s_0)X(s_0) \tag{97.7}$$

It may be verified by use of (97.3) and (97.6) that

$$M(s, s_0) = \begin{pmatrix} \dfrac{w}{w_0}\cos\psi - w_0'w\sin\psi\,, & ww_0\sin\psi \\ -\dfrac{1 + w_0w_0'ww'}{w_0w}\sin\psi - \left(\dfrac{w_0'}{w} - \dfrac{w'}{w_0}\right)\cos\psi, & \dfrac{w_0}{w}\cos\psi + w_0w\sin\psi \end{pmatrix} \tag{97.8}$$

where $w_0 = w(s_0)$, $\cdots$, and $\psi = \psi(s) - \psi(s_0)$.

If $s = s_0 + L$, then $w_0 = w$ and $w_0' = w'$ so that $M(s, s - L) \equiv M(s)$ is given by

$$M(s) = \begin{pmatrix} \cos\mu - w'w\sin\mu\,, & w^2\sin\mu \\ -\dfrac{1 + (ww')^2}{w^2}\sin\mu, & \cos\mu + w'w\sin\mu \end{pmatrix} \tag{97.9}$$

with

$$\mu = \psi(s) - \psi(s - L) = \int_{s-L}^{s} w^{-2}\,ds$$

Actually μ is independent of s, for if $s' \neq s$,

$$\begin{aligned} M(s') &= M(s', s)M(s)M(s, s') \\ &= M^{-1}(s, s')M(s)M(s, s') \end{aligned}$$

so that

$$\operatorname{tr} M(s') = \operatorname{tr} M(s) \qquad \text{or} \qquad \cos\mu' = \cos\mu$$

Therefore we write

$$\mu = \int_0^L w^{-2}\,ds \tag{97.10}$$

The motion can be regarded as a pseudoharmonic one with amplitude Aw and phase $\psi(s) = \int^s w^{-2}\,ds$. μ is the phase advance of the oscillation per period L, so the number of complete oscillations per circumference ν is

$$\nu = \frac{N\mu}{2\pi} \tag{97.11}$$

The condition for stability is that μ be real, hence that

$$|\operatorname{tr} M| = |2\cos\mu| \leq 2 \tag{97.12}$$

We assume the inequality sign to hold so that we are not on the boundary of the stable region.

To use the inequality (97.12) we must specify $n(s)$, i.e., consider some particular model. Following Courant and Snyder, we take the case of a machine with circular equilibrium orbits of radius R in which

$$\begin{aligned} n(s) &= n_1, & 0 < s < \frac{\pi R}{N} \\ n(s) &= -n_2, & \frac{\pi R}{N} < s < \frac{2\pi R}{N} \end{aligned} \tag{97.13}$$

there being N repetitions of this pattern around the machine. For vertical oscillations, then

$$M(0) = M\left(\frac{2\pi R}{N}, \frac{\pi R}{N}\right) M\left(\frac{\pi R}{N}, 0\right) \tag{97.14}$$

with

$$\begin{aligned} M\left(\frac{\pi R}{N}, 0\right) &= \begin{pmatrix} \cos\phi & (\sqrt{n_1}\,\kappa)^{-1}\sin\phi \\ -\sqrt{n_1}\,\kappa\sin\phi & \cos\phi \end{pmatrix} \\ M\left(\frac{2\pi R}{N}, \frac{\pi R}{N}\right) &= \begin{pmatrix} \cosh\psi & (\sqrt{n_2}\,\kappa)^{-1}\sinh\psi \\ \sqrt{n_2}\,\kappa\sinh\psi & \cosh\psi \end{pmatrix} \end{aligned} \tag{97.15}$$

where $\phi = \pi\sqrt{n_1}/N$, $\psi = \pi\sqrt{n_2}/N$. Equations (97.15) are obtained directly from the solution of the stable and unstable harmonic oscillator.

In this model the stability condition is

$$\tfrac{1}{2}|\operatorname{tr} M(0)| = \left|\cos\phi\cosh\psi - \frac{n_1 - n_2}{2(n_1 n_2)^{1/2}}\sin\phi\sinh\psi\right| < 1 \tag{97.16}$$

The region of stability is shown in Fig. 97-1. Not every point in this region is stable when account is taken of effects neglected here. The effect of imperfections with period C rather than L is to crosshatch the stability region with lines of instability, N in each direction. In addition, there are coupling resonances and resonances due to nonlinear effects which introduce instabilities. We cannot enter into these here.

Because the Hamiltonian G depends explicitly on the independent variable s there is no Jacobian integral of the motion. It is easily verified, however, that there is an integral of the motion given by

$$W = (X, UX) = A^2 \tag{97.17}$$

with

$$U = \begin{pmatrix} \dfrac{1 + (ww')^2}{w^2} & -ww' \\ -ww' & w^2 \end{pmatrix} \tag{97.18}$$

For each value of s and A, (97.17) defines an ellipse in phase space whose area is independent of s. [The area is proportional to the product of the principal semiaxes, hence to (product of eigenvalues)$^{-1/2}$ = (det U)$^{-1/2}$ = 1. See Appendix II.]

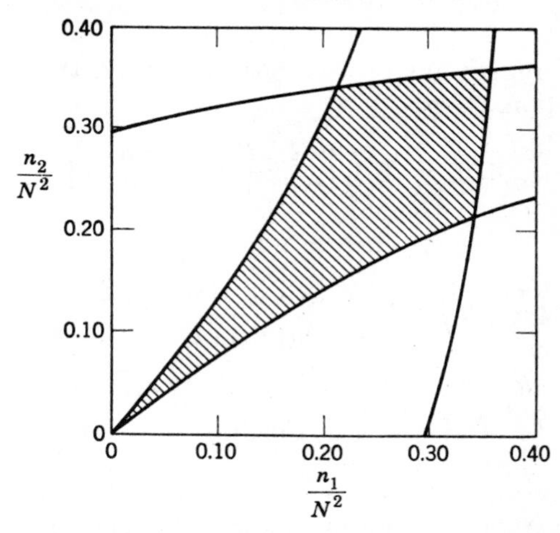

Fig. 97-1. Region of stability for radial and vertical oscillations. (Courtesy Courant and Snyder.)

The ellipse (97.17) has its greatest semiaxes in the z direction when $w' = 0$, $w = w_{\max}$. If we set the amplitude at this point equal to the effective aperture a of the accelerator for these oscillations, we see that the area $\mathfrak{A}$ of the ellipse is given by

$$\mathfrak{A} = \pi A^2 = \frac{\pi a^2}{w_{\max}^2} \tag{97.19}$$

$\mathfrak{A}$ is called the *admittance* of the accelerator for vertical oscillations. It is an important factor in determining the intensity of beam obtainable because the number of particles an injector can supply increases with the range of initial conditions acceptable by the accelerator.

98. ACCELERATION AND SYNCHROTRON OSCILLATIONS

So far we have considered only the motion in pure magnetic fields so that the energy of the particle is constant. We now consider some mechanisms of acceleration. In an FFAG accelerator or a cyclotron the magnetic field is constant in time and the acceleration must be produced by separately supplied electric fields oscillating with a frequency which is some harmonic of the frequency of rotation of a particle in an equilibrium orbit. This frequency ω_{rf} must change with time as the rotation frequency depends on the particle's energy. In nonfixed field accelerators the changing magnetic field itself induces an electric field in the correct direction to accelerate the contained particles. In the betatron this is the only accelerating field used.

Suppose the magnetic field to be constant along the orbit, the radius of which is then

$$r = \frac{pc}{eB} \tag{98.1}$$

If the field changes slowly we have

$$\frac{dr}{r} = \frac{dp}{p} - \frac{dB}{B}$$

Let us keep the orbit radius unchanged. There is a change $d\Phi$ in the flux enclosed by the orbit and a change dE in the particle energy:

$$dE = \frac{ev}{2\pi rc}\, d\Phi$$

Using the relativistic relations

$$\frac{v}{c} = \frac{pc}{E}, \qquad E^2 = m^2c^4 + p^2c^2$$

we obtain

$$d\Phi = 2\pi r^2\, dB \tag{98.2}$$

Thus if the orbit radius is to remain unchanged while the magnetic field is increased, the flux through the orbit must change by twice the amount that it would change if the field were uniform over the interior of the orbit. For high energy particles this means (1) there is a large volume of magnetic field, hence a large stored energy, and (2) the highest field attainable is not available at the orbit so the orbit radius must be large. This difficulty does not arise if separate accelerating fields are supplied. Then the magnetic field can be confined to the neighborhood of the orbit and the induced electric field is very small.

We now consider the acceleration to be produced entirely by oscillating electric fields acting across one or more gaps around the orbit. We assume the relative change of momentum per revolution to be very small so that the discrete impulses given to the particle can be averaged out and the acceleration can be considered continuous. The action of several gaps is now equivalent to the action of a single effective gap across which there is a voltage:

$$V(t) = V_0 \sin\left(\int_0^t \omega_{rf}(t)\,dt + \psi\right) \tag{98.3}$$

$V(t)$ is rapidly varying. It affects the particle only when the particle is in the gap, however, and if $\omega_{rf}(t)$ is nearly a harmonic of the rotation frequency of the particle in its orbit, the effect of V on the particle will change only a little from turn to turn. Thus if the value of $\int_0^t \omega_{rf}(t)\,dt + \psi$ when the particle is in the gap is ϕ, we may regard ϕ as a continuous, slowly varying function of the time. Then

$$\frac{d\phi}{dt} = \omega_{rf}(t) - h\omega(E) \tag{98.4}$$

where ω_{rf} is the hth harmonic of the frequency ω_s of a particle in the synchronous equilibrium orbit and $\omega(E)$ is the rotation frequency of a particle of energy E.

The energy gain is described by

$$\frac{dE}{dt} = \frac{\omega(E)}{2\pi} eV_0 \sin\phi \tag{98.5}$$

The pair of equations (98.4) and (98.5) specify ϕ and E as functions of t when $\omega_{rf}(t)$ is given. It is convenient to change the variable so that these equations become of Hamiltonian form. Define W by

$$W = \int_0^E \frac{dE'}{h\omega(E')} \tag{98.6}$$

Then we obtain

$$\begin{aligned} \frac{d\phi}{dt} &= \omega_{rf} - h\omega(W) \\ \frac{dW}{dt} &= \frac{eV_0}{2\pi h} \sin\phi \end{aligned} \tag{98.7}$$

which are derivable from the Hamiltonian

$$H(\phi, W, t) = -E(W) + W\omega_{rf} + \frac{eV_0}{2\pi h}\cos\phi \tag{98.8}$$

If ω_{rf} is constant, H is time-independent and describes oscillations in ϕ known as *synchrotron oscillations.* For sufficiently small amplitude these oscillations are harmonic. The existence of a stable phase about which oscillations can take place was first noted by Veksler and by McMillan.* This led to the development of all frequency-modulated accelerators, i.e., accelerators in which the momentum change is sufficiently large to affect the rotation frequency seriously.

From (98.7) we see that when ω_{rf} is constant there is a W which makes ϕ a constant. Assuming constant $\omega(W)$ means constant W (which is true everywhere except at the transition energy), the second of equations (98.7) then shows that $\phi = 0$ or π. If ω_{rf} changes adiabatically, then again there is an $\omega(W)$ which makes ϕ constant but which now implies a slowly changing W. Then ϕ is no longer 0 or π. For sufficiently small V_0, ϕ may differ considerably from these values without invalidating our adiabatic hypothesis. Whether the stable oscillations take place about $\phi_0 \gtrsim 0$ or about $\phi_1 \lesssim \pi$ depends upon the first two terms of H.

We may expand the first two terms of H about W_0, which is defined by

$$0 = \omega_{rf} - h\omega(W_0)$$

at some particular time. Then

$$-E(W) + W\omega_{rf} = -E(W_0) + h\omega(W_0)W_0 - \frac{1}{2}h\left(\frac{d\omega}{dW}\right)_0 (W - W_0)^2 \tag{98.9}$$

The constant terms are immaterial. If $(d\omega/dW)_0 > 0$ the stable oscillations take place about $\phi_0 \gtrsim 0$; above the transition energy where $(d\omega/dW)_0 < 0$ they take place about $\phi_1 \lesssim \pi$.

When the amplitude of oscillation of ϕ is small, the cosine in H may also be expanded. Below the transition energy, then, H becomes

$$H = \text{const} - \frac{1}{2}\left[h\left(\frac{d\omega}{dW}\right)_0 (W - W_0)^2 + \frac{eV_0}{\pi h}\sin\phi_0\,(\phi - \phi_0) + \frac{eV_0\cos\phi_0}{2\pi h}(\phi - \phi_0)^2\right] \tag{98.10}$$

* E. M. McMillan, *Phys. Rev.*, 68, 143 (1945); V. Veksler, *J. Phys. USSR*, 9, 153 (1945).

The frequency Ω of these oscillations is therefore given by

$$\Omega^2 = \left(\frac{d\omega}{dW}\right)_0 \frac{eV_0 \cos \phi_0}{2\pi} \tag{98.11}$$

$$= -\frac{\eta e V_0 \cos \phi_0}{2\pi E} {\omega_0}^2$$

where η is defined by (94.13) and $\omega_0 = \omega_s c/v$ is the frequency of a particle traversing the equilibrium orbit at the speed of light. Now $eV_0 \cos \phi_0 \ll E$ and in strong focusing accelerators $|\eta| \ll 1$ so that $\Omega^2 \ll {\omega_0}^2$. The synchrotron oscillations are of very low frequency compared to the betatron oscillations. There is therefore very little coupling between the two modes of oscillation.

Associated with the synchrotron oscillation is a radial oscillation because the particle executing the synchrotron oscillation has a variable momentum and is thus not in a single equilibrium orbit at all times. In weak focusing accelerators the amplitude of this radial oscillation or change of radius of the equilibrium orbit with energy may be large, requiring such machines to have much more radial than vertical aperture. In strong focusing machines the large momentum compaction makes this change of radius much smaller.

The variables ϕ, W are canonically conjugate. If they are taken as cartesian coordinates in a phase plane, then Liouville's theorem states that the density of phase points in the neighborhood of any given phase point cannot change provided that there is no coupling between the synchrotron and other degrees of freedom. The range of ϕ is restricted to 2π, say from $-\pi$ to π, while $0 \leq W \leq \infty$. This is illustrated in Fig. 98-1. The synchrotron oscillations discussed in the last few paragraphs are represented by small ellipses centered on the line $\phi = 0$ or on a parallel line. The acceleration process carries these ellipses slowly up the W axis until the final energy is reached.

In a conventional accelerator the range of W available to a particle at any one time is very narrow. In an FFAG accelerator the range of W available is essentially from $W = 0$ to $W = W_f$, the final value. There may be particles present represented by phase points anywhere in the shaded area of the phase plane shown in Fig. 98-2. Suppose there are particles present represented by phase points everywhere except for a narrow band at the top, from $W_f - \Delta$ to W_f. (These particles may have struck a target or have otherwise been lost.) We may now apply a radio-frequency voltage in such a way that a particle in region A of Fig. 98-2 would *lose* energy. The region A would then move down into the occupied part of the phase plane. Because of

Liouville's theorem some of the phase points below the line $W_f - \Delta$ must move above it, there being no other place to go. In this manner some particles outside of A are accelerated. The actual history of an accelerated particle may be complicated but the result is assured.

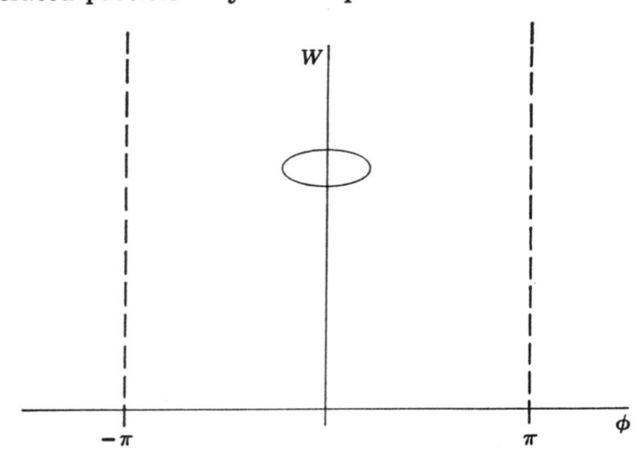

Fig. 98-1. Synchrotron oscillations in ϕ-W plane [W defined by (98.6)].

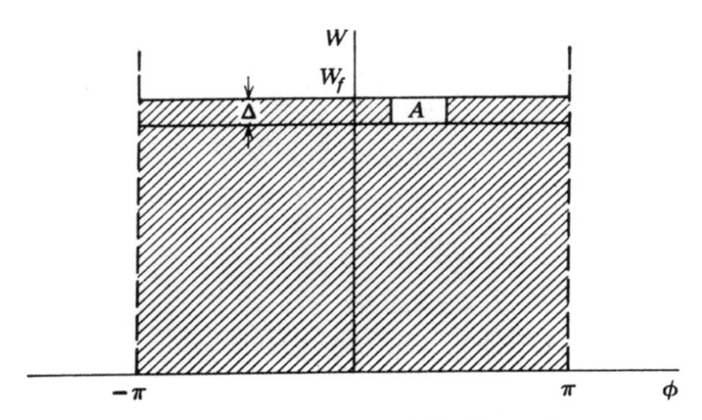

Fig. 98-2. ϕ-W plane for FFAG machine.

It is essential for this mechanism that the lower limit of W available in the accelerator be essentially zero. Otherwise particles could be lost out of the bottom of the diagram and none need be squeezed out the top. It is therefore not applicable to variable field accelerators.

APPENDIX I
RIEMANNIAN GEOMETRY

The configuration space of a dynamic system that is specified by f generalized coordinates is not in general a Euclidean space. By this we mean that there does not always exist a system of coordinates in terms of which the distance between any two neighboring points in the space is given simply by the sum of the squares of the coordinate differentials. A simple example of this occurs in the motion of a single particle on the surface of a fixed sphere. There exist no cartesian coordinates on the surface of a sphere and the space of two dimensions is said to be curved. Such spaces may be described in terms of Riemannian geometry rather than Euclidean geometry.

We denote by q^m $(m = 1 \cdots f)$ a set of coordinates which define a point in configuration space, and consider another set of coordinates $\bar{q}^n$ $(n = 1 \cdots f)$ which are known functions of the q^m:

$$\bar{q}^n = \bar{q}^n(q^1 \cdots q^f)$$

Infinitesimal changes dq^m in the q^m define infinitesimal changes $d\bar{q}^n$ in the $\bar{q}^n$ by

$$d\bar{q}^n = \frac{\partial \bar{q}^n}{\partial q^m} dq^m \qquad (m, n = 1 \cdots f) \tag{I.1}$$

(the summation convention being adopted). Quantities λ^m which transform according to the law

$$\bar{\lambda}^n = \frac{\partial \bar{q}^n}{\partial q^m} \lambda^m \tag{I.2}$$

are called the *contravariant components of a vector*. If the contravariant components of a vector are given in one coordinate system, then (I.2) gives them in any other coordinate system. However, there are components of vectors which do not transform according to (I.2). As an example, let

$$\lambda_m = \frac{\partial \phi}{\partial q^m}$$

where ϕ is a scalar function of the coordinates q^m. Then

$$\bar{\lambda}_n \equiv \frac{\partial \phi}{\partial \bar{q}^n} = \frac{\partial \phi}{\partial q^m} \frac{\partial q^m}{\partial \bar{q}^n}$$

or

$$\bar{\lambda}_n = \frac{\partial q^m}{\partial \bar{q}^n} \lambda^m \tag{I.3}$$

Quantities which transform according to (I.3) are called the *covariant components* of a vector. Contravariant components of a vector are labeled by superscripts and covariant components by subscripts.

If λ_m and μ^m are the covariant and contravariant components of two vectors respectively, the quantity $\lambda_m \mu^m$ is a scalar, or an invariant. For

$$\begin{aligned} \bar{\lambda}_n \bar{\mu}^n &= \frac{\partial q^m}{\partial \bar{q}^n} \lambda_m \frac{\partial \bar{q}^n}{\partial q^l} \mu^l \\ &= \lambda_m \mu^l \frac{\partial q^m}{\partial q^l} \\ &= \lambda_m \mu^l \delta_l{}^m \\ &= \lambda_m \mu^m \end{aligned} \tag{I.4}$$

In (I.4) the independence of q^m from q^l for $l \neq m$ has been used. The symbol $\delta_l{}^m$ is called a Kronecker delta; it has the value $+1$ when $m = l$, 0 otherwise. The summation convention is always applied to indices one of which is contravariant, the other covariant, because of the invariant significance of this combination.

The square of the length of a vector is a scalar. In particular let us consider the quantity $(dr)^2$. In cartesian coordinates this is given by

$$(dr)^2 \equiv ds^2 = (dx^1)^2 + (dx^2)^2 + (dx^3)^2$$

If generalized coordinates q^m are used, we have

$$(I.5) \qquad ds^2 = \left(\frac{\partial x^1}{\partial q^m}\,dq^m\right)^2 + \left(\frac{\partial x^2}{\partial q^m}\,dq^m\right)^2 + \left(\frac{\partial x^3}{\partial q^m}\,dq^m\right)^2$$

$$= g_{mn}\,dq^m\,dq^n$$

where

$$g_{mn} = \frac{\partial x^1}{\partial q^m}\frac{\partial x^1}{\partial q^n} + \frac{\partial x^2}{\partial q^m}\frac{\partial x^2}{\partial q^n} + \frac{\partial x^3}{\partial q^m}\frac{\partial x^3}{\partial q^n}$$

The quantity g_{mn} is called the *metric tensor*, which is here given in terms of its covariant components, as will appear immediately. Using other generalized coordinates $\bar{q}^r$, (I.5) is written

$$ds^2 = \bar{g}_{rs}\,d\bar{q}^r\,d\bar{q}^s$$

Now, using (I.1) and the invariance of ds^2, we obtain

$$\bar{g}_{rs}\frac{\partial \bar{q}^r}{\partial q^m}\frac{\partial \bar{q}^s}{\partial q^n}\,dq^m\,dq^n = g_{mn}\,dq^m\,dq^n$$

Identifying coefficients of products of coordinate differentials, we obtain

$$(I.6) \qquad \bar{g}_{rs}\frac{\partial \bar{q}^r}{\partial q^m}\frac{\partial \bar{q}^s}{\partial q^n} = g_{mn}$$

Thus each index on g is treated like the covariant index of a vector, resulting in the term covariant component of the metric tensor. In general, *quantities* a_{mn} *which transform according to* (I.6) *are called the covariant components of a tensor of the second order.* The generalization to an arbitrary order is obvious.

The square of the length of any vector with contravariant components λ^m is defined to be

$$(I.7) \qquad |\lambda|^2 = g_{mn}\lambda^m\lambda^n$$

So far we have given no connection between the covariant and the contravariant components of a vector. This is done by use of the metric tensor. We have seen by (I.4) that $\lambda_m\lambda^m$ is a scalar. This scalar we identify with the square of the length of the vector. Now let $\lambda^m = \mu^m + \nu^m$, $\lambda_m = \mu_m + \nu_m$, in which we assume that the connection between λ^m and λ_m is a linear one. Then

$$\lambda_m\lambda^m = g_{mn}\lambda^m\lambda^n$$

so that

$$(I.8) \qquad (\mu_m + \nu_m)(\mu^m + \nu^m) = g_{mn}(\mu^m + \nu^m)(\mu^n + \nu^n)$$

Equation (I.8) must hold for arbitrary μ^m and ν^m, and so we must have

$$\mu_m = g_{mn}\mu^n \tag{I.9}$$

This is the fundamental connection between the two kinds of components of a vector. In cartesian coordinates the g_{mn} are Kronecker deltas δ_{mn}, and the two kinds of components are equal.

Equation (I.9) may be solved for the μ^n in terms of the μ_m. We write this solution as

$$\mu^n = g^{nm}\mu_m \tag{I.10}$$

thereby defining the quantities g^{nm}. These are termed the contravariant components of the metric tensor. Clearly,

$$g^{nm}g_{ml} = \delta_l{}^n \tag{I.11}$$

The reader may verify the fact that g^{mn} transforms according to the law

$$\bar{g}^{rs} = g^{mn}\frac{\partial\bar{q}^r}{\partial q^m}\frac{\partial\bar{q}^s}{\partial q^n} \tag{I.12}$$

Quantities which transform according to (I.12) *are called the contravariant components of a second-order tensor.* The generalization to higher orders is clear, as is that to mixed tensors with some covariant and some contravariant indices. It is worth noting that the quantities $\delta_l{}^m$ are the mixed components of a tensor with the property that

$$\bar{\delta}_l{}^m = \delta_l{}^m \tag{I.13}$$

This is not the case for δ_{ml} or δ^{ml}, which are merely the particular forms of the metric tensor in cartesian coordinates.

The metric tensor can be used to *raise* or *lower* indices. Thus for any tensor $h_b{}^a$ we have

$$h_b{}^a = g_{bc}h^{ac} = g^{ac}h_{cb}$$

which is a straightforward generalization of (I.9) and (I.10), the rule for raising and lowering the indices on vectors.

We now apply the ideas developed in this appendix to the problem of specifying the velocity and acceleration of a particle in arbitrary generalized coordinates. Except in the case of cartesian coordinates, the coordinates themselves do not form the components of a vector since they do not transform like the components of a vector. The velocity, however, is a vector.

The generalized velocity components are taken to be $\dot{q}^j$. That these are the components of a vector is evident from their transformation properties. If the $\bar{q}^k$ are another set of coordinates,

$$\dot{\bar{q}}^k = \frac{\partial \bar{q}^k}{\partial q^l} \dot{q}^l \tag{I.14}$$

which is the transformation law for a vector given in terms of its contravariant components.

The case of the acceleration is more difficult, since the acceleration in curvilinear coordinates involves terms like the centripetal acceleration and the Coriolis acceleration. We now show how these appear automatically.

The generalized acceleration components cannot be taken to be the second time derivatives of the generalized coordinates because these quantities do not transform like the components of a vector. From (I.14)

$$\begin{aligned} \ddot{\bar{q}}^k &= \frac{d}{dt}\left(\frac{\partial \bar{q}^k}{\partial q^l} \dot{q}^l\right) \\ &= \frac{\partial \bar{q}^k}{\partial q^l} \ddot{q}^l + \frac{\partial^2 \bar{q}^k}{\partial q^m \, \partial q^l} \dot{q}^l \dot{q}^m \end{aligned} \tag{I.15}$$

The second term on the right spoils the transformation properties of the $\ddot{q}^l$.

There is another quantity which fails to transform like a tensor because of a term similar to that in (I.15). The proper combination of $\ddot{q}^l$ and this other quantity will then transform like a vector. To find this other quantity we use the transformation equation for the metric tensor:

$$g_{mn} = \bar{g}_{ij} \frac{\partial \bar{q}^i}{\partial q^m} \frac{\partial \bar{q}^j}{\partial q^n}$$

Then

$$\frac{\partial g_{mn}}{\partial q^r} = \frac{\partial \bar{g}_{ij}}{\partial \bar{q}^k} \frac{\partial \bar{q}^i}{\partial q^m} \frac{\partial \bar{q}^j}{\partial q^n} \frac{\partial \bar{q}^k}{\partial q^r} + \bar{g}_{ij}\left(\frac{\partial^2 \bar{q}^i}{\partial q^r \, \partial q^m} \frac{\partial \bar{q}^j}{\partial q^n} + \frac{\partial \bar{q}^i}{\partial q^m} \frac{\partial^2 \bar{q}^j}{\partial q^n \, \partial q^r}\right) \tag{I.16}$$

Two similar equations can be obtained from (I.16) by changing indices cyclically:

$$\frac{\partial g_{rm}}{\partial q^n} = \frac{\partial \bar{g}_{ki}}{\partial \bar{q}^j} \frac{\partial \bar{q}^i}{\partial q^m} \frac{\partial \bar{q}^j}{\partial q^n} \frac{\partial \bar{q}^k}{\partial q^r} + \bar{g}_{ij}\left(\frac{\partial^2 \bar{q}^i}{\partial q^n \, \partial q^r} \frac{\partial \bar{q}^j}{\partial q^m} + \frac{\partial \bar{q}^i}{\partial q^r} \frac{\partial^2 \bar{q}^j}{\partial q^m \, \partial q^n}\right)$$

$$\frac{\partial g_{nr}}{\partial q^m} = \frac{\partial \bar{g}_{jk}}{\partial \bar{q}^i} \frac{\partial \bar{q}^i}{\partial q^m} \frac{\partial \bar{q}^j}{\partial q^n} \frac{\partial \bar{q}^k}{\partial q^r} + \bar{g}_{ij}\left(\frac{\partial^2 \bar{q}^i}{\partial q^m \, \partial q^n} \frac{\partial \bar{q}^j}{\partial q^r} + \frac{\partial \bar{q}^i}{\partial q^n} \frac{\partial^2 \bar{q}^j}{\partial q^r \, \partial q^m}\right)$$

Now subtract (I.16) from the sum of the last two equations:

$$\text{(I.17)} \quad \left(\frac{\partial g_{rm}}{\partial g^n} + \frac{\partial g_{nr}}{\partial q^m} - \frac{\partial g_{mn}}{\partial q^r}\right)$$
$$= \left(\frac{\partial \bar{g}_{ki}}{\partial \bar{q}^j} + \frac{\partial \bar{g}_{jk}}{\partial \bar{q}^i} - \frac{\partial \bar{g}_{ij}}{\partial \bar{q}^k}\right) \frac{\partial \bar{q}^i}{\partial q^m} \frac{\partial \bar{q}^j}{\partial q^n} \frac{\partial \bar{q}^k}{\partial q^r} + 2\bar{g}_{ij} \frac{\partial \bar{q}^j}{\partial q^r} \frac{\partial^2 \bar{q}^i}{\partial q^m \, \partial q^n}$$

The quantity in parentheses fails to transform like a tensor by a term similar to the one in (I.15). These parenthesized quantities appear so often that they are given a name and a symbol. We define the symbol $[mn, r]$ by

$$\text{(I.18)} \qquad [mn, r] \equiv \frac{1}{2}\left(\frac{\partial g_{rm}}{\partial q^n} + \frac{\partial g_{nr}}{\partial q^m} - \frac{\partial g_{mn}}{\partial q^r}\right)$$

and call it a *Christoffel symbol* of the first kind. The Christoffel symbol of the second kind is defined by

$$\text{(I.19)} \qquad \left\{ {s \atop mn} \right\} = g^{sr}[mn, r] = \frac{1}{2} g^{sr} \left(\frac{\partial g_{rm}}{\partial q^n} + \frac{\partial g_{nr}}{\partial q^m} - \frac{\partial g_{mn}}{\partial q^r}\right)$$

Equation (I.17) may now be written as

$$\text{(I.20)} \qquad [mn, r] = \overline{[ij, k]} \frac{\partial \bar{q}^i}{\partial q^m} \frac{\partial \bar{q}^j}{\partial q^n} \frac{\partial \bar{q}^k}{\partial q^r} + \bar{g}_{ij} \frac{\partial^2 \bar{q}^i}{\partial q^m \, \partial q^n} \frac{\partial \bar{q}^j}{\partial q^r}$$

which may be solved for the second derivative, yielding

$$\text{(I.21)} \qquad \frac{\partial^2 \bar{q}^i}{\partial q^m \, \partial q^n} = \left\{ {r \atop mn} \right\} \frac{\partial \bar{q}^i}{\partial q^r} - \overline{\left\{ {i \atop jk} \right\}} \frac{\partial \bar{q}^j}{\partial q^m} \frac{\partial \bar{q}^k}{\partial q^n}$$

Inserting this into (I.15) we obtain, after some simplification,

$$\text{(I.22)} \qquad \ddot{\bar{q}}^k + \overline{\left\{ {k \atop mn} \right\}} \dot{\bar{q}}^m \dot{\bar{q}}^n = \left(\ddot{q}^l + \left\{ {l \atop rs} \right\} \dot{q}^r \dot{q}^s\right) \frac{\partial \bar{q}^k}{\partial q^l}$$

This shows that the quantities

$$\text{(I.23)} \qquad \ddot{q}^l + \left\{ {l \atop rs} \right\} \dot{q}^r \dot{q}^s \equiv a^l$$

are the contravariant components of a vector.

In cartesian coordinates the Christoffel symbols all vanish because the g_{mn} are constants in such a coordinate system. Thus the vector we have just obtained has the second time derivatives of the cartesian coordinates as components and is properly identified with the acceleration vector. In curvilinear coordinates the terms involving the Christoffel symbols give exactly such terms as the centripetal and Coriolis accelerations.

As an example, we consider the transformation from rectangular cartesian coordinates in two dimensions ($q^1 = x, q^2 = y$) to polar coordinates ($\bar{q}^1 = r, \bar{q}^2 = \theta$). Thus

$$ds^2 = g_{mn}\,dq^m\,dq^n = \bar{g}_{mn}\,d\bar{q}^m\,d\bar{q}^n$$

with

$$g_{mn} = \begin{pmatrix} 1 & 0 \\ 0 & 1 \end{pmatrix} = g^{mn}$$

$$\bar{g}_{ij} = \begin{pmatrix} 1 & 0 \\ 0 & r^2 \end{pmatrix}, \qquad \bar{g}^{ij} = \begin{pmatrix} 1 & 0 \\ 0 & r^{-2} \end{pmatrix}$$

All of the unbarred Christoffel symbols vanish and in the barred set of coordinates

$$\overline{[22, 1]} = -r, \qquad \overline{[21, 2]} = \overline{[12, 2]} = r$$

$$\overline{\begin{Bmatrix} 1 \\ 22 \end{Bmatrix}} = -r, \qquad \overline{\begin{Bmatrix} 2 \\ 12 \end{Bmatrix}} = \overline{\begin{Bmatrix} 2 \\ 21 \end{Bmatrix}} = \frac{1}{r}$$

all others vanishing.

The components of the acceleration vector are then given by (I.23):

$$\bar{a}^1 = \ddot{r} - r\dot{\theta}^2$$

$$\bar{a}^2 = \ddot{\theta} + 2\frac{\dot{r}\dot{\theta}}{r}$$

Similarly, if spherical coordinates $\bar{q}^1 = r, \bar{q}^2 = \theta, \bar{q}^3 = \phi$ are introduced to describe the motion of a particle in three dimensions, we have

$$ds^2 = dr^2 + r^2\,d\theta^2 + r^2 \sin^2\theta\,d\phi^2$$

so that

$$\bar{g}_{mn} = \begin{pmatrix} 1 & 0 & 0 \\ 0 & r^2 & 0 \\ 0 & 0 & r^2\sin^2\theta \end{pmatrix}, \qquad \bar{g}^{mn} = \begin{pmatrix} 1 & 0 & 0 \\ 0 & r^{-2} & 0 \\ 0 & 0 & r^{-2}\operatorname{cosec}^2\theta \end{pmatrix} \tag{I.24}$$

The only nonvanishing Christoffel symbols are those involving $\partial \bar{g}_{22}/\partial \bar{q}^1$, $\partial \bar{g}_{33}/\partial \bar{q}^1$, $\partial \bar{g}_{33}/\partial \bar{q}^2$, namely,

$$\overline{[22, 1]} = -r, \qquad \overline{\begin{Bmatrix} 1 \\ 22 \end{Bmatrix}} = -r$$

$$\overline{[12, 2]} = \overline{[21, 2]} = r, \qquad \overline{\begin{Bmatrix} 2 \\ 12 \end{Bmatrix}} = \overline{\begin{Bmatrix} 2 \\ 21 \end{Bmatrix}} = \frac{1}{r}$$

$$\overline{[33, 1]} = -r \sin^2 \theta, \qquad \overline{\begin{Bmatrix} 1 \\ 33 \end{Bmatrix}} = -r \sin^2 \theta$$

$$\overline{[13, 3]} = \overline{[31, 3]} = r \sin^2 \theta, \qquad \overline{\begin{Bmatrix} 3 \\ 13 \end{Bmatrix}} = \overline{\begin{Bmatrix} 3 \\ 31 \end{Bmatrix}} = \frac{1}{r}$$

$$\overline{[33, 2]} = -r^2 \sin \theta \cos \theta, \qquad \overline{\begin{Bmatrix} 2 \\ 33 \end{Bmatrix}} = -\sin \theta \cos \theta$$

$$\overline{[23, 3]} = \overline{[32, 3]} = r^2 \sin \theta \cos \theta, \qquad \overline{\begin{Bmatrix} 3 \\ 23 \end{Bmatrix}} = \overline{\begin{Bmatrix} 3 \\ 32 \end{Bmatrix}} = \cot \theta$$

and from (I.23) we obtain

$$\begin{aligned} \bar{a}^1 &= \ddot{r} - r\dot{\theta}^2 - r \sin^2 \theta \, \dot{\phi}^2 \\ \bar{a}^2 &= \ddot{\theta} + 2 \frac{\dot{r}\dot{\theta}}{r} - \sin \theta \cos \theta \, \dot{\phi}^2 \\ \bar{a}^3 &= \ddot{\phi} + \frac{2}{r} \dot{r}\dot{\phi} + 2 \cot \theta \, \dot{\theta}\dot{\phi} \end{aligned} \tag{I.25}$$

For a particle moving under the influence of no external forces, i.e., in a straight line with uniform velocity, these acceleration components all vanish, although in this case the quantities $\ddot{r}$, $\ddot{\theta}$, $\ddot{\phi}$, obtained by differentiating the coordinates twice with respect to the time, do not vanish.

We see, however, that even in the above simple cases the task of evaluating the Christoffel symbols is tedious. Fortunately, the method of Lagrange allows one to write down the equations of motion for a dynamic system in terms of generalized coordinates without requiring that the Christoffel symbols which characterize the space of those coordinates be evaluated.

To see this for a single particle, we note that the kinetic energy of a particle is $\frac{1}{2}m(ds/dt)^2$, where ds, the infinitesimal element of distance traveled, is given by (I.5):

$$ds^2 = g_{rs}\, dq^r\, dq^s$$

so that

$$T = \tfrac{1}{2} m g_{rs} \dot{q}^r \dot{q}^s$$

Hence

$$\text{(I.26)} \quad \frac{d}{dt}\left(\frac{\partial T}{\partial \dot{q}^r}\right) - \frac{\partial T}{\partial q^r} = m \left[\frac{d}{dt}\,(g_{rs}\dot{q}^s) - \frac{1}{2}\frac{\partial g_{ms}}{\partial q^r}\,\dot{q}^m\dot{q}^s\right]$$

$$= m\left[g_{rs}\ddot{q}^s + \left(\frac{\partial g_{rs}}{\partial q^m} - \frac{1}{2}\frac{\partial g_{ms}}{\partial q^r}\right)\dot{q}^m\dot{q}^s\right]$$

$$= m\left[g_{rs}\ddot{q}^s + \frac{1}{2}\left(\frac{\partial g_{rs}}{\partial q^m} + \frac{\partial g_{rm}}{\partial q^s} - \frac{\partial g_{ms}}{\partial q^r}\right)\dot{q}^m\dot{q}^s\right]$$

where in the last line the symmetry of $\dot{q}^m\dot{q}^s$ in m and s has been used to interchange m and s in the first term inside the parentheses. From (31.9) and (I.18) we see that (I.26) may be written thus

$$Q_r = m(g_{rs}\ddot{q}^s + [ms, r]\dot{q}^m\dot{q}^s)$$

or

$$\text{(I.27)} \qquad Q^l \equiv g^{lr}Q_r = m\left(\ddot{q}^l + \begin{Bmatrix} l \\ ms \end{Bmatrix} \dot{q}^m\dot{q}^s\right)$$

$$= ma^l$$

where the acceleration vector a^l is given by (I.23). Thus the generalized force vector Q^l is related to the acceleration of a single particle by a relation of the form force equals mass times acceleration, even when curvilinear coordinates are used. The Lagrangian formulation of the equations of motion thus eliminates the necessity of calculating the values of the Christoffel symbols explicitly. Alternatively, this formulation provides a convenient method of computing the Christoffel symbols of the second kind, if so desired, which is much simpler than the direct method used in (I.24), (I.25).

In discussing the motion of a number of particles of different mass, we have

$$T = \frac{1}{2}\sum_{\rho} m_\rho \left(\frac{ds_\rho}{dt}\right)^2 = \frac{1}{2}\sum_{\rho, i} m_\rho[\dot{x}_{i,\rho}\dot{x}_{i,\rho}]$$

where ds_ρ is the infinitesimal element of distance traveled by particle ρ. It is then convenient to define, for each particle, the vector $X_{i,\rho} = \sqrt{m_\rho}\, x_{i,\rho}$ in order that the individual masses no longer appear explicitly in the expression for the kinetic energy:

$$T = \frac{1}{2} \sum_{\rho,i} \dot{X}_{i,\rho} \dot{X}_{i,\rho}$$

The analysis then proceeds as for a single particle of unit mass.

We have considered in this appendix only those transformations which relate two coordinate systems *statically*, i.e., in which the $\bar{q}^n$ are functions solely of the q^m, and do not depend explicitly upon the time. We have seen in Sec. 43 that such explicit time dependence adds extra terms to (I.15), and these terms also spoil the transformation properties of $\ddot{q}^l$. It is possible to treat this situation formally by writing $q^{f+1} = t$, so that (I.1) is unaltered in form

$$d\bar{q}^n = \frac{\partial \bar{q}^n}{\partial q^m} dq^m \qquad (m, n = 1 \cdot \cdot \cdot f + 1)$$

but now refers to a transformation in $f + 1$ dimensions. Nonrelativistically, $\bar{q}^{f+1} = q^{f+1} = t$. However, the complete description of such transformations is possibly only within the framework of the general theory of relativity.

GENERAL REFERENCE

L. P. Eisenhart, *Riemannian Geometry*. Princeton University Press (1926)

APPENDIX II
LINEAR VECTOR SPACES

A vector in n-dimensional space is defined in a given cartesian coordinate system by n numbers in a definite order, the components of the vector. We shall denote such a vector by a letter such as x, or by a letter with subscript followed by a vertical line, as $x_{m|}$. Letters with subscripts that are not followed by a vertical line, as x_i, $x_{m|i}$, denote vector components. The summation convention will apply only to indices that label components, i.e., to subscripts not followed by a line.

The *sum* of two vectors is defined to be the vector whose components are the sums of the corresponding components of the individual vectors. Thus

$$(x + y)_i = x_i + y_i \tag{II.1}$$

The *product* of a number and a vector is defined to be the vector whose components are those of the original vector multiplied by the number. Thus

$$(\alpha x)_i = \alpha x_i \tag{II.2}$$

A space in which the sum of two vectors and the product of a number and a vector are defined in this way is called a *linear vector space.* We consider only those numbers and vectors which are real.

We introduce a set of n vectors e_i with components given by

$$\begin{aligned} e_1 &= (1, 0, 0, \cdots, 0) \\ e_2 &= (0, 1, 0, \cdots, 0) \\ &\vdots \\ e_n &= (0, 0, 0, \cdots, 1) \end{aligned} \tag{II.3}$$

These vectors are said to be a set of *basic vectors* for the space. According to the notation just adopted, these vectors should be denoted by $e_{i|}$ rather than by e_i. The index in this particular case does, however, refer to a particular coordinate direction, so here only we leave off the vertical line. This is necessary so that the summation convention will work out properly and enable us to retain the great simplicity gained by its use.

Any vector can be written as a sum of basic vectors with appropriate coefficients:

$$x = e_i x_i \tag{II.4}$$

The x_i are the components of x along the coordinate directions, as is seen from the form (II.3) of the e's.

The inner or scalar product of two vectors is defined by the equation

$$(x, y) = x_i y_i \tag{II.5}$$

This is a straightforward generalization of the scalar or dot product of two three-dimensional vectors (cf. Sec. 6). Thus the scalar product of two of the basic vectors is given by

$$(e_i, e_j) = \delta_{ij} \tag{II.6}$$

The length of a vector is defined to be

$$|x| = (x, x)^{1/2} = (x_i x_i)^{1/2} \tag{II.7}$$

The quantity $x_i x_i$ is, of course, nonnegative, and so the length defined in this way is always real and is taken to be positive. Some properties of the scalar product are

$$\begin{aligned} (x, y) &= (y, x) \\ (x, \alpha y) &= (\alpha x, y) = \alpha(x, y) \\ (x, y + z) &= (x, y) + (x, z) \end{aligned} \tag{II.8}$$

All of these follow from the definition (II.5) and the definition of a linear vector space.

The m component of a vector is the scalar product of the vector with the mth basic vector. Using (II.4) and (II.6), we obtain

$$\begin{aligned} (e_m, x) &= (e_m, e_i x_i) \\ &= (e_m, e_i) x_i \\ &= \delta_{mi} x_i \\ &= x_m \end{aligned} \tag{II.9}$$

The basic vectors e_i are linearly independent, by which we mean that a linear combination of the e_i vanishes only when the coefficients of the individual e_i vanish separately. In an n-dimensional space any $n + 1$ vectors are linearly dependent. Therefore any vector can be expressed in terms of any set of n linearly independent vectors. Let $a_{j|}$ be such a set and let x be a nonzero vector. Then

$$x - \sum_{j=1}^{n} a_{j|}x_j = 0 \tag{II.10}$$

for suitable x_j, not all of which vanish. The vectors $a_{j|}$ are said to *span* the space. They are in general not as convenient as the e_j because there is no simple expression such as (II.9) for the "components" x_j.

In three dimensions, if a and b are two vectors, the cosine of the angle θ between them may be written as

$$\cos\theta = \frac{a \cdot b}{|a||b|}$$

The cosine so defined is never greater than unity in absolute value. We wish to show that the n-dimensional generalization of this, namely

$$\cos\theta = \frac{(x, y)}{[(x, x)(y, y)]^{1/2}} \tag{II.11}$$

also is never greater than unity in absolute value. This result may be written as

$$(x, y)^2 \leq (x, x)(y, y) \tag{II.12}$$

in which form it is known as *Schwarz's inequality*. To prove this we note that

$$\begin{aligned}
(x, x)(y, y) - (x, y)^2 &= (y, y)\left[(x, x) - 2\frac{(x, y)^2}{(y, y)} + \frac{(x, y)^2}{(y, y)}\right] \\
&= (y, y)\left[x - \frac{(x, y)}{(y, y)}y\right]^2 \\
&\geq 0
\end{aligned}$$

The equality sign holds only when the components of x are proportional to those of y. In this case $\cos\theta = \pm 1$ and the vectors are parallel or antiparallel. If $(x, y) = 0$, then $\cos\theta = 0$ and the vectors are said to be *orthogonal*.

Another inequality follows from Schwarz's. It is the so-called *triangular inequality*, and it can be written

$$|x| + |y| \geq |x + y| \tag{II.13}$$

To prove this we evaluate $(x + y, x + y)$

$$\begin{aligned}(x + y, x + y) &= (x, x) + (y, y) + 2(x, y)\\ &\leq (x, x) + (y, y) + 2|x||y|\end{aligned}$$

where Schwarz's inequality has been used. When the square root of both sides is extracted, the result follows.

An *operator* in a vector space is a quantity which, applied to a given vector, yields another vector. An operator A is called *linear* when it satisfies the equations

$$\begin{aligned}A(x + y) &= Ax + Ay\\ A(\lambda x) &= \lambda Ax\end{aligned} \tag{II.14}$$

for any vectors x and y and for any number λ. Operators will be denoted by capital letters.

The sum of two operators is defined by the equation

$$(A + B)x = Ax + Bx \tag{II.15}$$

which must hold for every x. The product of two operators is defined by

$$(AB)x = A(Bx) \tag{II.16}$$

which must hold for every x. The order of addition does not matter, but the order of multiplication does. Equation (II.16) implies that the operator AB applied to x produces the same effect as first applying operator B and then applying operator A to the resulting vector. This is not necessarily the same as applying A and then B. Thus in general

$$ABx \neq BAx$$

If these two quantities are equal for every x, then A and B are said to *commute*. The product of operators is associative.

$$(AB)Cx = A(BC)x = ABCx$$

so that no parentheses are needed.

The operator I is defined by the equation

$$Ix = x \tag{II.17}$$

for all x. This operator is called the *identity operator*.

We say that two operators A and B are equal if $Ax = Bx$ for all x. This enables us to write down relations between operators without writing down the vector on which the operator operates each time.

If two operators A and B are such that

$$AB = I$$

then

$$BAB = B$$

and so

$$(BA)Bx = Bx$$

Any vector y may be written as Bx for suitable x. For let

$$a_{j|} = Be_j$$

so that

$$A \sum_i a_{j|}x_j = e_j x_j$$

If the $a_{j|}$ are linearly dependent, the left side vanishes for some nonvanishing set of x_j, which is impossible because the right side cannot vanish for nonvanishing x_j. The a_j are therefore linearly independent and span the space, so that any vector y may be expressed in terms of them, as was to be proved. Hence

$$BAy = y$$

for any y and so

$$BA = I$$

Two operators related in this way are said to be *reciprocals* of each other, or one is the *inverse* of the other. Not all operators have inverses, those not having them being called *singular*. The inverse of A is denoted by A^{-1}. A and A^{-1} commute.

We shall denote the *transpose* of an operator A by A^{T} and define it by the equation

$$(A^{\mathsf{T}}x, y) = (x, Ay) \tag{II.18}$$

for all x and y. The transpose of the product of two operators A and B is the product of their transposes in reverse order. For

$$((AB)^{\mathsf{T}}x, y) = (x, (AB)y)$$

and

$$(x, ABy) = (A^{\mathsf{T}}x, By) = (B^{\mathsf{T}}A^{\mathsf{T}}x, y)$$

so that

$$(AB)^{\mathsf{T}} = B^{\mathsf{T}}A^{\mathsf{T}} \tag{II.19}$$

A similar result holds for more factors. If $A = A^{\mathsf{T}}$, the operator is said to be *symmetric*.

An operator A is said to be *orthogonal* if

$$A^{\mathsf{T}}A = I \tag{II.20}$$

An orthogonal operator has the important property that

$$(Ax, Ay) = (x, y) \tag{II.21}$$

If A is orthogonal, then $A^{\mathsf{T}} = A^{-1}$ and A and A^{T} commute.

Let A be a linear operator, and let the vector resulting from the application of A to x be y. Thus

$$y = Ax \tag{II.22}$$

This vector equation may be written in terms of the components of the vectors x and y in a particular coordinate system with the unit vectors e_i as a basis. We obtain in this way a representation of the operator A. This representation will depend on the basic vectors chosen as well as on the operator itself. Let

$$y = e_j y_j, \qquad x = e_j x_j$$

Since A is a linear operator, (II.22) becomes

$$e_j y_j = A e_j x_j$$

Scalar multiplication of both sides by e_i yields

$$(e_i, e_j)y_j = y_i = (e_i, Ae_j)x_j \tag{II.23}$$

The quantity (e_i, Ae_j) does not depend on the vector x but only on the operator A and the basic vectors. It thus gives the representation of the operator we desire. We write

$$(e_i, Ae_j) = A_{ij} \tag{II.24}$$

so that equation (II.23) can be written as

$$y_i = A_{ij}x_j \tag{II.25}$$

Equation (II.25) is equivalent to (II.22) and is said to be (II.22) in the representation defined by the basic vectors e_i.

The numbers A_{ij} form an $n \times n$ array called a *matrix* with the first index labeling the rows and the second index labeling the columns; thus

$$\begin{pmatrix} A_{11} & A_{12} & A_{13} & A_{14} & \cdots & A_{1n} \\ A_{21} & A_{22} & A_{23} & A_{24} & \cdots & A_{2n} \\ \cdot & & & & & \\ \cdot & & & & & \\ \cdot & & & & & \\ A_{n1} & A_{n2} & A_{n3} & A_{n4} & \cdots & A_{nn} \end{pmatrix}$$

The numbers A_{ij} are called the *components* or *elements* of the matrix given above. We are thus led to consider a matrix as the *representative* of a linear operator in terms of a set of basic vectors. The elements of the matrix with two equal indices are called the *diagonal* elements. If only diagonal elements of a matrix are different from zero, the matrix is said to be in *diagonal form* or to be *diagonal*.

The representative of the identity operator is a diagonal matrix independent of the choice of basic vectors. Since by definition

$$x = Ix$$

we have

$$x_i = I_{ij}x_j, \qquad I_{ij} = \delta_{ij}$$

Thus the form of this matrix is

$$\begin{pmatrix} 1 & 0 & 0 & \cdots & 0 \\ 0 & 1 & 0 & \cdots & 0 \\ \cdot & & & & \\ \cdot & & & & \\ \cdot & & & & \\ 0 & 0 & 0 & \cdots & 1 \end{pmatrix}$$

for any set of basic vectors. This matrix is called the *unit matrix*.

The sum of two operators A and B was defined in (II.15). The representative of the sum of two operators can be found. If $A + B = C$,

$$Ce_j = (A + B)e_j = Ae_j + Be_j$$

Scalar multiplication by e_i on both sides gives

$$\begin{aligned} (e_i, Ce_j) = C_{ij} &= (e_i, Ae_j) + (e_i, Be_j) \\ &= A_{ij} + B_{ij} \end{aligned}$$

The matrix C_{ij} is called the *sum* of the matrices A_{ij} and B_{ij}. *The representative of the sum of two operators is the sum of their representatives.*

The product of two operators A and B was defined in (II.16). The representative of the product of two operators can be found. If $C = AB$, let

$$y = Bx, \qquad z = Ay = ABx$$

Then

$$y_j = B_{jk}x_k, \qquad z_i = A_{ij}y_j$$

so that

$$\begin{aligned} z_i &= A_{ij}B_{jk}x_k \\ &= (AB)_{ik}x_k \end{aligned}$$

The definition of the *product* of two matrices is that

$$(AB)_{ik} = A_{ij}B_{jk} \tag{II.26}$$

and so we see that *the representative of the product of two operators is the product of their representatives.*

The rule for multiplying matrices is unlike that for numbers. Each element of the product matrix is the sum of products of elements of the two matrices, those of a row in the first matrix and those of a column in the second matrix. Thus

$$\begin{pmatrix} a & b \\ c & d \end{pmatrix}\begin{pmatrix} e & f \\ g & h \end{pmatrix} = \begin{pmatrix} ae + bg & af + bh \\ ce + dg & cf + dh \end{pmatrix}$$

The order of multiplication is essential, as the product may depend on the order of the factors, so that the commutative law does not hold in general. If the order of two matrices does not affect their product, they are said to *commute*. The associative law of multiplication does hold for matrices.

Since an operator and its representative follow the same rules for multiplication and addition, which are the only manipulations we shall be making with operators, there is no need to have separate symbols for the two quantities. We shall therefore use capital letters to denote both linear operators and their representatives in terms of a particular set of basic vectors. If more than one set of basic vectors is in use, they will be distinguished by bars, and the representatives will be similarly distinguished.

The vector equation $y = Ax$, when written out in terms of a particular representation as in (II.25), is a set of n linear equations giving the y's in terms of the x's. Such a set of equations can be solved for the x's in terms of the y's provided that the determinant of the coefficients does not vanish. (This determinant is the Jacobian of the transformation from the x's to the y's; cf. Sec. 2.) The solution can be written as

$$x_i = A^{-1}{}_{ij}y_j \tag{II.27}$$

The matrix $A^{-1}{}_{ij}$ is that matrix which, when multiplied by the matrix A_{jk}, yields the unit matrix and is therefore the matrix inverse to A. The inverse matrix can be expressed in terms of the elements of A by use of the theory of determinants. A determinant can be expanded in terms of the minors of any row or column. Denoting the cofactor of the element A_{ij} by $\mathfrak{A}_{ij}$—the cofactor is the minor prefixed by $(-1)^{i+j}$— we have

$$A_{ij}\mathfrak{A}_{kj} = \det A \ \delta_{ik}$$

Thus

$$A_{ij} \frac{\alpha_{kj}}{\det A} = \delta_{ik}$$

so that, as $A_{ij}A^{-1}{}_{jk} = \delta_{ik}$, we may write the matrix inverse to A as

$$A^{-1}{}_{jk} = \frac{\text{cofactor of } A_{kj} \text{ in } \det A}{\det A} \tag{II.28}$$

If det A vanishes, there is no matrix inverse to A. Thus a singular operator is one whose representative has a zero determinant.

The transpose A^{T} of an operator A was defined by (II.18), namely by

$$(A^{\mathsf{T}}x, y) = (x, Ay)$$

When written out in a particular representation this becomes

$$y_i A^{\mathsf{T}}{}_{ij} x_j = x_j A_{ji} y_i$$

Thus

$$A^{\mathsf{T}}{}_{ij} = A_{ji} \tag{II.29}$$

and the representative of the transposed operator is the transposed matrix, i.e., the original matrix with its rows and columns interchanged.

An operator A is called symmetric if $A^{\mathsf{T}} = A$. The representative of a symmetric operator is a symmetric matrix, a matrix which is unchanged when the rows and columns are interchanged, so that $A_{ij} = A_{ji}$.

An operator A is called orthogonal if $A^{\mathsf{T}}A = I$. Written out in a particular representation, this becomes

$$\begin{aligned} (A^{\mathsf{T}}A)_{ik} &= A^{\mathsf{T}}{}_{ij}A_{jk} \\ &= A_{ji}A_{jk} \\ &= \delta_{ik} \end{aligned} \tag{II.30}$$

If each column of A is considered to be a vector, these n vectors are of unit length and are mutually orthogonal, according to (II.30). The same holds for the rows of A. Hence the names "orthogonal" matrix and "orthogonal" operator.

The basic vectors e_i form a complete, orthonormal set of vectors, i.e., a set of n mutually orthogonal vectors of unit length in terms of which any other vector may be described.

Any complete orthonormal set of basic vectors is equivalent to any other. Changing from one set to another is equivalent to making a rotation of the coordinate system.

Let us call the original orthonormal set of basic vectors e_i, and let us introduce a new orthonormal set of vectors $\bar{e}_j$. The conditions of orthogonality are

$$(e_i, e_j) = \delta_{ij}, \qquad (\bar{e}_i, \bar{e}_j) = \delta_{ij} \tag{II.31}$$

All vectors are now to be represented in terms of the new basic vectors $\bar{e}_j$. Therefore we write

$$e_i = \bar{e}_j S_{ji} \tag{II.32}$$

The coefficients S_{ji} form a matrix. Each column of this matrix consists of the components of one of the old basic vectors in the new representation, and so we should expect S to be an orthogonal matrix. We verify this by evaluating the scalar product

$$\begin{aligned}(e_i, e_j) = \delta_{ij} &= (\bar{e}_k S_{ki}, \bar{e}_m S_{mj}) \\ &= S_{ki} S_{mj} (\bar{e}_k, \bar{e}_m) \\ &= S_{ki} S_{kj} \\ &= S^{\mathsf{T}}{}_{ik} S_{kj}\end{aligned} \tag{II.33}$$

Thus (II.32) may also be written

$$\bar{e}_j = e_i S^{\mathsf{T}}{}_{ij} \tag{II.34}$$

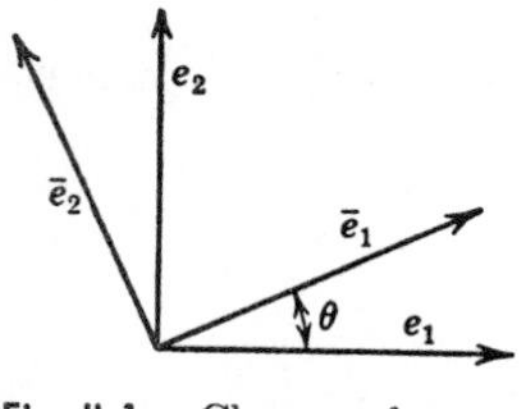

As an example of a change of representation (Fig. II-1) let $n = 2$. The old basic vectors are e_1, e_2, whereas the new ones are $\bar{e}_1$, $\bar{e}_2$. Then

$$\begin{aligned}e_1 &= \bar{e}_1 S_{11} + \bar{e}_2 S_{21} \\ &= \bar{e}_1 \cos\theta - \bar{e}_2 \sin\theta \\ e_2 &= \bar{e}_1 S_{12} + \bar{e}_2 S_{22} \\ &= \bar{e}_1 \sin\theta + \bar{e}_2 \cos\theta\end{aligned}$$

Fig. II-1. Change of representation.

so that $\bar{e}_1$, $\bar{e}_2$ are obtained from e_1, e_2 by a rotation through the angle θ in the direction from 1 to 2. The matrix S and its transpose are thus given by

$$S = \begin{pmatrix} \cos\theta & \sin\theta \\ -\sin\theta & \cos\theta \end{pmatrix}, \qquad S^{\mathsf{T}} = \begin{pmatrix} \cos\theta & -\sin\theta \\ \sin\theta & \cos\theta \end{pmatrix} \tag{II.35}$$

Matrix multiplication will show that $SS^{\mathsf{T}} = I = S^{\mathsf{T}}S$.

A vector x may be given in either representation. Denoting its components in the new representation by barred quantities, we have

$$x = \bar{e}_i\bar{x}_i = e_jx_j$$
$$= \bar{e}_iS_{ij}x_j$$

Thus

$$\bar{x}_i = S_{ij}x_j$$

or in matrix form

$$\bar{x} = Sx, \qquad x = S^{\mathsf{T}}\bar{x} = S^{-1}\bar{x} \tag{II.36}$$

The scalar product of two vectors is not changed when a change in basic vectors is made. As in (II.21)

$$(\bar{x}, \bar{y}) = (Sx, Sy)$$
$$= (x, S^{\mathsf{T}}Sy)$$
$$= (x, y)$$

An orthogonal transformation thus preserves the lengths of vectors and the angle between vectors, as it must if it is to represent a rotation of coordinates.

The matrix representing an operator depends on the basic vectors defining the representation as well as on the operator. We may find the connection between two representatives of the operator:

$$\bar{A}_{ij} = (\bar{e}_i, A\bar{e}_j)$$
$$= (e_kS^{\mathsf{T}}{}_{ki}, Ae_mS^{\mathsf{T}}{}_{mj})$$
$$= S^{\mathsf{T}}{}_{ki}(e_k, Ae_m)S^{\mathsf{T}}{}_{mj}$$
$$= S_{ik}A_{km}S^{\mathsf{T}}{}_{mj}$$

Hence

$$\bar{A} = SAS^{\mathsf{T}} = SAS^{-1} \tag{II.37}$$

For consistency in the theory it must follow that two successive changes in representation are equivalent to a single change. First we show that, if the transformation matrices are S and U respectively, the transformation matrix for the single equivalent transformation is US. We then show that this product matrix is also orthogonal. Let

$$e_i = \bar{e}_jS_{ji}, \qquad \bar{e}_j = \bar{\bar{e}}_kU_{kj}$$

Then

$$e_i = \bar{\bar{e}}_kU_{kj}S_{ji}$$
$$= \bar{\bar{e}}_k(US)_{ki}$$

That the product of two orthogonal matrices is again orthogonal follows from the rule for taking the transpose of a product (II.19). If $S^{\mathsf{T}}S = I$ and $U^{\mathsf{T}}U = 1$,

$$(\text{II.38}) \qquad \begin{aligned} (US)^{\mathsf{T}}(US) &= S^{\mathsf{T}}U^{\mathsf{T}}US \\ &= S^{\mathsf{T}}S \\ &= I \end{aligned}$$

Every orthogonal matrix has an inverse, since every matrix has a transpose and the transpose of an orthogonal matrix is its inverse. This result can be seen in another way. The rule for the multiplication of two $n \times n$ determinants is formally the same as that for two n-dimensional matrices, and the value of a determinant is not affected by interchanging rows and columns. Thus

$$\begin{aligned} \det (SS^{\mathsf{T}}) &= \det S \det S^{\mathsf{T}} \\ &= (\det S)^2 \end{aligned}$$

But

$$\begin{aligned} \det (SS^{\mathsf{T}}) &= \det I \\ &= 1 \end{aligned}$$

Therefore

$$(\text{II.39}) \qquad \det S = \pm 1$$

If the plus sign holds, the transformation is called *proper;* if the minus sign holds, it is called *improper*. Hence an orthogonal matrix cannot be singular and therefore has an inverse.

There are two scalar quantities of importance associated with the representative of a linear operator which are invariant under an orthogonal transformation. (Actually there are n such quantities, but usually only two of them are of interest individually.) These are the determinant of the matrix representing the operator, and the sum of the diagonal elements of the matrix. The latter is called the *spur* or *trace* of the matrix and is denoted by tr A. If (II.37) gives $\bar{A}$ in terms of A and S, then

$$(\text{II.40}) \qquad \begin{aligned} \det \bar{A} &= \det (SAS^{\mathsf{T}}) \\ &= \det S \det A \det S^{\mathsf{T}} \\ &= \det SS^{\mathsf{T}} \det A \\ &= \det A \end{aligned}$$

Also

$$\text{(II.41)} \qquad \begin{aligned} \operatorname{tr} \bar{A} &= \bar{A}_{ii} \\ &= S_{ik} A_{km} S^{\mathsf{T}}{}_{mi} \\ &= S_{ik} S_{im} A_{km} \\ &= \delta_{km} A_{km} \\ &= A_{kk} \\ &= \operatorname{tr} A \end{aligned}$$

Thus both det A and tr A are invariant under orthogonal transformations.

We have introduced orthogonal transformations as coordinate transformations. The vectors such as x are not affected by such a transformation, but their components are changed. Orthogonal transformations can be looked at in another way: All the vectors in the n-dimensional space are rotated by application of the operator, the coordinate system remaining unchanged. This is the viewpoint we adopt when discussing the displacements of a rigid body. There the body rotates, carrying with it vectors fixed in the body. The relation between these vectors in the original and final positions of the body is that given by an orthogonal transformation. There is a difference in the two viewpoints, however, because a rotation of the coordinate system through an angle θ about an axis has the same effect as a rotation of all the vectors through an angle $-\theta$ about the same axis.

The result of operating on a vector with a linear operator is in general to change the length and direction of the vector. Corresponding to a given operator A there are certain vectors, however, which are not changed in direction by application of the operator although they are generally changed in length. For such a vector x we have the equation

$$\text{(II.42)} \qquad Ax = \alpha x$$

The vector x is called an *eigenvector* of the operator A, and the number α is called the *eigenvalue* corresponding to this eigenvector.

Writing (II.42) out in terms of a particular representation, we obtain

$$A_{ij} x_j = \alpha x_i$$

or

$$\text{(II.43)} \qquad (A_{ij} - \alpha \delta_{ij}) x_j = 0$$

This is a set of homogeneous linear equations for the components x_j and can be solved only if the determinant of the coefficients vanishes:

$$\text{(II.44)} \qquad \det (A_{ij} - \alpha \delta_{ij}) = \det (A - \alpha I) = 0$$

α must be a root of an nth degree algebraic equation:

$$\begin{vmatrix} A_{11}-\alpha & A_{12} & A_{13} & \cdots & A_{1n} \\ A_{21} & A_{22}-\alpha & A_{23} & \cdots & A_{2n} \\ \cdot & & & & \\ \cdot & & & & \\ \cdot & & & & \\ A_{n1} & A_{n2} & A_{n3} & \cdots & A_{nn}-\alpha \end{vmatrix} = 0$$

The roots of (II.43) may be complex even if all the elements of A are real. So, although the operator A has n eigenvalues and n corresponding eigenvectors, these may not be the kind of vectors we have been considering, because we have restricted ourselves to real quantities. We can, however, show that *all the eigenvalues of a real symmetric operator are real,* and, since real symmetric operators are the ones in whose eigenvalues we shall be primarily interested, this is sufficient. Suppose that α is a complex eigenvalue of the real symmetric operator A and x the corresponding eigenvector with complex components. Then, denoting the complex conjugate of a quantity by *, we have

$$Ax = \alpha x$$

$$Ax^* = \alpha^* x^*$$

because A is real. Scalar multiplication of the first with x^* and the second with x gives

$$(x^*, Ax) = \alpha(x^*, x)$$

$$(x, Ax^*) = \alpha^*(x, x^*)$$

Because A is symmetric, the left-hand sides of these equations are equal, and $(x^*, x) = (x, x^*)$, so that

$$\alpha = \alpha^*$$

and α is real. All the components of the eigenvectors may also be taken as real.

There are n eigenvalues and eigenvectors associated with every linear operator. These n eigenvalues may not all be distinct. If there are multiple eigenvalues, the operator is said to be *degenerate.* The various eigenvectors of an operator will be denoted by $x_{l|}$, where l runs from 1 to n, and the corresponding eigenvalues will be denoted by $\alpha_{l|}$. Thus

$$Ax_{l|} = \alpha_{l|} x_{l|} \tag{II.45}$$

Note that according to our notation the index l is not summed!

The eigenvectors of a symmetric operator corresponding to distinct eigenvalues are orthogonal to each other. Let

$$Ax_{l|} = \alpha_{l|}x_{l|}$$
$$Ax_{m|} = \alpha_{m|}\, x_{m|}$$

Then

$$(x_{m|}, Ax_{l|}) = \alpha_{l|}(x_{m|}, x_{l|})$$
$$(x_{l|}, Ax_{m|}) = \alpha_{m|}(x_{l|}, x_{m|}) \tag{II.46}$$

Because A is symmetric, the left sides of these equations are equal. Thus

$$(\alpha_{l|} - \alpha_{m|})(x_{l|}, x_{m|}) = 0$$

and, if the first factor is different from zero, the second factor must vanish. This proves the theorem.

Equations such as (II.42) do not determine the length of the vector x, but only its direction. We are free to choose this length in the most convenient manner, which is to make the vector a unit vector. An eigenvector with length unity is said to be *normalized.*

If a symmetric operator has all distinct eigenvalues, its normalized eigenvectors form a complete orthonormal set. If there exist multiple eigenvalues, the eigenvectors can be so chosen that they again form a complete orthonormal set. Suppose that there are m identical eigenvalues. Then of the n homogeneous linear equations (II.43) with this value of α there are only $n - m$ which are independent. These together with the normalization condition give $n - m + 1$ independent conditions on the n components. There is thus an $(m - 1)$-fold infinity of solutions. We pick one solution arbitrarily, say $x_{1|}$. If we now require that any other solution of the equations corresponding to the same eigenvalue be orthogonal to $x_{1|}$, there is only an $(m - 2)$-fold infinity of solutions. We pick one, $x_{2|}$, and then require the next solution corresponding to this eigenvalue to be orthogonal to $x_{1|}$ and $x_{2|}$. Proceeding in this way we get m normalized and mutually orthogonal eigenvectors corresponding to the eigenvalue of m-fold degeneracy. All these vectors are orthogonal to any eigenvector corresponding to a different eigenvalue. Hence we have an orthonormal set. This set is not a unique one because of the arbitrary choices involved in finding the eigenvectors corresponding to multiple eigenvalues.

Since the eigenvectors of a symmetric operator can be chosen to form a complete orthonormal set, they can also be chosen as the basic set of vectors of a cartesian coordinate system. In this coordinate system the matrix of the operator assumes an especially simple form.

Let the linear operator A, which is symmetric so that its eigenvalues are all real, have the n real eigenvectors $x_{i|}$, and let these eigenvectors be normalized and, if the operator A is degenerate, also orthogonalized. Then

$$(x_{i|}, x_{j|}) = \delta_{ij}$$

We now choose these $x_{i|}$ as a new basic set of vectors:

$$\bar{e}_i = x_{i|}$$

The transformation from the old to the new coordinate system is given by the orthogonal operator S:

$$e_i = \bar{e}_j S_{ji}, \qquad \bar{e}_i = e_j S^{\mathsf{T}}{}_{ji} = S_{ij} e_j$$

Since

$$x_{i|} = x_{i|j} e_j$$

where $x_{i|j}$ denotes the jth component of the ith eigenvector, we have

$$S_{ij} = x_{i|j}, \qquad S^{\mathsf{T}}{}_{ij} = x_{j|i}$$

The rows of S are the eigenvectors in the original coordinate system.

The operator A has a matrix $\bar{A}_{ij}$ in the barred coordinate system which, because of (II.46) is given by

$$\bar{A}_{ij} = (x_{i|}, \alpha_{j|} x_{j|}) = \alpha_{j|} \delta_{ij} \tag{II.47}$$

The representative of A is thus a diagonal matrix when the eigenvectors are chosen as the basic vectors, and the elements appearing along the diagonal are the eigenvalues of the operator. We speak of this as a representation in which A is diagonal.

The problem of finding the eigenvalues and eigenvectors of an operator is equivalent to finding a representation in which the operator is diagonal. For, if the eigenvectors are known, the transformation matrix S can be constructed with these vectors as rows, and the new representation gained is the one in which A is diagonal, whereas, if the representation in which A is diagonal is known, the diagonal elements are the eigenvalues and the coordinate directions are the directions of the eigenvectors.

The two invariants under orthogonal transformations (II.40), (II.41) are now seen to be connected with the eigenvalues of the operator A. Det A is the product of the n eigenvalues, and tr A is their sum.

Tr A is the coefficient of $(-\alpha)^{n-1}$ in (II.44), and det A is the constant in this equation. The coefficients of all the other powers of α are also invariant. Since the coefficient of α^n is unity, there are n nontrivial invariants.

The representative of a linear operator A is known when all the scalar products (e_i, Ae_j) are known, the e_i being any set of basic vectors. This representation is equally well known if the scalar product (y, Ax) is known for all vectors x and y, for these vectors may be expressed as combinations of the basic vectors. The scalar product (y, Ax) is a bilinear form in the components of x and y

$$(y, Ax) = A_{ij}y_ix_j$$

and so we may say that the representative of a linear operator is associated with a bilinear form, knowledge of one giving the other. If the operator A is symmetric, it is sufficient to know the scalar products

$$(x, Ax)$$

for all x. Then

$$(x, Ax) = A_{ij}x_ix_j$$

is a quadratic form in the components of x. We frequently meet with symmetric operators given by a quadratic form, such as the kinetic energy of a system.

The algebraic process of diagonalizing a symmetric matrix can be given a geometrical meaning with the aid of quadratic forms. The equation

$$(x, Ax) = 1 \tag{II.48}$$

is the equation of a quadratic surface, an ellipsoid or a hyperboloid, in n dimensions. If the representation of A is diagonal, only squared terms appear in (II.48), the cross terms having zero coefficients, and the quadratic surface is then said to be referred to *principal axes.* If the surface is an ellipsoid, the coordinate axes of the representation in which A is diagonal meet the surface at right angles. In two dimensions the ellipsoid would be an ellipse, and the principal axes would be the lines along the major and minor axes respectively. The eigenvectors of an operator A lie along the principal axes of the quadratic surface defined by (II.48).

The eigenvalues of the operator are connected with the semiaxes of the surface. Written in standard form, the equation of such a surface is

$$\left(\frac{x_1}{a_1}\right)^2 + \left(\frac{x_2}{a_2}\right)^2 + \cdots + \left(\frac{x_n}{a_n}\right)^2 = 1$$

where for definiteness all the terms have been taken as positive, giving an ellipsoid. The hyperboloid can be treated similarly. Now, when A is diagonal, (II.48) can be written

$$\alpha_{i|}x_ix_i = 1$$

Thus

$$\alpha_{i|} = \frac{1}{(a_i)^2}$$

where the a_i are the semiaxes of the quadric.

If two of the eigenvalues of A are equal so that A is degenerate, two of the semiaxes are equal and the quadric is one of revolution. This gives a geometrical interpretation of the freedom of choice of the eigenvectors in such a situation. Any pair of mutually orthogonal lines in the plane of a pair of equal semiaxes forms a pair of semiaxes. Thus any such pair may be chosen as the directions of eigenvectors. If more than two eigenvalues are equal, the freedom of choice is greater yet.

A quadratic form is called *positive definite* if the coefficients are such that the value of the form is positive for all nonzero vectors x. It is called *indefinite* if the value can have either sign. An operator with only positive eigenvalues is associated with a positive definite form, and the corresponding quadric is an ellipsoid. If the operator has eigenvalues of both signs, the form is indefinite and the quadric is a hyperboloid. If the eigenvalue zero appears, the quadric is cylindrical.

From the point of view of quadratic forms a new definition of the eigenvalues of an operator can be given. This definition is useful for the computation of eigenvalues when solution of the determinantal equation (II.44) is difficult. Consider a representation in which A is diagonal. Then, if the eigenvalues of A are numbered in a decreasing sequence, we have

$$\begin{aligned}(x, Ax) &= \alpha_{i|}x_ix_i \\ &\leq \alpha_{1|}x_ix_i \\ &= \alpha_{1|}(x, x)\end{aligned}$$

since $\alpha_{1|}$ is the largest eigenvalue. The equality sign holds only when the vector x is $x_{1|}$. We may thus define $\alpha_{1|}$ by the equation

$$\alpha_{1|} = \max_x \frac{(x, Ax)}{(x, x)} \tag{II.49}$$

where the symbol $\max_x$ means maximum as x is varied.

The kth eigenvalue can be obtained in a similar way, if use is made of the fact that $x_{k|}$ is orthogonal to all the eigenvectors with smaller values of k. Thus

$$\begin{aligned}(x, Ax) &= \alpha_{i|}x_ix_i \\ &\leq \sum_{i=1}^{k-1} \alpha_{i|}x_i^2 + \alpha_{k|}\sum_{i=k}^{n} x_i^2 \\ &= \sum_{i=1}^{k-1} (\alpha_{i|} - \alpha_{k|})x_i^2 + \alpha_{k|}x_ix_i\end{aligned}$$

so that

$$\frac{(x, Ax)}{(x, x)} \leq \alpha_{k|} + \sum_{i=1}^{k-1} \frac{(\alpha_{i|} - \alpha_{k|})x_i^2}{(x, x)} \tag{II.50}$$

The maximum of the quantity on the left is greater than or equal to $\alpha_{k|}$, being of course $\alpha_{1|}$. If, however, we maximize this quantity subject to the conditions that x be orthogonal to $k - 1$ vectors $y_{j|}$, then this maximum will in general be less than $\alpha_{1|}$. The smallest value of the maximum will occur when the $y_{j|}$ are linear combinations of the first $k - 1$ eigenvectors, since then $x_i = 0$ for $i < k$ and all the terms in the sum on the right of (II.50) will vanish. Thus we may write

$$\alpha_{k|} = \min_{y_{i|}} \ \max_x \frac{(x, Ax)}{(x, x)} \tag{II.51}$$

$$(x, y_{i|}) = 0 \qquad (i = 1, 2, \cdots, k - 1)$$

APPENDIX III GROUP THEORY AND MOLECULAR VIBRATIONS

In this appendix we discuss the degeneracy inherent in the eigenvibrations of systems possessing symmetries. The systems of interest are polyatomic molecules.

Consider a polyatomic molecule containing some identical atoms. There may exist rotations of the molecule or reflections of the molecule whose effects are to replace each atom by an identical atom. The equilibrium configuration of the molecule is unchanged by such operations. They are called *symmetry operations* and are denoted by S_α $(\alpha = 1, 2, \cdots, h)$. The symmetry operations of a molecule constitute a group. We are concerned here only with those molecules which have symmetry operations in addition to the identity operation.

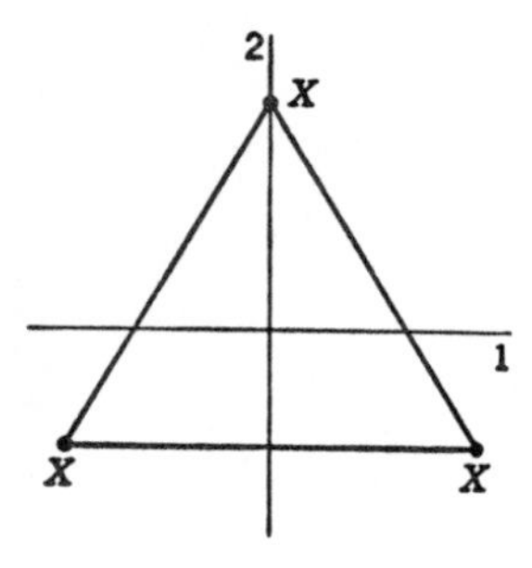

Fig. III-1. The molecule X_3.

As an example take the molecule X_3 consisting of three X atoms which, at equilibrium are at the corners of an equilateral triangle as shown in Fig. III-1. This molecule admits six symmetry operations if we exclude reflections in the plane of the molecule. There are rotations about an axis through the center of the triangle normal to the plane of the molecule through the angles 0, $2\pi/3$, and $4\pi/3$. In addition, there are reflections in lines through the center of the triangle and through each of the corners in turn.

Let the atoms of the molecule be displaced a small amount from their equilibrium positions. This displacement is described by the vector x in the configuration space of the molecule. If a symmetry operation S_α is now carried out, the atoms will be near an equilibrium configuration of the molecule. The displacement of the atoms from equilibrium after the symmetry operation is described by the vector y in configuration space. x and y are linearly related. We write this as

$$y = D(\alpha)x \tag{III.1}$$

In this way we obtain a *representation* of the symmetry group by a set of linear transformations. If a coordinate system is introduced, (III.1) can be written out in components and the $D(\alpha)$ become matrices. The $D(\alpha)$ are real orthogonal transformations. If rectangular coordinates are used, the matrices resulting are real orthogonal matrices.

Successive symmetry operations result in successive linear transformations, whose resultant is described by a matrix which is the matrix product of the matrices of the individual linear transformations. Thus the product of two symmetry group elements corresponds to the product of the representative matrices. If $S_\gamma = S_\beta S_\alpha$, then $D(\gamma) = D(\beta)D(\alpha)$.

If in the space R of the vectors x and y (the configuration space of the molecule) it is possible to introduce a subspace R_{I} and its complement R_{II} (so that $R = R_{\text{I}} + R_{\text{II}}$) in such a way that no symmetry operation applied to a vector in R_{I} gives it a part lying in R_{II} or applied to a vector in R_{II} gives it a part lying in R_{I}, then the representation D in the space R is *completely reducible*. If a coordinate system is introduced in such a way that the basis vectors lie entirely in R_{I} or in R_{II}, then the matrices $D(\alpha)$ will split into two blocks along the diagonal. The representation is then *reduced* and the subspaces R_{I} and R_{II} are *invariant* subspaces. If no such subspaces (except $R_{\text{I}} = R$ or $R_{\text{II}} = R$) exist, the representation is *irreducible*.

Two irreducible representations of a group are *equivalent* if the matrices constituting the two representations can be made identical by an appropriate choice of coordinate systems. The basis vectors of the coordinate systems which make the matrices identical define *equivalent directions* in the spaces of the two representations.

The number of inequivalent irreducible representations $D_\lambda(\alpha)$ of a group depends only on the group and is finite. Any representation can be reduced; i.e., a coordinate system can be introduced in the vector space of the representation $D(\alpha)$ in such a way that the matrices $D(\alpha)$ all split into blocks along the diagonal, each block corresponding to some irreducible representation $D_\lambda(\alpha)$. In this reduction an irreducible representation may occur more than once. The lth occurrence of the λth irreducible representation is labeled by λl. We choose coordinates so that all the matrices of equivalent irreducible representations are alike. Thus the matrix $D_{\lambda l}(\alpha)$ has elements independent of l.

A displacement vector x may be written as a sum of vectors, one in each invariant subspace:

$$x = \sum_{\lambda} \sum_{l} x_{\lambda l} \tag{III.2}$$

The potential energy due to such a displacement then may be written

$$V = \frac{1}{2} \sum_{\lambda,\mu} \sum_{l,m} V_{\lambda l i \mu m j} x_{\lambda l i} x_{\mu m j} \tag{III.3}$$

The summation convention holds for i, j, which have ranges depending on λ, μ.

The potential energy is not affected by a symmetry operation, which changes no relative distances within the molecule. Thus

$$V = \frac{1}{2} \sum_{\lambda,\mu} D_{\lambda}(\alpha)_{ri} D_{\mu}(\alpha)_{sj} \sum_{l,m} V_{\lambda l r \mu m s} x_{\lambda l i} x_{\mu m j} \tag{III.4}$$

Equating coefficients of the x's in (III.3) and (III.4), we obtain

$$D_{\lambda}(\alpha)_{ri} D_{\mu}(\alpha)_{sj} V_{\lambda l r \mu m s} = V_{\lambda l i \mu m j} \tag{III.5}$$

This may be written out in matrix notation. The matrices $D_{\lambda}(\alpha)$, $D_{\mu}(\alpha)$ are square but not necessarily of the same number of dimensions. The matrix $V_{\lambda l \mu m}$ has as many rows as $D_{\lambda}(\alpha)$ and as many columns as $D_{\mu}(\alpha)$. Equation (III.5) becomes

$$D^{\mathsf{T}}{}_{\lambda}(\alpha) V D_{\mu}(\alpha) = V \tag{III.6}$$

or, because the matrices $D_{\lambda}(\alpha)$ are orthogonal,

$$V D_{\mu}(\alpha) = D_{\lambda}(\alpha) V \tag{III.7}$$

where the indices λ, l, μ, m on V are suppressed.

Schur's lemma in the theory of groups states that any matrix which commutes with all the matrices of an irreducible representation is a multiple of the unit matrix, and that any matrix V which satisfies (III.7) for $\lambda \neq \mu$ vanishes.

From Schur's lemma we can immediately conclude that: (1) the potential energy matrix contains no elements connecting displacements in inequivalent irreducible subspaces; (2) the potential energy matrix connects only displacements in equivalent directions in equivalent irreducible subspaces, and all displacements in such directions are coupled with equal coefficients. The same argument applies to the kinetic energy matrix. Thus, only oscillations involving displacements in equivalent directions in equivalent irreducible subspaces can be coupled in the Lagrangian of the molecule, and all couplings between displacements in equivalent irreducible subspaces are equal.

From this statement it follows that:

1. A given eigenvibration involves only displacements in equivalent directions in a set of equivalent irreducible subspaces.
2. There are as many eigenvibrations of a given frequency as there are dimensions in one of the corresponding set of irreducible subspaces, because of the equality of the coupling between displacements in all such subspaces.
3. The determinantal equation (42.12) for the eigenfrequencies breaks into factor determinants. There are as many distinct factor determinants as there are sets of nonequivalent irreducible subspaces in configuration space. Each factor determinant appears once for each dimension of the set of equivalent subspaces.
4. Each factor determinant has as many rows and columns as there are members of the corresponding set of equivalent subspaces.

The problem now remains to reduce the representation of the symmetry group which is obtained in the configuration space of the molecule. It is possible to discover the number of degeneracies and their multiplicity without actually finding a coordinate system which reduces the representation. This is accomplished by finding how often each irreducible representation of the symmetry group occurs in the representation obtained in configuration space. We state without proof the relevant theorem (III.9) from the theory of groups and illustrate the method by applying it to the molecule X_3Y whose prototype is ammonia.

Two elements S_α, S_β of a group are in the same *class* if there exists an element S_γ of the group such that

$$S_\alpha = S_\gamma S_\beta S_\gamma^{-1} \tag{III.8}$$

If the group is abelian, i.e., commutative, each element is in a class by itself. The identity element is always alone in its class.

The trace of a representative matrix $D(\alpha)$ is denoted by $\chi(\alpha)$. $\chi(\alpha)$ is independent of the coordinate system used to determine the matrix $D(\alpha)$, and is the same for all $D(\alpha)$ representing elements in the same class. The set of numbers $\chi(\alpha)$ is called the *character* of the representation. The character of a reducible representation is the sum of the characters of the irreducible representations contained within it. This is obviously true in the coordinate which reduces the representation, but the character is independent of the coordinate system.

The characters of two irreducible representations $D_\lambda(\alpha)$, $D_\mu(\alpha)$ of a group satisfy an orthogonality relation

$$\sum_\alpha \chi_\lambda(\alpha)\chi_\mu(\alpha) = h\delta_{\lambda\mu} \tag{III.9}$$

where h is the order of the group, i.e., the number of elements in the group. [Equation (III.9) holds only for real orthogonal representations. In the more general case of representation by unitary transformations the left side must be written $\sum_\alpha \chi_\lambda{}^*(\alpha)\chi_\mu(\alpha)$.]

If the reducible representation $D(\alpha)$ contains the λth irreducible representation n_λ times, then for the character of this reducible representation we may write

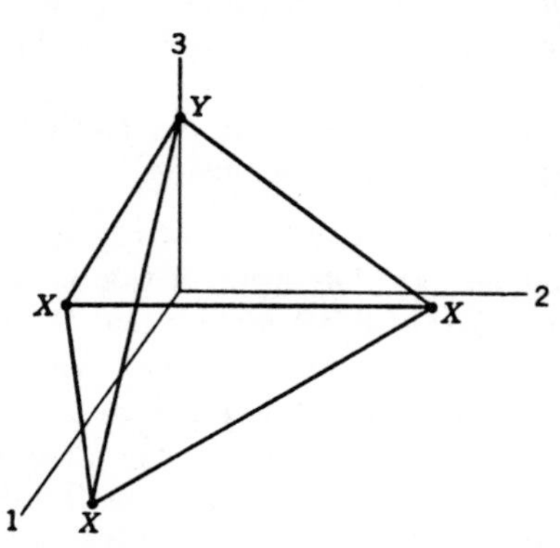

Fig. III-2. The X_3Y molecule.

$$\chi(\alpha) = \sum_\lambda n_\lambda \chi_\lambda(\alpha) \tag{III.10}$$

Applying (III.9), then,

$$n_\mu = \frac{1}{h}\sum_\alpha \chi_\mu(\alpha)\chi(\alpha) \tag{III.11}$$

By this means we may determine how often each irreducible representation occurs in any given representation, and thus we can find the number of degenerate eigenvibrations of a molecule and the multiplicities of the degeneracies.

As an example consider the molecule X_3Y whose structure is shown in Fig. III-2 and whose symmetry group is that of the molecule X_3 mentioned earlier. The group has six elements. Its multiplication table is readily determined and is as follows:

	S_1	S_2	S_3	S_4	S_5	S_6
S_1	S_1	S_2	S_3	S_4	S_5	S_6
S_2	S_2	S_3	S_1	S_5	S_6	S_4
S_3	S_3	S_1	S_2	S_6	S_4	S_5
S_4	S_4	S_6	S_5	S_1	S_3	S_2
S_5	S_5	S_4	S_6	S_2	S_1	S_3
S_6	S_6	S_5	S_4	S_3	S_2	S_1

This group is identical with the permutation group on three things, the rotations being cyclic permutations, the reflections being interchanges of pairs. The irreducible representations are well known. There are three of them because there are three classes, hence three independent values of $\chi(\alpha)$ for a given representation. There can thus be at most three orthogonal sets of χ's.

The three irreducible representations may be chosen as

	$\lambda = 1$	$\lambda = 2$	$\lambda = 3$
$D_\lambda(S_1)$	(1)	(1)	$\begin{pmatrix} 1 & 0 \\ 0 & 1 \end{pmatrix}$
$D_\lambda(S_2)$	(1)	(1)	$\begin{pmatrix} -\frac{1}{2} & -\sqrt{3}/2 \\ \sqrt{3}/2 & -\frac{1}{2} \end{pmatrix}$
$D_\lambda(S_3)$	(1)	(1)	$\begin{pmatrix} -\frac{1}{2} & \sqrt{3}/2 \\ -\sqrt{3}/2 & -\frac{1}{2} \end{pmatrix}$
$D_\lambda(S_4)$	(1)	(-1)	$\begin{pmatrix} 1 & 0 \\ 0 & -1 \end{pmatrix}$
$D_\lambda(S_5)$	(1)	(-1)	$\begin{pmatrix} -\frac{1}{2} & \sqrt{3}/2 \\ \sqrt{3}/2 & \frac{1}{2} \end{pmatrix}$
$D_\lambda(S_6)$	(1)	(-1)	$\begin{pmatrix} -\frac{1}{2} & -\sqrt{3}/2 \\ -\sqrt{3}/2 & \frac{1}{2} \end{pmatrix}$

The corresponding characters are

$$\chi_1(\alpha) = 1;\ 1, 1;\ 1, 1, 1$$

$$\chi_2(\alpha) = 1;\ 1, 1;\ -1, -1, -1$$

$$\chi_3(\alpha) = 2;\ -1, -1;\ 0, 0, 0$$

The semicolons separate the various classes.

The representation of the symmetry group in the molecule's configuration space of twelve dimensions may be chosen as follows:

$$D(1) = \begin{pmatrix} R_1 & 0 & 0 & 0 \\ 0 & R_1 & 0 & 0 \\ 0 & 0 & R_1 & 0 \\ 0 & 0 & 0 & R_1 \end{pmatrix}, \quad R_1 = \begin{pmatrix} 1 & 0 & 0 \\ 0 & 1 & 0 \\ 0 & 0 & 1 \end{pmatrix}$$

$$D(2) = \begin{pmatrix} 0 & 0 & R_2 & 0 \\ R_2 & 0 & 0 & 0 \\ 0 & R_2 & 0 & 0 \\ 0 & 0 & 0 & R_2 \end{pmatrix}, \quad R_2 = \begin{pmatrix} \cos 2\pi/3 & -\sin 2\pi/3 & 0 \\ \sin 2\pi/3 & \cos 2\pi/3 & 0 \\ 0 & 0 & 1 \end{pmatrix}$$

$$D(3) = \begin{pmatrix} 0 & R_3 & 0 & 0 \\ 0 & 0 & R_3 & 0 \\ R_3 & 0 & 0 & 0 \\ 0 & 0 & 0 & R_3 \end{pmatrix}, \quad R_3 = \begin{pmatrix} \cos 2\pi/3 & \sin 2\pi/3 & 0 \\ -\sin 2\pi/3 & \cos 2\pi/3 & 0 \\ 0 & 0 & 1 \end{pmatrix}$$

$$D(4) = \begin{pmatrix} R_4 & 0 & 0 & 0 \\ 0 & 0 & R_4 & 0 \\ 0 & R_4 & 0 & 0 \\ 0 & 0 & 0 & R_4 \end{pmatrix}, \quad R_4 = \begin{pmatrix} 1 & 0 & 0 \\ 0 & -1 & 0 \\ 0 & 0 & 1 \end{pmatrix}$$

$$D(5) = \begin{pmatrix} 0 & R_5 & 0 & 0 \\ R_5 & 0 & 0 & 0 \\ 0 & 0 & R_5 & 0 \\ 0 & 0 & 0 & R_5 \end{pmatrix}, \quad R_5 = \begin{pmatrix} \cos 2\pi/3 & \sin 2\pi/3 & 0 \\ \sin 2\pi/3 & -\cos 2\pi/3 & 0 \\ 0 & 0 & 1 \end{pmatrix}$$

$$D(6) = \begin{pmatrix} 0 & 0 & R_6 & 0 \\ 0 & R_6 & 0 & 0 \\ R_6 & 0 & 0 & 0 \\ 0 & 0 & 0 & R_6 \end{pmatrix}, \quad R_6 = \begin{pmatrix} \cos 2\pi/3 & -\sin 2\pi/3 & 0 \\ -\sin 2\pi/3 & -\cos 2\pi/3 & 0 \\ 0 & 0 & 1 \end{pmatrix}$$

Here the *submatrix* R_α describes the rotation or reflection of the displacement vector of a given atom, and the *supermatrix* on the left describes the interchange of atoms among themselves. The first three rows and columns of the supermatrix involve the X atoms, the fourth involves the single Y atom.

The character of the configuration space representation is seen to be

$$\chi(\alpha) = \{12;\ 0, 0;\ 2, 2, 2\}$$

Application of (III.11) now yields

$$n_1 = 3, \qquad n_2 = 1, \qquad n_3 = 4$$

Included in all the above considerations have been the rigid displacements of the molecule as well as displacements involved in vibration only. The former could have been excluded by using a six-dimensional configuration space obtained by constraining the molecule to have no translation of the center of mass or rotation about it. It is easier to investigate the behavior of the rigid motions under the symmetry operations. The translations and the rotations do not get mixed by symmetry operations and can therefore be discussed separately.

The three components of the rigid translation form a polar vector and thus transform under all the symmetry operations according to the matrices R_α. The three components of an infinitesimal rotation form an axial vector and therefore transform under S_1, S_2, and S_3 according to R_1, R_2, and R_3, but under S_4, S_5, and S_6 according to

$-R_1$, $-R_2$, and $-R_3$. The character for translations is thus

$$X_t(\alpha) = \{3;\ 0, 0;\ 1, 1, 1\}$$

and for rotations

$$X_r(\alpha) = \{3;\ 0, 0;\ -1, -1, -1\}$$

The translations thus account for one irreducible representation with $\lambda = 1$ and one with $\lambda = 3$; the rotations account for the representation $\lambda = 2$ and one with $\lambda = 3$. The vibrations therefore are described by displacements which lie in two one-dimensional irreducible subspaces $\lambda = 1$ and in two two-dimensional subspaces $\lambda = 3$. Of the six eigenvibrations, then, two are nondegenerate, and there are two pairs which are doubly degenerate.

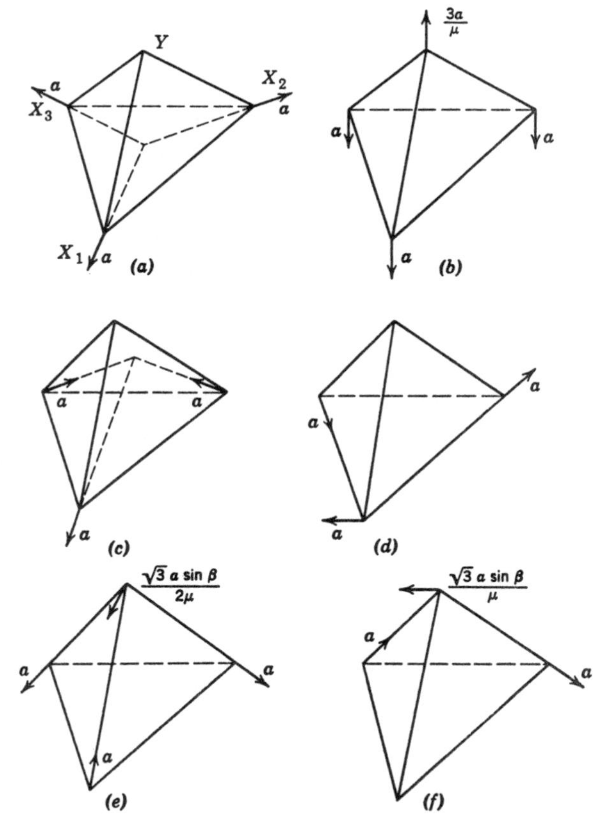

Fig. III-3. Eigenvibrations of X_3Y.

Six possible displacements of the atoms of this molecule which do not contain any rigid motion are shown in Fig. III-3. Those illustrated in (*a*) and (*b*) are invariant under all symmetry operations and occupy the two subspaces of irreducible representation $D_1(\alpha)$. Those illustrated in (*c*) and (*d*) become mixed with each other, as do those shown in (*e*) and (*f*), under symmetry operations. They are displacements in the two irreducible subspaces of two dimensions. The displacements (*c*) and (*d*) get mixed with the same coefficients as (*e*) and (*f*) so that (*c*) and (*e*) are in equivalent directions in their respective subspaces, as are (*d*) and (*f*).

We may now apply the rules of p. 367 to this example and see geometrically what is involved.

1. A given eigenvibration may involve linear combinations of displacements (*a*) and (*b*) or (*c*) and (*e*) or (*d*) and (*f*).
2. Eigenvibrations which are linear combinations of (*a*) and (*b*) are nondegenerate. Eigenvibrations which are linear combinations of (*c*) and (*e*) are degenerate with eigenvibrations which are the *same* linear combinations of (*d*) and (*f*).
3. There are two distinct factor determinants in the determinantal equation, one appearing once, one twice.
4. Each factor determinant has two rows and two columns.

It is possible by group theoretic means, therefore, to obtain considerable information about degeneracies without actually solving the small oscillation problem provided symmetries exist.

GENERAL REFERENCES

E. Wigner, *Göttingen Nachr.*, p. 133 (1930).

J. E. Rosenthal and G. M. Murphy, *Rev. Mod. Phys.*, 8, 317 (1936).

E. Wigner, *Gruppen Theorie.* Vieweg und Sohn, Braunschweig (1931), and Edwards Bros., Ann Arbor (1946). *Group Theory.* Academic Press, New York (1959).

G. Hertzberg, *Infrared and Raman Spectra of Polyatomic Molecules.* Van Nostrand, New York (1945).

APPENDIX IV QUATERNIONS AND PAULI SPIN MATRICES

We have seen in Secs. 47 and 48 how the rotation of a rigid body about a point may be specified by an orthogonal operator or by the Euler angles of the rotation. An alternative method of representing rotations by means of quaternions was developed by Sir W. R. Hamilton, and although it is not used widely in classical mechanics it is closely related to that introduced by W. Pauli for describing the spin of a particle in quantum mechanics.

The radius vector $\mathbf{r}$ of a point in a rigid body is carried over into the vector $\bar{\mathbf{r}}$ by a rotation of the body through the angle θ about an axis passing through the origin in the direction of the unit vector $\mathbf{n}$. θ is positive if $\mathbf{n}$ is in the direction of advance of a right-hand screw rotated through the angle θ about $\mathbf{n}$. We seek an expression for $\bar{\mathbf{r}}$ in terms of $\mathbf{r}$, $\mathbf{n}$, θ.

We break the vector $\mathbf{r}$ up into three parts as shown in Fig. IV-1. The first part $\mathbf{a}$ is parallel to the axis of rotation. The second part $\mathbf{b}$ is perpendicular to the axis of rotation and to $\bar{\mathbf{r}} - \mathbf{r}$. The third part $\mathbf{c}$ is parallel to $\bar{\mathbf{r}} - \mathbf{r}$. We have

$$\mathbf{a} = (\mathbf{r} \cdot \mathbf{n})\mathbf{n}$$

$$\text{(IV.1)} \qquad \mathbf{b} = \cos^2 \frac{\theta}{2} (\mathbf{n} \times \mathbf{r}) \times \mathbf{n} + \cos \frac{\theta}{2} \sin \frac{\theta}{2} (\mathbf{n} \times \mathbf{r})$$

$$\mathbf{c} = -\cos \frac{\theta}{2} \sin \frac{\theta}{2} (\mathbf{n} \times \mathbf{r}) + \sin^2 \frac{\theta}{2} (\mathbf{n} \times \mathbf{r}) \times \mathbf{n}$$

Then

$$\text{(IV.2)} \qquad \begin{aligned} \mathbf{r} &= \mathbf{a} + \mathbf{b} + \mathbf{c} \\ \bar{\mathbf{r}} &= \mathbf{a} + \mathbf{b} - \mathbf{c} \end{aligned}$$

It is readily verified that the first of equations (IV.2) holds with the values given by (IV.1). The displaced vector $\bar{\mathbf{r}}$ is given by

$$\text{(IV.3)} \quad \begin{aligned} \bar{\mathbf{r}} &= (\mathbf{r} \cdot \mathbf{n})\mathbf{n} + \left(\cos^2 \frac{\theta}{2} - \sin^2 \frac{\theta}{2}\right) (\mathbf{n} \times \mathbf{r}) \times \mathbf{n} \\ &\qquad + 2 \cos \frac{\theta}{2} \sin \frac{\theta}{2} (\mathbf{n} \times \mathbf{r}) \\ &= (1 - \cos \theta)(\mathbf{r} \cdot \mathbf{n})\mathbf{n} + \cos \theta \, \mathbf{r} + \sin \theta \, (\mathbf{n} \times \mathbf{r}) \end{aligned}$$

If $\mathbf{r}$ is parallel to $\mathbf{n}$, then (IV.3) shows that $\bar{\mathbf{r}} = \mathbf{r}$, which is true also if θ is zero. If $\mathbf{r}$ is perpendicular to $\mathbf{n}$, then $\bar{\mathbf{r}} = \cos \theta \mathbf{r} + \sin \theta (\mathbf{n} \times \mathbf{r})$,

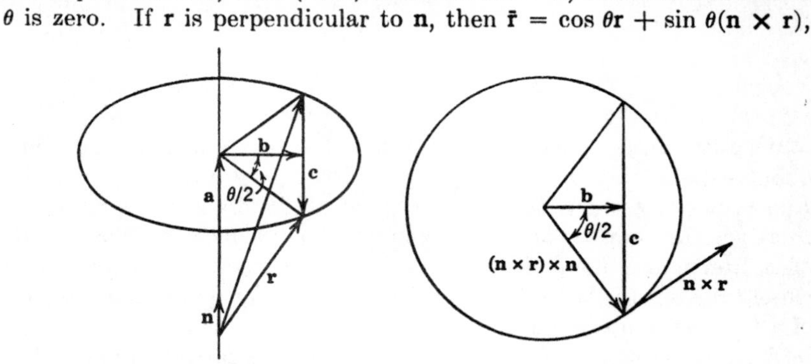

Fig. IV-1. Effect on radius vector of rotation about an axis.

which is the well-known formula for rotations in a plane. In terms of components ($\mathbf{n} = e_3$) it is

$$\text{(IV.4)} \qquad \begin{aligned} \bar{x}_1 &= \cos \theta \, x_1 - \sin \theta \, x_2 \\ \bar{x}_2 &= \sin \theta \, x_1 + \cos \theta \, x_2 \\ \bar{x}_3 &= x_3 \end{aligned}$$

Equation (IV.4) for rotations in the 1-2 plane can be expressed in complex form. Let

$$z = x_1 + ix_2$$

Then

$$\text{(IV.5)} \qquad \begin{aligned} e^{i\theta}z &= (\cos \theta + i \sin \theta)(x_1 + ix_2) \\ &= (\cos \theta \, x_1 - \sin \theta \, x_2) + i(\sin \theta \, x_1 + \cos \theta \, x_2) \\ &= \bar{x}_1 + i\bar{x}_2 \end{aligned}$$

where (IV.4) has been used. Thus we may write

$$\bar{z} = e^{i\theta}z = e^{i(\theta/2)}ze^{i(\theta/2)} \tag{IV.6}$$

where this last form of writing is used to make this equation resemble one we shall derive later—(IV.16).

A rotation in the plane is specified by a single parameter θ, and its effect on a vector is described by a complex number of unit modulus. To specify a rotation in space requires three parameters, and we attempt to describe the effect of a spatial rotation on a vector by a *hypercomplex number* of unit norm. A hypercomplex number is one of the form

$$a_0 + a_i e_i$$

a_0 and the a_i are real numbers; the e_i are generalizations of the square root of -1. If there are three such e's, so that $i = 1, 2, 3$, then the hypercomplex number is called a *quaternion*, and the e_i obey the rules of multiplication:

$$\begin{aligned} e_i^2 &= -1 \qquad (i = 1, 2, 3) \\ e_i e_j &= \epsilon_{ijk} e_k \qquad (i \neq j) \end{aligned} \tag{IV.7}$$

where the ϵ_{ijk} are the antisymmetric coefficients introduced in Sec. 6. These hypercomplex numbers do not obey all the laws of the algebra of complex numbers. They add like complex numbers, but their multiplication is not commutative.

The product of two quaternions, p and q, is given by

$$\begin{aligned} pq &= (p_0 + p_i e_i)(q_0 + q_j e_j) \\ &= p_0 q_0 - p_i q_i + p_0 q_i e_i + q_0 p_i e_i + \epsilon_{ijk} p_i q_j e_k \end{aligned} \tag{IV.8}$$

The product qp differs from pq only in the sign of the last term, so that, if $p_i = \alpha q_i$, where α is any real number, p and q commute.

The form of a quaternion suggests consideration of the quaternion as a combination of a scalar q_0 and a vector with components q_i. If we write

$$q = q_0 + \mathbf{q} \tag{IV.9}$$

then

$$r = pq = p_0 q_0 - \mathbf{p} \cdot \mathbf{q} + p_0 \mathbf{q} + q_0 \mathbf{p} + \mathbf{p} \times \mathbf{q} \tag{IV.10}$$

so that, if $r = r_0 + \mathbf{r}$, we have

$$\text{(IV.11)} \qquad \begin{aligned} r_0 &= p_0 q_0 - \mathbf{p} \cdot \mathbf{q} \\ \mathbf{r} &= p_0 \mathbf{q} + q_0 \mathbf{p} + \mathbf{p} \times \mathbf{q} \end{aligned}$$

The analogy to ordinary complex numbers is enhanced by this view. A general quaternion may be written

$$\text{(IV.12)} \qquad q = q_0 + q_1 \mathbf{n}$$

where $\mathbf{n}$ is a unit vector. q_0 is the analog of the real part; q_1, that of the imaginary part. The product of $\mathbf{n}$ with itself must be taken in the sense of quaternions, and since $\mathbf{n}$ has no scalar part, $n^2 = -\mathbf{n} \cdot \mathbf{n} = -1$, so that $\mathbf{n}$ is analogous to the imaginary $\sqrt{-1}$.

With every quaternion, q, we associate another, q^*, its conjugate, defined by

$$\text{(IV.13)} \qquad q^* = q_0 - q_1 \mathbf{n} = q_0 - q_i e_i$$

The product

$$\text{(IV.14)} \qquad \begin{aligned} q^* q &= (q_0 - q_1 \mathbf{n})(q_0 + q_1 \mathbf{n}) \\ &= q_0^2 + q_1^2 \\ &= q q^* \end{aligned}$$

is a real positive number called the *norm* of the quaternion.

A rotation about an axis in the direction of $\mathbf{n}$ through the angle θ we now describe by a quaternion whose vector lies in the direction of $\mathbf{n}$, and whose scalar and vector parts are suggested by the last part of (IV.6):

$$\text{(IV.15)} \qquad q = \cos \frac{\theta}{2} + \sin \frac{\theta}{2} \mathbf{n}$$

Hence $qq^* = 1$, and q is a quaternion of unit norm. If the vector $\mathbf{r}$ of (IV.3) is now taken to be a quaternion r with zero scalar, we find that

$$\text{(IV.16)} \qquad \begin{aligned} q r q^* &= \cos \theta \mathbf{r} + (1 - \cos \theta)(\mathbf{n} \cdot \mathbf{r})\mathbf{n} + \sin \theta (\mathbf{n} \times \mathbf{r}) \\ &= \bar{r} \end{aligned}$$

the last step coming from (IV.3). The effect of a rotation is thus completely specified by the quaternion q.

The components of the quaternion, q_0, q_1, q_2, q_3, which are subject to the condition that the sum of their squares is unity, are the homogeneous parameters of Euler. In a cartesian coordinate system they are given by

$$
\begin{aligned}
q_0 &= \cos\frac{\theta}{2} \\
q_1 &= \sin\frac{\theta}{2}\cos\alpha \\
q_2 &= \sin\frac{\theta}{2}\cos\beta \\
q_3 &= \sin\frac{\theta}{2}\cos\gamma
\end{aligned}
\tag{IV.17}
$$

where $\cos\alpha$, $\cos\beta$, $\cos\gamma$ are the direction cosines of the axis of rotation.

That the quaternions constitute a representation of the rotation group follows from the equation of transformation (IV.16). For, if p and q are two quaternions of unit norm representing two rotations, and if q is applied first, then

$$
\begin{aligned}
\bar{r} &= qrq^* \\
\bar{\bar{r}} &= p\bar{r}p^* \\
&= pqrq^*p^* \\
&= (pq)r(pq)^*
\end{aligned}
\tag{IV.18}
$$

where the identity $(pq)^* = q^*p^*$, the proof of which is left to the reader, has been used. Also if $qq^* = 1$, $pp^* = 1$, then

$$(pq)(pq)^* = pqq^*p^* = pp^* = 1$$

so that pq has norm unity. The unit quaternion exists, every quaternion of unit norm has an inverse of unit norm, and quaternions obey the associative law of multiplication. The representation is double-valued, for both q and $-q$ represent the same rotation.

For some purposes it is convenient to introduce the quantities σ_j in place of the e_j, the σ's being defined by

$$\sigma_j = ie_j \tag{IV.19}$$

where $i = \sqrt{-1}$. The advantage lies in the fact that

$$\sigma_j^2 = +1 \qquad (j = 1, 2, 3)$$

without the minus sign of (IV.7). The rest of (IV.7) becomes

$$\sigma_k\sigma_l = i\epsilon_{klm}\sigma_m \qquad (k \neq l) \tag{IV.20}$$

This shows that the σ's anticommute:

$$\sigma_k \sigma_l = -\sigma_l \sigma_k \tag{IV.21}$$

The σ's are known as the *Pauli spin operators.* They are used to describe the spin of the electron in quantum mechanics.

The rotation of a rigid body with time is represented by having the homogeneous parameters functions of time. The quaternion q is written $q(t)$, and the value of $r(t)$ at any time t is expressed as

$$r(t) = q(t) r_0 q^*(t) \tag{IV.22}$$

where r_0 is the initial value of r, and $q(0) = 1$.

To find the velocities of particles composing the body we consider the displacement that takes place in the very short time dt, where powers of dt higher than the first are neglected. In the time dt, q will depart very little from unity and can be written, in view of (IV.15), as

$$q = 1 + \frac{d\theta}{2}\mathbf{n} \tag{IV.23}$$

Thus

$$\begin{aligned} r(t + dt) &= r + dr = \left(1 + \frac{d\theta}{2}\mathbf{n}\right) r \left(1 - \frac{d\theta}{2}\mathbf{n}\right) \\ &= r + d\theta(\mathbf{n} \times \mathbf{r}) \end{aligned} \tag{IV.24}$$

or

$$\begin{aligned} d\mathbf{r} &= d\theta(\mathbf{n} \times \mathbf{r}) \\ \dot{\mathbf{r}} &= \frac{d\theta}{dt}(\mathbf{n} \times \mathbf{r}) = \boldsymbol{\omega} \times \mathbf{r} \end{aligned} \tag{IV.25}$$

The vector $\boldsymbol{\omega}$ has the magnitude $d\theta/dt$ and lies along the direction of the axis of rotation. This axis is the *instantaneous axis of rotation* of the body. Its direction is not necessarily constant. The vector character of $\boldsymbol{\omega}$ follows from (IV.22), for, if two successive small displacements about nonparallel axes are made, then from (IV.22)

$$\begin{aligned} r + dr &= \left(1 + \frac{d\theta'}{2}\mathbf{n}'\right)\left(1 + \frac{d\theta}{2}\mathbf{n}\right) r \left(1 - \frac{d\theta}{2}\mathbf{n}\right)\left(1 - \frac{d\theta'}{2}\mathbf{n}'\right) \\ &= r + (d\theta\mathbf{n} + d\theta'\mathbf{n}') \times \mathbf{r} \\ \dot{\mathbf{r}} &= (\boldsymbol{\omega} + \boldsymbol{\omega}') \times \mathbf{r} \end{aligned} \tag{IV.26}$$

This result is in contrast with that for finite rotations. The difference lies in the fact that the quaternions representing infinitesimal rotations commute, since the noncommutative part involves differentials of order higher than the first, whereas quaternions representing finite rotations do not necessarily commute.

The angular velocity vector $\boldsymbol{\omega}$ may be found as a function of q and $\dot{q}$. Differentiating (IV.22) with respect to the time we have, since $\dot{r}_0$ vanishes,

$$\dot{r} = \dot{q}r_0q^* + qr_0\dot{q}^* \tag{IV.27}$$

Now

$$r_0 = q^*rq \tag{IV.28}$$

so that

$$\begin{aligned} \dot{r} &= \dot{q}q^*r + rq\dot{q}^* \\ &= \dot{q}q^*r - r\dot{q}q^* \end{aligned} \tag{IV.29}$$

Now the only parts of r and $\dot{q}q^*$ which do not commute are the vector parts entering the vector product. Hence, denoting the vector part of the quaternion $\dot{q}q^*$ by $\overrightarrow{\dot{q}q^*}$, we have

$$\begin{aligned} \dot{\mathbf{r}} &= (\overrightarrow{\dot{q}q^*}) \times \mathbf{r} - \mathbf{r} \times (\overrightarrow{\dot{q}q^*}) \\ &= 2(\overrightarrow{\dot{q}q^*}) \times \mathbf{r} \end{aligned} \tag{IV.30}$$

Thus, comparing (IV.30) with (IV.25), we see that

$$\boldsymbol{\omega} = 2(\overrightarrow{\dot{q}q^*}) \tag{IV.31}$$

The three rotations described in Sec. 48 to specify the Eulerian angles may be represented by three quaternions

$$\begin{aligned} q_\phi &= \cos\frac{\phi}{2} + \sin\frac{\phi}{2}\,e_3 \\ q_\theta &= \cos\frac{\theta}{2} + \sin\frac{\theta}{2}\,(q_\phi e_1 q_\phi{}^*) \\ &= q_\phi\left(\cos\frac{\theta}{2} + \sin\frac{\theta}{2}\,e_1\right)q_\phi{}^* \\ q_\psi &= q_\theta q_\phi\left(\cos\frac{\psi}{2} + \sin\frac{\psi}{2}\,e_3\right)q_\phi{}^*q_\theta{}^* \end{aligned} \tag{IV.32}$$

Hence

$$q = q_\psi q_\theta q_\phi = \left(\cos\frac{\phi}{2} + \sin\frac{\phi}{2}e_3\right)\left(\cos\frac{\theta}{2} + \sin\frac{\theta}{2}e_1\right)\left(\cos\frac{\psi}{2} + \sin\frac{\psi}{2}e_3\right) \tag{IV.33}$$

or

$$\begin{aligned} q_0 &= \cos\frac{\theta}{2}\cos\frac{\phi+\psi}{2} \\ q_1 &= \sin\frac{\theta}{2}\cos\frac{\phi-\psi}{2} \\ q_2 &= \sin\frac{\theta}{2}\sin\frac{\phi-\psi}{2} \\ q_3 &= \cos\frac{\theta}{2}\sin\frac{\phi+\psi}{2} \end{aligned} \tag{IV.34}$$

These components are the same in both coordinate systems because the vector part of the quaternion lies along the axis of rotation.

By use of (IV.19) a quaternion may be written as a 2×2 matrix. The customary representation of the Pauli spin operators is by the matrices

$$\sigma_1 = \begin{pmatrix} 0 & 1 \\ 1 & 0 \end{pmatrix}, \qquad \sigma_2 = \begin{pmatrix} 0 & -i \\ i & 0 \end{pmatrix}, \qquad \sigma_3 = \begin{pmatrix} 1 & 0 \\ 0 & -1 \end{pmatrix}$$

Thus we may write

$$\begin{aligned} q &= q_0 - iq_k\sigma_k \\ &= \begin{pmatrix} q_0 - iq_3 & -iq_1 - q_2 \\ -iq_1 + q_2 & q_0 + iq_3 \end{pmatrix} \\ &= \begin{pmatrix} \alpha & \beta \\ -\beta^* & \alpha^* \end{pmatrix} \end{aligned} \tag{IV.35}$$

The hypercomplex number q is specified by two complex numbers, α and β.

From (IV.35) we see that

$$\begin{aligned} \det(q) &= q_0^2 + \vec{q}^{\,2} \\ &= qq^* \\ &= \alpha\alpha^* + \beta\beta^* \end{aligned}$$

Thus if q represents a rotation, the parameters α and β must satisfy

$$\alpha\alpha^* + \beta\beta^* = 1$$

In this case the parameters α and β are called the Cayley-Klein parameters of the rotation. Insertion of (IV.34) in (IV.35) yields the connection between them and the Euler angles.

INDEX

Abelian group, 367
Absolute integral invariant, 235
  generalized, 242
Acceleration, 12, 29
  in cylindrical coordinates, 14
  generalized, 339
  variational principle involving, 266
Accelerators, high energy, 314 f.
  van de Graaff, 112
Action, 155, 171, 189, 323
  as generator of contact transformation, 185
Action and angle variables, 188 f., 193
  in a Coulomb field, 206
  in perturbation theory, 245
Action principle, 170
Adiabatic change, 195
  damping, 323
  invariance, 195, 323
  perturbation, 50
Admittance of an accelerator, 328
Ammonia-like molecule, 368
Angle variable, 190
  see also Action and angle variables
Angular momentum, 4, 36, 40, 85, 108
  in electromagnetic field, 313
  as generator of contact transformation, 218, 225, 227, 261
  integral, 90, 92, 96
  Poisson bracket relations for, 225
  orbital, 42, 91
  spin, 42, 146, 258, 304
  tensor in relativity theory, 298
  total, 202
Angular velocity, 38, 142, 146, 379
Anharmonic oscillator, 47 f., 192
Anomalous magnetic moment, 307
Approximation methods, *see* Perturbation methods
Apse, 64
Apsidal distance, 95
Atmospheric drag, 71
Axial vector, 21
Axis, principal, 147
  of rotation, 137, 378

Barn, 101
Basic vectors, 346
Beta-ray spectrometer, 111
Betatron oscillations, 314, 316, 318
Body coordinate system, 136, 150
Bohr correspondence principle, 206
Bohr orbits, 98, 109
Boltzmann equation, 268
Bulk viscosity, 271
Burn-out point, 98

CO_2-like molecule, 120
Canonical equations, *see* Hamilton's equations
Canonical transformation, *see* Contact transformations
Canonically conjugate variables, *see* Conjugate variables
Cartesian coordinates, 6
Cayley-Klein parameters, 139, 381
Center of mass, 30
Center-of-mass coordinate system, 62

Central motion, 63, 90, 200
equivalent Hamiltonian (θ motion), 203
equivalent potential (radial motion), 96, 202
Hamilton-Jacobi equation, 200
Centrifugal potential, 93
Centripetal acceleration, 13, 146, 162
Character of representation, 367
Chasles' theorem, 139
Child-Langmuir law, 54
Christoffel symbols, 340
Class of group elements, 367
Closed-loop transfer function, 133
Commuting operators, 352
Complete orthonormal set, 359
Configuration space, 7, 10, 32, 166, 325
of molecule, 364
of rigid body, 136
Conical pendulum, 104
Conjugate quaternion, 376
Conjugate variables, 156, 232
Conservation laws, angular momentum, 30, 218
energy, 35, 279
momentum, 30, 217
see also Generator, Integral of motion
Conservative field, 34
Constraints, 10, 77
Contact transformations, 172, 183, 187
extended, 241
generated by action, 185
Hamiltonian, 217
momentum, 217, 224
orbital angular momentum, 218, 225, 261
total angular momentum, 260
infinitesimal, 215
Continuous medium, 273
Contravariant components, 335
Coordinates, body, 136, 150
cartesian, 6
cyclic, ignorable, 83, 159, 188, 225
cylindrical, 13, 159
elliptical, 16, 25, 207
normal, 116
parabolic, 16, 25
polar, 7, 341
quasi, 144, 254
rotating, 88, 108, 161, 262
Coordinates, spherical, 8, 160, 341
Coriolis acceleration, 13, 14, 89, 146, 162
Correspondence principle, 206
Coulomb scattering, 100 f.
see also Central motion, Kepler orbit
Coupled oscillators, 115
Covariant components, 336
Critically damped oscillator, 75
Cross section, 100, 101
Cyclic coordinate, 83, 159, 188, 225
Cylindrical coordinates, 13, 159
Cylindrical magnetron, 110

Damped oscillator, 74
Degenerate eigenvibration, 368
Degenerate system, 118, 205, 248
Degrees of freedom, 10, 136, 230
Deuteron, 65, 262
Differential cross section, 101
Dilatation viscosity, 271
Dipole, motion of dipole in field of, 260
particle in field of, 209
Dirac delta function, 286
Dirac theory, 258, 301
Dissipation function, 89
Dissipative systems, 74, 131
Double pendulum, 117
Drag, atmospheric, 71
Dust particle, motion of, 73

Effective magnetic moment, 309
Eigenvalue, 357
Eigenvector, 357
Eigenvibration, 116, 367
Electric dipole moment of moving magnet, 305
Electromagnetic field, Lorentz transformation for, 293
motion of particle in, 85, 163, 223, 297
Elliptic integral, 52
Elliptical coordinates, 16, 25, 207
reduction to, parabolic coordinates, 210
plane polar coordinates, 209
Energy conservation, 35, 279
see also Conservation laws, Jacobian integral
Energy density, 275
Energy flux, 275
Energy integral, 243

Energy-momentum tensor, 279
Ensemble, 235
Equation of state, 165
Equations of motion, 29, 186
  alternative forms for, 179
  with arbitrary independent variable, 240
  for continuous medium, 275
  Euler's, 149
  in field theory, 285
  Hamilton-Jacobi, 154, 184
  Hamilton's, 156, 183, 232, 241
  Lagrange's, 77, 167, 220
  in momentum space, 180
  in phase space, 233
  in quasi coordinates, 255
  in relativity theory, 302
Equilibrium orbit, 315
Equivalence principle, 45
Euclidean space, 335
Euler's angles, 139, 140, 142, 257, 379, 381
Euler's equations of motion, 149
Euler's homogeneous parameters, 377
Event, 290
Extended contact transformations, 241
Extended phase space, 241

Field, dipole, 209, 260
  irrotational, 34
  lamellar, 34, 80, 81
Field index, 318
Field theory, 273
  Hamiltonian form, 284
First integral, *see* Integral of motion
Fixed constraint, 10
Fixed field alternating gradient synchrotron, 317
Force, 28
  generalized, 78
  Lorentz, 297
  many-body, 29, 35
  noncentral, 258
  parallelogram law, 29
Four-velocity, 294
Free rotator, 148
Frenet formulas, 319

Galilean transformation, 288
Gauge transformation, 304
General theory of relativity, 6, 45, 344
Generalized acceleration, 339
  coordinates, 8, 10, 13, 337
  force, 79
  integral invariant, 242
  momentum, 155
  Poisson bracket, 241
Generating function, 183, 191
Generator, 174
  of contact transformation, 172, 302
  of infinitesimal contact transformations, 216, 233
    angular momentum as, 218, 225, 227, 261
    momentum as, 217, 224
Gradient, 20, 34
Gravitation, two centers of, 206
Gravitational constant, 5, 96
Gravitational mass, 43
Group, abelian, 367
  of contact transformations, 176
  Lorentz, 303
  permutation, 369
  of rotations, 139
  of symmetry operations, 364

Hamiltonian, 156
  in field theory, 284
  for particle, betatron oscillations, 320
    central plus constant force, 249
    Coulomb plus magnetic field, 248
    electromagnetic field, 163, 223
    rotating coordinates, 161
    spherical coordinates, 160
    synchrotron oscillations, 331
    two magnetic dipoles, 260
  in relativity theory, 298
  in terms of arbitrary independent variable, 240, 320
  in terms of complex variables, 193
Hamilton-Jacobi equation, 154, 183
  for central field, 201
  for Coulomb and uniform fields, 210
  for dipole field, 209
  for elliptic coordinates, 208
  for parabolic coordinates, 211
  in perturbation theory, 251
  in relativity theory, 300
Hamilton's equations, 156, 183, 232, 241
  in field theory, 285

Hamilton's equations, in quasi coordinates and momenta, 256
in relativity theory, 302
Hamilton's principal function, 168, 184
Hamilton's principle, 168
for continuous media, 274
in relativity theory, 296
Harmonic oscillator, 89, 194, 242
two dimensions, 84
Heisenberg uncertainty principle, 157
Holonomic constraint, 10
Homogeneous Lorentz transformation, 303
Homogeneous parameters of Euler, 377
Hypercomplex number, 375

Identity operator, 348
Ignorable coordinate, 83, 159, 188, 225
Impact parameter, 101
Improper Lorentz transformation, 291
Independent variable, change of, 238
Inertia tensor, 40, 146, 151, 258
Inertial mass, 43
Inertial system, 146, 150, 304
Infinitesimal rotation, 218, 370
Inhomogeneous Lorentz group, 303
Inner product, 18, 346
Instantaneous axis, 378
Integral invariants, 235, 242
Integral of motion, angular momentum, 90, 92, 96
energy, 243
momentum, 83
see also Jacobian integral
International prototype meter, 23
kilogram, 28
Interval, 290
Invariant subspace, 365
Inverted pendulum, 67
Irreducible representation, 365
Irrotational field, 34
Isochronous pendulum, 46
Isolated system, 38

Jacobian determinant, 9, 233
of a contact transformation, 235
Jacobian integral, 82, 84, 89, 110, 156, 163, 327
Jacobi's identity, 221, 228

Kepler orbit, 93, 248
see also Central motion
Kepler's laws, 96
Kinetic energy, 31
relative, 61
rigid body, 40, 146, 151
spin, 258
string, 124, 275
in various coordinates, cylindrical, 159
elliptical, 207
generalized, 343
parabolic, 210
rotating, 161
spherical, 160
Kinetic momentum, 163, 223
Kronecker delta, 336

Laboratory and center-of-mass coordinates, 62
Lagrange brackets, 237 f.
Lagrange multipliers, 88
Lagrange's equations, 77 f., 167, 220
with arbitrary independent variable, 240
for continuous medium, 275
in momentum space, 180
in phase space, 233
in quasi coordinates, 255
for rigid body, 151
see also Equations of motion
Lagrangian, 81 f.
particle, in accelerator, 315
in electromagnetic field, 85, 297
with spin, 258
Lagrangian density, 274
Lamellar field, 34
Langmuir-Child law, 54
Laplace transforms, 131
Larmor frequency, 108
Larmor's theorem, 107
Least action principle, 170
Left-handed axes, 19, 21, 22
see also Reflection
Length, 23
Libration, 52, 192, 197
Linear coordinate transformation, 179
Linear operator, 348
Linear vector space, 345 f.
Liouville's theorem, 233, 236, 268, 322

Lorentz force, 297
Lorentz transformation, 288

Magnetic field, charged particle in, 110, 248, 260
  magnet in, 262
Magnetic moment, 228, 258, 304
  anomalous, 307
  effective, 309
Magnetic quantum number, 91
Magnetron, 110
Many-body forces, 29, 35
Mass, 27
Mathieu's equation, 67, 68
Maxwell's equations, 86, 229, 289, 293
Mean pressure, 271
Meteoritic dust particle, 73
Michelson-Morley experiment, 288
Minimum energy orbit, 99
Missile, motion of, 98
Molecular oscillations, 364
  CO_2, 120
  NH_3, 368
Moment of a vector, 36, 149
Momental ellipsoid, 147
Moments of inertia, 40, 147
Momentum, 28, 83, 87
  conservation of, 30, 217
  in field theory, 284
  as generator of contact transformation, 217, 224
  integral, 83
  quasi, 256
  in relativity theory, 296
Momentum compaction, 317
Momentum density, 275
Momentum space, 157
Moving constraint, 10, 12
Multiply-periodic systems, 196, 245

NH_3-like molecule, 368
Natural system, 83, 156, 185
Natural units, 28, 51
Navier-Stokes equation, 268
Newton's laws, Law of Gravitation, 43
  First Law, 29
  Second Law, 28
  Third Law, 30
Noncentral forces, 258 f.
Nondegenerate system, 248
Nonholonomic constraint, 11
Norm of quaternion, 376
Normal modes, 116, 367
  for CO_2-like molecule, 120
  damped, 134
  for double pendulum, 117
  for masses on a string, 124
  for NH_3-like molecule, 368
  for triple pendulum, 118
Normalized eigenvector, 359
Null-vector, 294
Nutation, 149

Open-loop transfer function, 133
Operator, 348
Orbit, Bohr, 98, 109
  equilibrium, 315
  Kepler, 94, 248
  minimum energy, 99
  positive energy, 214
  satellite, 98
Orbital angular momentum, 42, 91
  see also Angular momentum
Orthogonal, operator, 350
  transformation, 357
  vectors, 347
Oscillations, betatron, 314, 316, 318
  molecular, 120, 368
  small, 113 f.
  synchrotron, 314, 331
Oscillator, 47
  anharmonic, 192
  critically damped, 75
  damped, 74
  simple harmonic, 89, 194, 242
  two-dimensional, 84
Oscillators, coupled, 115
Outer product, 22

Parabolic coordinates, 16, 25
Parallelogram law, 29
Particle-antiparticle pair, 301
Pauli spin matrices, 140, 373, 380
Pendulum, conical, 104
  double, 117
  inverted, 67 f.
  isochronous, 46
  plane, 50
  spherical, 103
  triple, 118
Periodic boundary conditions, 127

Periodic systems, 48, 188, 196, 245
Permutation group, 369
Perturbation methods, 105, 244
  stationary state, 245
  time-dependent, 251
Perveance, 56
Phase space, 156, 192
  extended, 241
  Lagrangian in, 233
Plane polar coordinates, 7, 341
Poinsot's representation, 149
Poisson brackets, 220 f.
  of angular momentum components, 225
  in field theory, 285
  generalized, 241
  invariance under contact transformations, 221
  involving vectors, 227
  in quantum mechanics, 230
  of quasi variables, 256
  relation to Lagrange brackets, 238
  in simplified notation, 233
  of spin components, 256
  of velocity components, 223
Potential energy, 32
  matrix, 366
Poynting-Robertson effect, 73
Precession, of free rotator, 149
  Larmor, 107
  of spherical pendulum, 104
  of spin, 263
  Thomas, 310
Pressure, thermodynamic, 272
Principal axes, 147
Principle, equivalence, 45
  Hamilton's, 168, 274, 296
  stationary action, 170, 315
  uncertainty, 157
  variational, 166, 265
Products of inertia, 40
Proper Lorentz transformation, 291
Proper time, 290
Proper transformation, 356

Q of an oscillator, 75
Quadratic form, 361
Quantum electrodynamics, 2
Quantum field theory, 2, 286
Quantum theory, action and angle variables, 193
Quantum theory, angular momentum, 91
  Hamilton-Jacobi equation, 214
  initial conditions, 49
  perturbation theory, 245
  Poisson brackets, 231
  spin, 373
Quasi coordinates, 144, 254
Quasi momenta, 256
Quaternions, 373 f.

Rayleigh dissipation function, 89
Reciprocal, 349
Reduced mass, 59
Reflection, in a mirror, 21
  of a molecule, 364
  in a moving point, 313
  in a point, 26
Relative integral invariant, 235
  generalized, 242
Relativity theory, 240, 286, 287 f.
  general, 6, 45, 344
Representation, by quaternion, 378
  of symmetry group, 365
Representative, 351, 367
Rest energy, 295, 300
Riemannian geometry, 335
Right-handed axes, 19, 21, 22
Rigid body, 39, 136, 373
Rigid rotator, 139, 148
Rocket motion, 70
Rotating coordinates, 88, 108, 161
  for spin resonance, 262
Rotation (motion), 192, 197
  axis of, 378
Rotation, coupled space and spin, 260
  infinitesimal, 218, 370
  of molecule, 364
  of rigid body, 373
  specification of, by contact transformation, 218
    by Euler's angles, 139
    in field theory, 283
    by orthogonal matrix, 137, 354
    by quaternions, 373
    by rotations about orthogonal axes, 144
Routh's function, 181
Rutherford scattering, 100 f.

Satellite, motion of, 98
Scalar potential, 86, 292

Scalar product, 18, 346
Schrödinger's equation, 214
Schur's lemma, 366
Schwarz's inequality, 347
Singular operator, 349
Small oscillations, 113 f.
Space-charge limitation, 54
Space-like vector, 294
Space-time, 288
Special theory of relativity, *see* Relativity theory
Spherical coordinates, 8, 160, 341
Spherical pendulum, 103
Spin, 42, 146, 258, 304
  see also Angular momentum
Standing wave, 127
Stark effect, 210, 214
Stationary action principle, 170, 315
Statistical mechanics, 235, 268
Stokes' law, 72
Stokes' relation, 272
Stress-energy tensor, 283
Strings, oscillations of, 124, 273
Strong focusing accelerators, 324
Submatrix, 370
Subspace, 365
Supermatrix, 370
Symmetric operator, 349
Symmetric rotator, 151
Symmetry, related to integrals of the motion, 224
Symmetry operation, 364
Symplectic matrices, 234, 326
Synchrotron oscillations, 314, 331
Systems, degenerate, 118, 205, 248
  dissipative, 74, 131
  isolated, 31
  natural, 83, 156, 185
  periodic, 188, 196, 245

Tensor, 22, 337
  angular momentum, 298
  energy-momentum, 279
  inertia, 40, 146, 151, 258
  stress-energy, 283
Thermodynamic pressure, 272
Thomas precession, 304
Time-like vector, 294
Top, motion of, 153
Total scattering cross section, 101
Trace, 361
Trajectory, 157, 247, 315
Transfer function, 133
Transformation, contact, 172, 183, 187
  Galilean, 288
  gauge, 304
  infinitesimal contact, 215
    extended, 241
  Laplace, 131
  linear coordinate, 179
  Lorentz, 288, 291, 303
  orthogonal, 357, 365
  proper, 356
  theory, 302
Transition energy, 318, 331
Transpose, 349
Traveling waves, 127, 275
Triangular inequality, 348
Triple pendulum, 118
Triple product, 19
Two centers of gravitation, 206

Uncertainty principle, 157

Van de Graaff accelerator, 112
Variational derivative, 285
Variational principles, 166
  in rocket motion, 265
Vector, 17
  axial, 21
  basic, 346
  in configuration space, 226
  covariant and contravariant components, 335
  in n-dimensional space, 345
  polar, 21
Vector potential, 86, 292
  point dipole, 260
  uniform magnetic field, 107
Vector product, 18
Velocity, 11
  in cylindrical coordinates, 13
  generalized, 11, 338
  in spherical coordinates, 15
Virial function, 165
Virial theorem, 164
Virtual states, 214
Viscosity, 72, 271

Weak focusing accelerators, 323
Work done by a force, 31
  in small displacement, 79
Workless constraint, 10

A CATALOG OF SELECTED

DOVER BOOKS

IN SCIENCE AND MATHEMATICS

A CATALOG OF SELECTED

DOVER BOOKS

IN SCIENCE AND MATHEMATICS

Astronomy

BURNHAM'S CELESTIAL HANDBOOK, Robert Burnham, Jr. Thorough guide to the stars beyond our solar system. Exhaustive treatment. Alphabetical by constellation: Andromeda to Cetus in Vol. 1; Chamaeleon to Orion in Vol. 2; and Pavo to Vulpecula in Vol. 3. Hundreds of illustrations. Index in Vol. 3. 2,000pp. 6⅛ x 9¼.
23567-X, 23568-8, 23673-0 Three-vol. set

THE EXTRATERRESTRIAL LIFE DEBATE, 1750–1900, Michael J. Crowe. First detailed, scholarly study in English of the many ideas that developed from 1750 to 1900 regarding the existence of intelligent extraterrestrial life. Examines ideas of Kant, Herschel, Voltaire, Percival Lowell, many other scientists and thinkers. 16 illustrations. 704pp. 5⅜ x 8½.
40675-X

A HISTORY OF ASTRONOMY, A. Pannekoek. Well-balanced, carefully reasoned study covers such topics as Ptolemaic theory, work of Copernicus, Kepler, Newton, Eddington's work on stars, much more. Illustrated. References. 521pp. 5⅜ x 8½.
65994-1

AMATEUR ASTRONOMER'S HANDBOOK, J. B. Sidgwick. Timeless, comprehensive coverage of telescopes, mirrors, lenses, mountings, telescope drives, micrometers, spectroscopes, more. 189 illustrations. 576pp. 5⅝ x 8¼. (Available in U.S. only.)
24034-7

STARS AND RELATIVITY, Ya. B. Zel'dovich and I. D. Novikov. Vol. 1 of *Relativistic Astrophysics* by famed Russian scientists. General relativity, properties of matter under astrophysical conditions, stars, and stellar systems. Deep physical insights, clear presentation. 1971 edition. References. 544pp. 5⅝ x 8¼. 69424-0

Chemistry

CHEMICAL MAGIC, Leonard A. Ford. Second Edition, Revised by E. Winston Grundmeier. Over 100 unusual stunts demonstrating cold fire, dust explosions, much more. Text explains scientific principles and stresses safety precautions. 128pp. 5⅜ x 8½.
67628-5

THE DEVELOPMENT OF MODERN CHEMISTRY, Aaron J. Ihde. Authoritative history of chemistry from ancient Greek theory to 20th-century innovation. Covers major chemists and their discoveries. 209 illustrations. 14 tables. Bibliographies. Indices. Appendices. 851pp. 5⅜ x 8½.
64235-6

CATALYSIS IN CHEMISTRY AND ENZYMOLOGY, William P. Jencks. Exceptionally clear coverage of mechanisms for catalysis, forces in aqueous solution, carbonyl- and acyl-group reactions, practical kinetics, more. 864pp. 5⅜ x 8½.
65460-5

THE HISTORICAL BACKGROUND OF CHEMISTRY, Henry M. Leicester. Evolution of ideas, not individual biography. Concentrates on formulation of a coherent set of chemical laws. 260pp. 5⅜ x 8½. 61053-5

A SHORT HISTORY OF CHEMISTRY, J. R. Partington. Classic exposition explores origins of chemistry, alchemy, early medical chemistry, nature of atmosphere, theory of valency, laws and structure of atomic theory, much more. 428pp. 5⅜ x 8½. (Available in U.S. only.) 65977-1

GENERAL CHEMISTRY, Linus Pauling. Revised 3rd edition of classic first-year text by Nobel laureate. Atomic and molecular structure, quantum mechanics, statistical mechanics, thermodynamics correlated with descriptive chemistry. Problems. 992pp. 5⅜ x 8½. 65622-5

Engineering

DE RE METALLICA, Georgius Agricola. The famous Hoover translation of greatest treatise on technological chemistry, engineering, geology, mining of early modern times (1556). All 289 original woodcuts. 638pp. 6¾ x 11. 60006-8

FUNDAMENTALS OF ASTRODYNAMICS, Roger Bate et al. Modern approach developed by U.S. Air Force Academy. Designed as a first course. Problems, exercises. Numerous illustrations. 455pp. 5⅜ x 8½. 60061-0

DYNAMICS OF FLUIDS IN POROUS MEDIA, Jacob Bear. For advanced students of ground water hydrology, soil mechanics and physics, drainage and irrigation engineering and more. 335 illustrations. Exercises, with answers. 784pp. 6⅛ x 9¼. 65675-6

ANALYTICAL MECHANICS OF GEARS, Earle Buckingham. Indispensable reference for modern gear manufacture covers conjugate gear-tooth action, gear-tooth profiles of various gears, many other topics. 263 figures. 102 tables. 546pp. 5⅜ x 8½. 65712-4

MECHANICS, J. P. Den Hartog. A classic introductory text or refresher. Hundreds of applications and design problems illuminate fundamentals of trusses, loaded beams and cables, etc. 334 answered problems. 462pp. 5⅜ x 8½. 60754-2

MECHANICAL VIBRATIONS, J. P. Den Hartog. Classic textbook offers lucid explanations and illustrative models, applying theories of vibrations to a variety of practical industrial engineering problems. Numerous figures. 233 problems, solutions. Appendix. Index. Preface. 436pp. 5⅜ x 8½. 64785-4

STRENGTH OF MATERIALS, J. P. Den Hartog. Full, clear treatment of basic material (tension, torsion, bending, etc.) plus advanced material on engineering methods, applications. 350 answered problems. 323pp. 5⅜ x 8½. 60755-0

A HISTORY OF MECHANICS, René Dugas. Monumental study of mechanical principles from antiquity to quantum mechanics. Contributions of ancient Greeks, Galileo, Leonardo, Kepler, Lagrange, many others. 671pp. 5⅜ x 8½. 65632-2

METAL FATIGUE, N. E. Frost, K. J. Marsh, and L. P. Pook. Definitive, clearly written, and well-illustrated volume addresses all aspects of the subject, from the historical development of understanding metal fatigue to vital concepts of the cyclic stress that causes a crack to grow. Includes 7 appendixes. 544pp. 5⅜ x 8½. 40927-9

STATISTICAL MECHANICS: Principles and Applications, Terrell L. Hill. Standard text covers fundamentals of statistical mechanics, applications to fluctuation theory, imperfect gases, distribution functions, more. 448pp. 5⅜ x 8½. 65390-0

THE VARIATIONAL PRINCIPLES OF MECHANICS, Cornelius Lanczos. Graduate level coverage of calculus of variations, equations of motion, relativistic mechanics, more. First inexpensive paperbound edition of classic treatise. Index. Bibliography. 418pp. 5⅜ x 8½. 65067-7

THE VARIOUS AND INGENIOUS MACHINES OF AGOSTINO RAMELLI: A Classic Sixteenth-Century Illustrated Treatise on Technology, Agostino Ramelli. One of the most widely known and copied works on machinery in the 16th century. 194 detailed plates of water pumps, grain mills, cranes, more. 608pp. 9 x 12. 28180-9

ORDINARY DIFFERENTIAL EQUATIONS AND STABILITY THEORY: An Introduction, David A. Sánchez. Brief, modern treatment. Linear equation, stability theory for autonomous and nonautonomous systems, etc. 164pp. 5⅜ x 8¼. 63828-6

ROTARY WING AERODYNAMICS, W. Z. Stepniewski. Clear, concise text covers aerodynamic phenomena of the rotor and offers guidelines for helicopter performance evaluation. Originally prepared for NASA. 537 figures. 640pp. 6⅛ x 9¼. 64647-5

INTRODUCTION TO SPACE DYNAMICS, William Tyrrell Thomson. Comprehensive, classic introduction to space-flight engineering for advanced undergraduate and graduate students. Includes vector algebra, kinematics, transformation of coordinates. Bibliography. Index. 352pp. 5⅜ x 8½. 65113-4

HISTORY OF STRENGTH OF MATERIALS, Stephen P. Timoshenko. Excellent historical survey of the strength of materials with many references to the theories of elasticity and structure. 245 figures. 452pp. 5⅜ x 8½. 61187-6

ANALYTICAL FRACTURE MECHANICS, David J. Unger. Self-contained text supplements standard fracture mechanics texts by focusing on analytical methods for determining crack-tip stress and strain fields. 336pp. 6⅛ x 9¼. 41737-9

Mathematics

HANDBOOK OF MATHEMATICAL FUNCTIONS WITH FORMULAS, GRAPHS, AND MATHEMATICAL TABLES, edited by Milton Abramowitz and Irene A. Stegun. Vast compendium: 29 sets of tables, some to as high as 20 places. 1,046pp. 8 x 10½. 61272-4

FUNCTIONAL ANALYSIS (Second Corrected Edition), George Bachman and Lawrence Narici. Excellent treatment of subject geared toward students with background in linear algebra, advanced calculus, physics and engineering. Text covers introduction to inner-product spaces, normed, metric spaces, and topological spaces; complete orthonormal sets, the Hahn-Banach Theorem and its consequences, and many other related subjects. 1966 ed. 544pp. 6⅛ x 9¼. 40251-7

ASYMPTOTIC EXPANSIONS OF INTEGRALS, Norman Bleistein & Richard A. Handelsman. Best introduction to important field with applications in a variety of scientific disciplines. New preface. Problems. Diagrams. Tables. Bibliography. Index. 448pp. 5⅜ x 8½. 65082-0

FAMOUS PROBLEMS OF GEOMETRY AND HOW TO SOLVE THEM, Benjamin Bold. Squaring the circle, trisecting the angle, duplicating the cube: learn their history, why they are impossible to solve, then solve them yourself. 128pp. 5⅜ x 8½. 24297-8

VECTOR AND TENSOR ANALYSIS WITH APPLICATIONS, A. I. Borisenko and I. E. Tarapov. Concise introduction. Worked-out problems, solutions, exercises. 257pp. 5⅝ x 8¼. 63833-2

THE ABSOLUTE DIFFERENTIAL CALCULUS (CALCULUS OF TENSORS), Tullio Levi-Civita. Great 20th-century mathematician's classic work on material necessary for mathematical grasp of theory of relativity. 452pp. 5⅝ x 8¼. 63401-9

AN INTRODUCTION TO ORDINARY DIFFERENTIAL EQUATIONS, Earl A. Coddington. A thorough and systematic first course in elementary differential equations for undergraduates in mathematics and science, with many exercises and problems (with answers). Index. 304pp. 5⅜ x 8½. 65942-9

FOURIER SERIES AND ORTHOGONAL FUNCTIONS, Harry F. Davis. An incisive text combining theory and practical example to introduce Fourier series, orthogonal functions and applications of the Fourier method to boundary-value problems. 570 exercises. Answers and notes. 416pp. 5⅜ x 8½. 65973-9

COMPUTABILITY AND UNSOLVABILITY, Martin Davis. Classic graduate-level introduction to theory of computability, usually referred to as theory of recurrent functions. New preface and appendix. 288pp. 5⅜ x 8½. 61471-9

ASYMPTOTIC METHODS IN ANALYSIS, N. G. de Bruijn. An inexpensive, comprehensive guide to asymptotic methods–the pioneering work that teaches by explaining worked examples in detail. Index. 224pp. 5⅜ x 8½ 64221-6

ESSAYS ON THE THEORY OF NUMBERS, Richard Dedekind. Two classic essays by great German mathematician: on the theory of irrational numbers; and on transfinite numbers and properties of natural numbers. 115pp. 5⅜ x 8½. 21010-3

APPLIED COMPLEX VARIABLES, John W. Dettman. Step-by-step coverage of fundamentals of analytic function theory–plus lucid exposition of five important applications: Potential Theory; Ordinary Differential Equations; Fourier Transforms; Laplace Transforms; Asymptotic Expansions. 66 figures. Exercises at chapter ends. 512pp. 5⅜ x 8½. 64670-X

INTRODUCTION TO LINEAR ALGEBRA AND DIFFERENTIAL EQUATIONS, John W. Dettman. Excellent text covers complex numbers, determinants, orthonormal bases, Laplace transforms, much more. Exercises with solutions. Undergraduate level. 416pp. 5⅜ x 8½. 65191-6

MATHEMATICAL METHODS IN PHYSICS AND ENGINEERING, John W. Dettman. Algebraically based approach to vectors, mapping, diffraction, other topics in applied math. Also generalized functions, analytic function theory, more. Exercises. 448pp. 5⅝ x 8¼. 65649-7

CALCULUS OF VARIATIONS WITH APPLICATIONS, George M. Ewing. Applications-oriented introduction to variational theory develops insight and promotes understanding of specialized books, research papers. Suitable for advanced undergraduate/graduate students as primary, supplementary text. 352pp. 5⅜ x 8½. 64856-7

COMPLEX VARIABLES, Francis J. Flanigan. Unusual approach, delaying complex algebra till harmonic functions have been analyzed from real variable viewpoint. Includes problems with answers. 364pp. 5⅜ x 8½. 61388-7

AN INTRODUCTION TO THE CALCULUS OF VARIATIONS, Charles Fox. Graduate-level text covers variations of an integral, isoperimetrical problems, least action, special relativity, approximations, more. References. 279pp. 5⅜ x 8½. 65499-0

CATASTROPHE THEORY FOR SCIENTISTS AND ENGINEERS, Robert Gilmore. Advanced-level treatment describes mathematics of theory grounded in the work of Poincaré, R. Thom, other mathematicians. Also important applications to problems in mathematics, physics, chemistry and engineering. 1981 edition. References. 28 tables. 397 black-and-white illustrations. xvii + 666pp. 6⅛ x 9¼. 67539-4

INTRODUCTION TO DIFFERENCE EQUATIONS, Samuel Goldberg. Exceptionally clear exposition of important discipline with applications to sociology, psychology, economics. Many illustrative examples; over 250 problems. 260pp. 5⅜ x 8½. 65084-7

NUMERICAL METHODS FOR SCIENTISTS AND ENGINEERS, Richard Hamming. Classic text stresses frequency approach in coverage of algorithms, polynomial approximation, Fourier approximation, exponential approximation, other topics. Revised and enlarged 2nd edition. 721pp. 5⅜ x 8½. 65241-6

INTRODUCTION TO NUMERICAL ANALYSIS (2nd Edition), F. B. Hildebrand. Classic, fundamental treatment covers computation, approximation, interpolation, numerical differentiation and integration, other topics. 150 new problems. 669pp. 5⅜ x 8½. 65363-3

Physics

OPTICAL RESONANCE AND TWO-LEVEL ATOMS, L. Allen and J. H. Eberly. Clear, comprehensive introduction to basic principles behind all quantum optical resonance phenomena. 53 illustrations. Preface. Index. 256pp. 5⅜ x 8½. 65533-4

ULTRASONIC ABSORPTION: An Introduction to the Theory of Sound Absorption and Dispersion in Gases, Liquids and Solids, A. B. Bhatia. Standard reference in the field provides a clear, systematically organized introductory review of fundamental concepts for advanced graduate students, research workers. Numerous diagrams. Bibliography. 440pp. 5⅜ x 8½. 64917-2

QUANTUM THEORY, David Bohm. This advanced undergraduate-level text presents the quantum theory in terms of qualitative and imaginative concepts, followed by specific applications worked out in mathematical detail. Preface. Index. 655pp. 5⅜ x 8½. 65969-0

ATOMIC PHYSICS (8th edition), Max Born. Nobel laureate's lucid treatment of kinetic theory of gases, elementary particles, nuclear atom, wave-corpuscles, atomic structure and spectral lines, much more. Over 40 appendices, bibliography. 495pp. 5⅜ x 8½. 65984-4

AN INTRODUCTION TO HAMILTONIAN OPTICS, H. A. Buchdahl. Detailed account of the Hamiltonian treatment of aberration theory in geometrical optics. Many classes of optical systems defined in terms of the symmetries they possess. Problems with detailed solutions. 1970 edition. xv + 360pp. 5⅜ x 8½. 67597-1

THIRTY YEARS THAT SHOOK PHYSICS: The Story of Quantum Theory, George Gamow. Lucid, accessible introduction to influential theory of energy and matter. Careful explanations of Dirac's anti-particles, Bohr's model of the atom, much more. 12 plates. Numerous drawings. 240pp. 5⅜ x 8½. 24895-X

ELECTRONIC STRUCTURE AND THE PROPERTIES OF SOLIDS: The Physics of the Chemical Bond, Walter A. Harrison. Innovative text offers basic understanding of the electronic structure of covalent and ionic solids, simple metals, transition metals and their compounds. Problems. 1980 edition. 582pp. 6⅛ x 9¼. 66021-4

HYDRODYNAMIC AND HYDROMAGNETIC STABILITY, S. Chandrasekhar. Lucid examination of the Rayleigh-Benard problem; clear coverage of the theory of instabilities causing convection. 704pp. 5⅜ x 8¼. 64071-X

INVESTIGATIONS ON THE THEORY OF THE BROWNIAN MOVEMENT, Albert Einstein. Five papers (1905–8) investigating dynamics of Brownian motion and evolving elementary theory. Notes by R. Fürth. 122pp. 5⅜ x 8½. 60304-0

THE PHYSICS OF WAVES, William C. Elmore and Mark A. Heald. Unique overview of classical wave theory. Acoustics, optics, electromagnetic radiation, more. Ideal as classroom text or for self-study. Problems. 477pp. 5⅜ x 8½. 64926-1

METHODS OF THERMODYNAMICS, Howard Reiss. Outstanding text focuses on physical technique of thermodynamics, typical problem areas of understanding, and significance and use of thermodynamic potential. 1965 edition. 238pp. 5⅜ x 8½. 69445-3

TENSOR ANALYSIS FOR PHYSICISTS, J. A. Schouten. Concise exposition of the mathematical basis of tensor analysis, integrated with well-chosen physical examples of the theory. Exercises. Index. Bibliography. 289pp. 5⅜ x 8½. 65582-2

RELATIVITY IN ILLUSTRATIONS, Jacob T. Schwartz. Clear nontechnical treatment makes relativity more accessible than ever before. Over 60 drawings illustrate concepts more clearly than text alone. Only high school geometry needed. Bibliography. 128pp. 6⅛ x 9¼. 25965-X

THE ELECTROMAGNETIC FIELD, Albert Shadowitz. Comprehensive undergraduate text covers basics of electric and magnetic fields, builds up to electromagnetic theory. Also related topics, including relativity. Over 900 problems. 768pp. 5⅝ x 8¼. 65660-8

GREAT EXPERIMENTS IN PHYSICS: Firsthand Accounts from Galileo to Einstein, edited by Morris H. Shamos. 25 crucial discoveries: Newton's laws of motion, Chadwick's study of the neutron, Hertz on electromagnetic waves, more. Original accounts clearly annotated. 370pp. 5⅜ x 8½. 25346-5

RELATIVITY, THERMODYNAMICS AND COSMOLOGY, Richard C. Tolman. Landmark study extends thermodynamics to special, general relativity; also applications of relativistic mechanics, thermodynamics to cosmological models. 501pp. 5⅜ x 8½. 65383-8

LIGHT SCATTERING BY SMALL PARTICLES, H. C. van de Hulst. Comprehensive treatment including full range of useful approximation methods for researchers in chemistry, meteorology and astronomy. 44 illustrations. 470pp. 5⅜ x 8½. 64228-3

STATISTICAL PHYSICS, Gregory H. Wannier. Classic text combines thermodynamics, statistical mechanics and kinetic theory in one unified presentation of thermal physics. Problems with solutions. Bibliography. 532pp. 5⅜ x 8½. 65401-X